THE EVOLUTION OF THE INTERSTELLAR MEDIUM

A SERIES OF BOOKS ON RECENT DEVELOPMENTS IN ASTRONOMY AND ASTROPHYSICS

© Copyright 1990 Astronomical Society of the Pacific
390 Ashton Avenue, San Francisco, California 94112

All rights reserved

Printed by BookCrafters, Inc.

First published 1990

Library of Congress Catalog Card Number: 90-84615
ISBN 0-937707-31-7

D. Harold McNamara, Managing Editor of Conference Series
293 ESC Brigham Young University
Provo, UT 84602

ASTRONOMICAL SOCIETY OF THE PACIFIC

CONFERENCE SERIES

Volume 12

THE EVOLUTION OF THE INTERSTELLAR MEDIUM

Edited by

Leo Blitz

A SERIES OF BOOKS ON RECENT DEVELOPMENTS IN ASTRONOMY AND ASTROPHYSICS

Vol. 1-Progress and Opportunities in Southern Hemisphere Optical Astronomy: The CTIO 25th Anniversary Symposium
ed. V. M. Blanco and M. M. Phillips   ISBN 0-937707-18-X

Vol. 2-Proceedings of a Workshop on Optical Surveys for Quasars
ed. P. S. Osmer, A. C. Porter, R. F. Green, and C. B. Foltz   ISBN 0-937707-19-8

Vol. 3-Fiber Optics in Astronomy
ed. S.C. Barden   ISBN 0-937707-20-1

Vol. 4-The Extragalactic Distance Scale: Proceedings of the ASP 100th Anniversary Symposium
ed. S. van den Bergh and C. J. Pritchet   ISBN 0-937707-21-X

Vol. 5-The Minnesota Lectures on Clusters of Galaxies and Large-Scale Structure
ed. J. M. Dickey   ISBN 0-937707-22-8

Vol. 6-Synthesis Imaging in Radio Astronomy: A Collection of Lectures from the Third NRAO Synthesis Imaging Summer School
ed. R. A. Perley, F. R. Schwab, and A. H. Bridle   ISBN 0-937707-23-6

Vol. 7-Properties of Hot Luminous Stars: Boulder-Munich Workshop
ed. C. D. Garmany   ISBN 0-937707-24-4

Vol. 8-CCDs in Astronomy
ed. George H. Jacoby   ISBN 0-937707-25-2

Vol. 9-Cool Stars, Stellar Systems, and the Sun
ed. G. Wallerstein   ISBN 0-937707-27-9

Vol. 10-The Evolution of the Universe of Galaxies, The Edwin Hubble Centennial Symposium
ed. Richard G. Kron   ISBN 0-937707-28-7

Inquiries concerning these volumes should be directed to the:
  Astronomical Society of the Pacific
  CONFERENCE SERIES
  390 Ashton Avenue
  San Fransico, CA   94112-1722

This book is dedicated to the memory of Munio Blitz

TABLE OF CONTENTS

Preface ... v

Conference Registrants ... vii

SECTION I - The Three Phase Model Revisited

The Three Phase Model of the Interstellar Medium:
Where Does it Stand Now?
Christopher F. McKee ... 3

SECTION II - The Contents of the Interstellar Medium

Properties of the ISM: Gas in the Halo
Blair D. Savage ... 33

Dense Gas in the Galaxy
N.Z. Scoville ... 49

Interstellar Dust and Extinction
John S. Mathis ... 63

The High-Energy Component of the ISM: Cosmic Rays
and Gamma Rays
Hans Bloemen ... 79

The Roles of Cosmic Rays in Interstellar Dynamics
T.W. Hartquist ... 99

SECTION III - Evolutionary Processes in the Interstellar Medium

Turbulent Stripping of Interstellar Clouds by Interaction
with Supernova Remnants
Richard I. Klein, Christopher F. McKee, Philip Colella ... 117

Winds from Hot Stars
John H. Bieging ... 137

The Total Rate of Mass Return to the Interstellar Medium from
Red Giants and Planetary Nebulae
G.R. Knapp, K.P. Rauch, and E.M. Wilcots ... 151

Photodissociation Regions
David J. Hollenbach ... 167

Bipolar Outflows: Evolutionary and
Global Considerations
Luis F. Rodriguez ... 183

SECTION IV - Evolution of Dust, Gas, and Chemistry

Evolution of Interstellar Dust
Bruce T. Draine ... 193

The Chemistry of the Diffuse Interstellar Gas
Ewine F. van Dishoeck ... 207

Evolution of the Chemistry in Dense Clouds
                                Lucy M. Ziurys   229

Theories of Molecular Cloud Formation
                                Bruce G. Elmegreen   247

The Evolution of Galactic Giant Molecular Clouds
                                Leo Blitz   273

Formation of High Mass Stars   *Wm. J. Welch*   291

SECTION V - Posters

Observational Constraints on an Embedded Cloud Model
for the Soft X-ray Diffuse Background   *D.N. Burrows*   303

Probing the ISM with Neutron Stars
                                Dieter Hartmann   305

Low Frequency Observations of Galactic SNRS and the
Distribution of Low Density Ionized Gas
in the ISM   *Namir E. Kassim*   307

Far-IR Observations of the N/O Ratio in Interstellar Gas
                   J.P. Simpson, S.W.J. Colgan,
        E.F. Erickson, M.R. Haas, and R.H. Rubin   311

IRAS Observations of a Large Circumstellar Dust
Shell Around W Hydrae
                              George W. Hawkins   313

High Resolution ^{12}CO (J=1→0)
Observations of NGC 7027
              David J. Wilner, John H. Bieging,
              and *Harley A. Thronson, Jr.*   315

Synthesis Observations of J=1-0 HCO$^+$ In DR21
 Rognvald P. Garden, Dan Grolemund, and John Carlstrom   317

The 3.3 Micron Feature, H_2, and Ionized Gas in
the Orion Bar
             K. Sellgren, A.T. Tokunaga, and Y. Nakada   321

Observations of the Galactic Plane by the Zodiacal
Infrared Project
           L.J. Rickard, S.W. Stemwedel, and S.D. Price   323

Stars and Interstellar Matter in Ophiuchus/Scorpius
                                E.J. de Geus   325

High Spectral and Spatial Resolution Imaging of the
Shock Waves in Orion
       Michael Burton, Joss Bland, D. Axon, P. Brand,
      R. Garden, T. Geballe, D. Hollenbach, J. Hough,
                I. McLean, and A. Moorhouse   327

Galactic Structure in the Fourth Quadrant Deduced
  from Interstellar Lines, H II Regions,
  and Molecular Hydrogen   *J.J. Rickard*   329

Comparison of NH_3 and CS Distribution in NGC 2071
       Shudong Zhou, Neal J. Evans II, and Lee G. Mundy   331

Methanol Masers and Star Formation: VLA Observations
  of the NGC 6334 Region
                       K.M. Menten and M.J. Reid   333

Stellar Mass Loss Rates in Magellanic Cloud Stars
                            Catharine D. Garmany   335

Diffuse Ionized Gas in the Andromeda Galaxy
               Rene A.M. Walterbos, and Robert Braun   337

Dense Gas in the Starburst Galaxy NGC 253   *J.E. Carlstrom*   339

High Resolution Observations of NGC 7538 IRS 1
       Preethi Pratap, Wolfgang Batrla, and Lewis Snyder   343

Evolution of a Superbubble Blastwave in a Magnetized Medium
     Katia M. Ferriere, Ellen G. Zweibel, and Mordecai-Mark Maclow   345

Preface

The celebration of the 100th anniversary of the Astronomical Society of the Pacific included three scientific symposia: *The Evolution of the Solar System, The Evolution of the Interstellar Medium,* and *The Evolution of the Universe.* This volume is the proceedings of the second of these symposia.

I have remarked on more than one occasion that the highest compliment that one can pay to a contemporary work of theoretical astrophysics is that it doesn't matter whether the theory is wrong. The point is that subsequent observations may ultimately determine whether any theory is right or wrong, but a robust theory will stimulate other work to lead astronomers to a correct description of the Universe. In the mid 1970s Chris McKee, Jerry Ostriker and Len Cowie wrote a seminal series of papers on the energization of the interstellar medium by supernovae, culminating in the famous McKee-Ostriker three phase model of the interstellar medium. Although the paper was admittedly incomplete, in that it ignored the cold molecular phase of the ISM, the paper served as a benchmark, as one of the first comprehensive works that discussed in detail how the diffuse phases of the ISM evolve from one another. At the Berkeley meeting, Chris McKee commented that I wouldn't have addressed my remark to the McKee-Ostriker model if I thought the theory was correct (provoking not a small amount of laughter), but the model has indeed stimulated a great deal of interest and activity. It is therefore appropriate that this volume begins with Chris's evaluation of how the theory has stood up over the past decade.

Since that time, there has been a great deal of other work on the evolution of the ISM, and the three day meeting in Berkeley, from June 21-23 1989 attempted to assess the progress that had been made in this area. After all, in the astronomical cycle that leads from diffuse gas to dense molecular clouds to stars and then back into the interstellar medium, it is the evolution from the diffuse gas back into stars that is the least well understood.

Although the contents of this volume are organized a bit differently from that of the symposium itself, the flavor of the meeting is retained in the contributions. After the McKee paper, the second section of the book contains contributions that assess the state of the interstellar medium. The third section details the processes involved in determining the evolution of the ISM. The fourth section contains the papers that deal most directly with the evolution of the dust, the denser phases of the ISM, and the chemical evolution of the gas. Most of the contributions are extensively referenced and serve as both excellent and comprehensive reviews of the subjects they address; there is much material in the book that is not found elsewhere.

The meeting brought out a large number of poster contributions and a few very lively poster sessions. Because many of the authors worked so hard at them, I thought that short brief descriptions of many of the papers presented would provide a taste of the session itself. Consequently, the last section of the book is devoted to many of the poster presentations that were shown at the meeting. I have not attempted to intersperse them within the chapters that contain the invited contributions because most of the subject matter did not fit neatly into the catagories I used to organize the book.

A number of people deserve special mention. The organizing committee for the meeting consisted, besides myself, of Chris McKee, Frank Shu, Charlie

Lada, Ewine van Dishoeck, Mike Jura, and Jill Knapp. An enormous debt of gratitude goes to Harold Weaver and Andy Fraknoi who were responsible for so much of the organization of the meeting. They allowed me to have to worry only about the scientific content of the meeting. Everyone who has been involved in organizing a meeting knows that the hardest part of the job was done by Harold and Andy. Dana Mediate helped me with much of the work of actually putting the book together. Eugène de Geus helped with some of the technical aspects of making the book look more presentable. To all of them, I give my most sincere thanks.

Leo Blitz
College Park

LIST OF PARTICIPANTS

William Agnew	Unaffiliated
Mary Barsony	U.C. Berkeley
Mitchell Begelman	Colorado
Manfred Bester	U.C. Berkeley
John Bieging	U.C. Berkeley
John Black	Steward Observatory
Leo Blitz	Maryland
Hans Bloemen	Leiden
David Burrows	Penn State
Harold Butner	Texas
Michael Burton	U.C. Irvine
Richard Buss, Jr.	NASA/Ames
John Carlstrom	U.C. Berkeley
Roger Chevalier	Virginia
Rafael Costero	UNAM, Mexico
Nancy Cox	East Bay Astronomical Society
Nahide Craig	Livermore
Eugene de Geus	Maryland
Bruce Draine	Princeton
Ewine van Dishoeck	Caltech
Bruce Elmegreen	IBM Watson Research Center
Katia Ferriere	Colorado
William Forrest	Rochester
Margaret Frerking	Jet Propulsion Laboratory
Rognvald Garden	U.C. Irvine
Catharine Garmany	Colorado
Roy Garstang	Colorado
John Gaustad	Swarthmore
Peter Goldreich	Caltech
James Green	Colorado
Michael Haas	NASA/Ames
Pieter Hartmann	Livermore
Tom Hartquist	Max Planck/Garching
George Hawkins	U.C.L.A.
Isabel Hawkins	U.C. Berkeley
David Hollenbach	NASA/Ames
Namir Kassim	Naval Research Lab
Steven Kilston	Lockheed Research Labs
Richard Klein	U.C. Berkeley
Jill Knapp	Princeton
Bruce Lafferty	Unaffiliated
Charles Lada	U.C. Berkeley
G.E. Langer	Colorado College
Chun Ming Leung	Rensselaer
Kurt Liffman	NASA/Ames
Simon Lilly	Hawaii
K.Y. Lo	Illinois

Felix Lockman	NRAO
Julie Lutz	Washington State
Mordecai-Mark MacLow	NASA/Ames
Ronald Maddalena	NRAO
Loris Magnani	Arecibo Observatory
Jerome Malenfant	Mississippi
John Mathis	Wisconsin
Chris McKee	U.C. Berkeley
Gary Melnick	Center for Astrophysics
Karl Menten	Center for Astrophysics
Lee Mundy	Maryland
David Neufeld	U.C. Berkeley
Charlie Oostdyk	Orange County Astronomers
Patrick Palmer	Chicago
Dick Plambeck	U.C. Berkeley
Daniel Popper	U.C.L.A.
Preethi Pratap	Illinois
Richard Puchalsky	Maryland
Jim Rickard	Unaffiliated
Lee Rickard	Naval Research Labs
Luis Rodriguez	UNAM, Mexico
Steven Ruden	U.C. Berkeley
Alexander Rudolph	Maryland
Michael Rupen	Center for Astrophysics
Edwin Salpeter	Cornell
Blair Savage	Wisconsin
Richard Schwartz	Santa Barbara Astronomy Group
Nick Scoville	Caltech
Kris Sellgren	Hawaii
Frank Shu	U.C. Berkeley
Michael Shull	Colorado
Jan Simpson	NASA/Ames
Michael Sitko	Cincinnati
Robert Smith	U.C. Irvine
Sally Stemwedel	Applied Research Corp.
Charles Townes	U.C. Berkeley
John Vallerga	U.C. Berkeley
Peter Vedder	U.C. Berkeley
Christopher Walker	Caltech
Rene Walterbos	U.C. Berkeley
Zhong Wang	Caltech
Jack Welch	U.C. Berkeley
David Wilner	U.C. Berkeley
Douglas Wood	Center for Astrophysics
Ann Orel Woodin	Livermore
Gareth Wynn-Williams	Hawaii
Theodore Yu	U.C. Berkeley
Shudong Zhou	Texas
Lucy Ziurys	Arizona State

Section I

The Three Phase Model Revisited

THE THREE PHASE MODEL OF THE INTERSTELLAR MEDIUM: WHERE DOES IT STAND NOW?

CHRISTOPHER F. MCKEE
Departments of Physics and Astronomy
University of California, Berkeley, CA 94720

ABSTRACT The structure of the interstellar medium (ISM) is governed by energy injection by stars. Both warm neutral gas and ionized gas extend well above the Galactic plane; the weight of this gas is supported primarily by turbulent motions. In the absence of this turbulence, the scale height of the ISM would be substantially smaller than observed. Interstellar turbulence is generated primarily by supernovae: Random supernovae generate a three-phase medium in which hot, low-density gas surrounds warm and cold clouds and determines their pressure; supernovae in large associations of massive stars generate superbubbles, in which the hot gas is dominant. The observed level of turbulence is shown to require a large filling factor for supernova remnants, supporting the three-phase model of the ISM. The observational status of the three phase model is reviewed, with a summary of both its successes and its problems. The Appendix gives a discussion of the Galactic supernova rate; Galactic evidence gives a conservative estimate for this rate of 0.022 SN yr^{-1}, consistent with evidence from extragalactic SN.

I. INTRODUCTION

The interstellar medium plays a crucial role in modern astrophysics because of its central importance in star formation and in the formation and evolution of galaxies. The ISM is quite complex, with atomic, molecular, and ionized gas, dust grains, cosmic rays, magnetic fields, and radiation all interacting through a variety of physical and chemical processes. The overall energetics of the ISM are governed largely by the outflow of energy from stars—in the form of radiation, winds, and supernova explosions—through the ISM to the intergalactic medium. Much of this energy is from massive stars, which are continually forming out of the ISM and which return most of their mass back to the ISM. The low-mass stars which also form do not return so much of their mass to the ISM, and they gradually deplete the ISM of its gas and dust.

The ISM is observed to be highly inhomogeneous, with much of the mass concentrated in clouds—both atomic and molecular—which occupy a small fraction of the volume, whereas much of the volume is filled with warmer intercloud gas. Field, Goldsmith, and Habing (1969; hereafter FGH) provided a theoretical explanation of how such multiphase behavior is possible: the temperature dependence of the cooling rate of interstellar gas is such that cold ($T \sim 10^2$ K), neutral (H I) clouds can coexist in pressure equilibrium with warm ($T \sim 10^4$ K) intercloud H I only over a limited range of pressures, close to those observed in the ISM. Subsequently, Cox and Smith (1974) argued that supernova explosions in the Galaxy are sufficiently frequent that supernova remnants tend to overlap, forming a network of hot tunnels in the ISM with a filling factor of about 10%; Smith (1977) simulated this model numerically. McKee and Ostriker (1977; hereafter MO) argued that the filling factor of the hot gas is substantially larger than this, so that it forms the background medium in which the two phases of the FGH model are embedded. The ISM thus consists of three phases: the cold neutral medium (CNM), taken to have a temperature of 80 K; the the warm medium at $T \sim 8000$ K, which has both neutral (WNM) and ionized (WIM) parts; and the hot ionized intercloud medium (HIM) at a calculated temperature $T \sim 5 \times 10^5$ K. The clouds are assumed to have a spectrum of masses extending up to the point that they become self-gravitating, estimated at about 10^4 $M_\odot$. Most molecular clouds are observed to be more massive than this and are indeed self-gravitating (Larson 1981), so they are not explicitly included in the theory; the gas they contain is allowed for by including it in the mean density of the medium. The molecular clouds do not constitute a fourth phase because they are self-gravitating, and so are not in pressure equilibrium with the rest of the ISM. The three-phase model differs from the two-phase model in several important respects: The most obvious difference is that a large fraction of the volume of the ISM is predicted to be almost empty, filled with low-density, hot gas; as a result, mass exchange between the hot phase and the other phases, due to thermal evaporation and condensation, is crucial in determining the equilibrium (see Begelman and McKee 1990 for an analysis of multiphase equilibria with mass exchange). The thermal pressure in the ISM is determined dynamically by the evolution of supernova remnants (SNRs) in the ISM, permitting prediction of the distribution of the pressures in the ISM. A third feature of the three-phase model is that the ionized component of the warm medium, the WIM, is predicted to be an important component of the diffuse ISM, in agreement with observation.

In the years since it was proposed, the three-phase model has been reasonably successful in accounting for a number of observed properties of the ISM and in providing several testable predictions. Work on the model since then has included numerical simulations of SNR expansion (Cowie, McKee, and Ostriker 1981), interstellar bubbles (Wolff and Durisen 1987), and the time dependent evolution of a three-phase ISM (Habe, Ikeuchi, and Tanaka 1981; Ikeuchi, Habe, and Tanaka 1984). Attention has also focused on the injection of energy from the disk into the halo in a galactic fountain (Shapiro and Field 1976; Chevalier and Oegerle 1979; Bregman 1980; Cowie 1987; Corbelli and Salpeter 1988); however, Cox (1981) has argued that such a fountain must be weak. A variant of the fountain model in which the energy injection into the halo is localized to large stellar associations (the "chimney model") has been developed by Ikeuchi (1987) and extended by Norman and Ikeuchi (1989). Remarkably enough, the most striking prediction of the three-phase model—the large filling factor for the hot gas—remains in dispute. It is now generally agreed that the Sun is located in a bubble of hot gas (Cox and Reynolds 1987), but it has proven extremely difficult to measure the amount of hot gas outside this local bubble. Indeed, Cox (1988) has advanced arguments that the filling factor for the hot gas in the Galactic disk is small, perhaps only about 10% (Cox and Slavin 1989). The current status of our knowledge of hot gas in the Galaxy, both theoretical and observational, is treated in a comprehensive review by Spitzer (1990).

Observations in the last decade have shown that the ISM extends much farther from the plane of the Galaxy than originally believed: Lockman (1984) has found a component of HI with a scale height of about 500 pc, Reynolds (1989a) has inferred a scale height for ionized gas of 1–2.5 kpc from observations of pulsar dispersion measures, and a number of investigators have studied halo gas far from the plane through its absorption lines (Savage 1987). The three-phase model is a local theory which predicts the conditions in the Galactic plane, where the mean density of hydrogen nuclei is $\bar{n}_H \simeq 1$ cm^{-3}. MO were able to neglect the finite scale height of the gas disk because in their model supernova remnants expand to radii comparable to, but not greater than, the disk scale height. In fact, however, any model for the ISM in a disk galaxy is directly coupled to the vertical structure of the ISM since, in equilibrium, the pressure in the plane equals the weight of the gas above it (e.g., Spitzer 1978). The observations which show that the gas disk is thicker than previously thought at the same time imply that the weight of the ISM is greater than previously believed. Does this enhanced weight crush the hot phase virtually out of existence? Cox (1988) has argued that it does: he infers that the magnetic field must be rather large, $B \simeq 5$ μG, and that this large field prevents SNRs from filling much of the ISM. Here we shall adopt the opposite viewpoint: the processes which energize the ISM and determine its pressure force the gas to greater heights, increasing its weight (McKee 1990). We shall explore this idea after reviewing the current state of our knowledge of the vertical structure of the ISM.

II. VERTICAL STRUCTURE OF THE ISM: OBSERVATIONS

The mass and scale height of the various components of the ISM have been reviewed by Bloemen (1987), Spitzer (1990), and McKee (1990). Table I, taken from McKee (1990), is an inventory of the density distributions of the various components in the solar vicinity. The densities are given in terms of $\bar{n}_H(z)$, the volume averaged density of hydrogen nuclei as a function of distance from the Galactic plane, z. The surface densities Σ include the mass of helium, which we take to be 10% by number. The Table begins with the coldest component, the molecular gas. The surface density of this gas is changing rapidly within a kiloparsec of the Sun, so this value is necessarily somewhat uncertain. The values in Table I are based on the survey of Bronfman et al. (1988), and are compatible with those found by Scoville and Sanders (1987).

The distribution of neutral atomic hydrogen is complex: Lockman (1984) has identified three distinct components. Kulkarni and Heiles (1987) identify the component with the lowest scaleheight with the HI clouds (the CNM) and the remaining two components with the intercloud HI (which we label as WNM$_1$ and WNM$_2$). One of the components (WNM$_2$) extends into the halo, as has been confirmed by ultraviolet observations of high latitude stars (Lockman, Hobbs, and Shull 1986). The mean densities and scale heights of the two WNM components listed in Table I are taken from Lockman (1984). His method of analysis is insensitive to opaque HI, so the surface density of the CNM has been adjusted to give the total HI column observed by Heiles (1976); the scale height is taken from Falgarone and Lequeux (1973).

Ionized hydrogen near the Galactic plane is in two forms, photoionized (the WIM) and collisionally ionized (the HIM). The density and scale height of the WIM can be inferred from observations of pulsar dispersion measures (e.g., Kulkarni and Heiles 1987), which imply an electron distribution

$$\bar{n}_e = 0.015\exp-(|z|/70\text{ pc}) + 0.025\exp-(|z|/H_w) \quad \text{cm}^{-3}, \tag{1}$$

with a scale height $H_w \simeq 1$ kpc. Data on the pulsars recently discovered in globular clusters led Reynolds (1989a) to suggest 1 kpc $\lesssim H_w \lesssim 2.5$ kpc, with a best estimate of $H_w = 1.5$ kpc. Subsequent discoveries (e.g., the pulsars in M13 and M53—Anderson et al. 1989a,b) favor a somewhat lower value: the median scale height inferred from globular cluster pulsars more than 1 kpc from the Galactic plane is about 1.0 kpc. Allowing for the roughly 4% of the electrons contributed by helium (Mathis 1986), a correction more important in principle than in practice, gives the mean hydrogen density listed in Table I. Observations of the WIM in emission lines (reviewed in Reynolds 1989b) also imply a thick disk of warm, ionized gas. Such a WIM disk is not unique to the Galaxy: a thicker, more massive disk of WIM has been observed in NGC 891, which has more active star formation than the Galaxy (Rand, Kulkarni, and Hester 1990).

TABLE I Vertical Mass Distribution of the ISM

Component	$\bar{n}_H(z)$ (cm^{-3})	Σ ($M_\odot$ pc^{-2})
H$_2$	$0.54\exp-\frac{1}{2}(z/60)^2$	2.80
HI: CNM	$0.39\exp-\frac{1}{2}(z/135)^2$	4.54
WNM$_1$	$0.093\exp-\frac{1}{2}(z/254)^2$	2.04
WNM$_2$	$0.053\exp-(|z|/480)$	1.75
HII:[a] WIM	$0.96[0.015\exp-(|z|/70) + 0.025\exp-(|z|/H_w)]$	1.72/1.36
HIM	$0 / 0.0015\exp-(|z|/3000)$	0/0.31
Total		12.8

[a] Two cases are considered for the hot component of the ISM (HIM): (1) no HIM / (2) HIM based on MO. The second case accounts for some of the electrons observed in the dispersion measure of high altitude pulsars, thereby reducing the scale height of the WIM from 1 kpc for Case 1 to 780 pc for Case 2.

The characteristics of the HIM are more controversial, as indicated above. To gauge the effects of the HIM on the vertical structure of the ISM, we consider two cases: (1) The HIM is negligible, in the sense that its mean pressure is small compared to that of the other components; a corona supported by cosmic rays (Chevalier and Fransson 1984; Hartquist and Morfill 1986) is consistent with this case. (2) The HIM is similar to that envisioned by MO: in the plane, it has a temperature of about 5×10^5 K, a density $n_H \simeq 3\times 10^{-3}$ cm^{-3}, and a filling factor

f_h of about 0.5, so that the mean density in the plane is $\bar{n}_\mathrm{H} \simeq 1.5 \times 10^{-3}$ cm^{-3}. We adopt a scale height $H_h = 3$ kpc for the mean density, which is compatible with the adopted temperature (Bloemen 1987) and also with the observations of halo gas (Savage 1987). This model for the HIM contributes about 20% of the dispersion measure of the high altitude pulsars (see Table I; note that $\bar{n}_e = 1.2 \bar{n}_\mathrm{H}$ in the fully ionized HIM). As a result, since the density of the electrons in the WIM is assumed to obey equation (1), the scaleheight of the WIM is reduced accordingly, to $H_w \simeq 800$ pc. It should be noted that the density of the HIM adopted here is almost an order of magnitude less than the minimum value considered by Bloemen (1987) for $H_h = 3$ kpc; indeed, most of the densities he considered for this scale height have been eliminated by the observations of the globular cluster pulsars.

The accuracy of this inventory of the mass distribution is difficult to assess, but it is consistent with the available data on the column densities and scale heights of H$_2$, HI, and HII; the total column density of all the components in this model is 13 $M_\odot$ pc^{-2}, the same as that quoted by Kulkarni and Heiles (1987). Furthermore, the total midplane density is 1.1 cm^{-3}, in good agreement with the value 1.2 cm^{-3} found from an analysis of extinction data by Spitzer (1978). Our inventory of the mass is in reasonably good agreement with that of Bloemen (1987): he found a total neutral column of 10.8 $M_\odot$ pc^{-2} (using our He/H ratio), compared to the 11.1 $M_\odot$ pc^{-2} found here, but his estimate of the H$_2$ column was about 1 $M_\odot$ pc^{-2} higher, and that of the CNM column about 1 $M_\odot$ pc^{-2} lower, than ours. He did not consider the WIM, since he regarded its properties as too uncertain at that time. As noted above, many of the conditions he considered for the HIM have been eliminated by the recent observations of pulsars in globular clusters.

The large scale heights for some of the components of the ISM imply that the ISM is indeed heavy. Using Kuijken and Gilmore's (1989a) estimate for the gravitational acceleration above the disk of the Galaxy, McKee (1990) found that the total weight of the ISM is about 3×10^{-12} dyne cm^{-2}. If the ISM is in hydrostatic equilibrium, this must be balanced by a pressure of the same magnitude. This appears to be the case: From the data of Kulkarni and Fich (1985) and of Reynolds (1985), he inferred that the turbulent pressure in the Galactic plane is about 1.8×10^{-12} dyne cm^{-2}, corresponding to a mean one-dimensional velocity dispersion $\sigma \simeq 8$ km s^{-1}. Cosmic rays contribute a pressure 0.5×10^{-12} dyne cm^{-2}. The magnetic field was taken to have a uniform component in the plane of 2.5 μG and a random, isotropic component of the same magnitude, corresponding to an rms field of 3.5 μG. The random component of the field exerts less pressure than the uniform component (Parker 1969), so the effective field strength is about 3.0 μG and the magnetic pressure is only about 10% of the total. The strength of the interstellar magnetic field must be determined by observation, not theory, but this model shows that it is not necessary to assume a large value of the interstellar magnetic field in order to reconcile the observations of the pressure of the ISM with its weight. Bahcall (1984) has constructed a variety of models for the gravitational field of the Galaxy, all of which yield higher accelerations than the Kuijken and Gilmore model; had these been used, the agreement between the weight and the pressure would not have been as good, but it still would have been within the errors. In fact, observations show that some of the HI at the poles is falling towards the plane (Weaver 1974), so precise agreement between the weight and the pressure is not expected.

III. VERTICAL STRUCTURE OF THE ISM: ISOTHERMAL MODELS

In view of the inordinate complexity of the observed ISM, it is instructive to pull back and develop simple toy models to see what the ISM would be like if only a few processes were operating. Here we shall consider isothermal models for the vertical structure of the ISM. Let $C \equiv (P/\rho)^{1/2}$ be the isothermal sound speed for the ISM, where P is the total pressure and ρ the density. In order that the model not be too unrealistic, we shall include a magnetic pressure and a pressure due to cosmic rays in P. Let $g(z)$ be the magnitude of the z-component of the gravitational acceleration, and let $\bar{g}(z)$ be the mean value,

$$z\bar{g}(z) \equiv \int_0^z g(z')dz'. \tag{2}$$

Then the solution of the equation of hydrostatic equilibrium is

$$\rho(z) = \rho_0 \exp - \left[\frac{z\bar{g}(z)}{C^2}\right]. \tag{3}$$

If the gas disk is sufficiently thin and has a small fraction of the total mass, then the acceleration will be approximately a linear function of height, $g(z) \simeq z g_0'$, with the midplane gradient g_0' constant. In this case the density is a Gaussian,

$$\rho(z) = \rho_0 \exp -\frac{1}{2}\left(\frac{z^2}{H^2}\right), \tag{4}$$

where the scale height is

$$H = \frac{C}{g_0'^{1/2}} = 116 C_6 \quad \text{pc}, \tag{5}$$

with $C_6 \equiv C/(10^6 \text{ cm s}^{-1})$. The numerical value for the scale height is based on an acceleration gradient $g_0' = 2.4 \times 10^{-11}$ cm s^{-2} pc^{-1}; g_0' is proportional to the total density in the Galactic plane, and the adopted value is intermediate between those corresponding to densities of 0.185 $M_\odot$ pc^{-3} (Bahcall 1984) or 0.1 $M_\odot$ pc^{-3} (Kuijken and Gilmore 1989b).

a) One Phase: The Dead ISM

We begin with the simplest case, in which the rate of energy injection into the ISM is so low that the gas is all cold, say at a temperature of 80 K. If we assume that the thermal, magnetic, and cosmic ray pressures are all equal, then the isothermal sound speed in the medium is 1.25 km s^{-1}, and the scale height of the gas is only about 15 pc! In the absence of self-gravity, such an ISM would be lifeless, with a scale height which is small even compared to that of the molecular gas in the Galaxy. If the surface density were adequate for self-gravity to become important, structure would form in the disk, but the scale height would not grow significantly; hence, this model can be rejected outright.

b) Two Phases: The Quiescent ISM

Consider now a higher heating rate, so that some of the gas is at a temperature of order 10^4 K. It is generally possible for interstellar gas to have two phases, one warm and one cold, coexisting at the same pressure: The warm phase can be in equilibrium for thermal pressures less than $P_{th,max}$, whereas the cold phase can be in equilibrium for thermal pressures greater than $P_{th,min}$; a two-phase medium is possible provided $P_{th,max} > P_{th} > P_{th,min}$ (FGH). A recent calculation of the heating and cooling of the interstellar gas including heating due to the ejection of photoelectrons from grains and the damping of hydromagnetic waves found $\tilde{P}_{th,max} \simeq 2000$ cm^{-3} K (we use $\tilde{P} \equiv P/k$ for convenience; 2000 cm^{-3} K corresponds to 2.8×10^{-13} dyne cm^{-2}) and $\tilde{P}_{th,min} \simeq 150$ cm^{-3} K (Ferriere, Zweibel, and Shull 1989). However, since the typical value of the observed thermal pressure is about 3600 cm^{-3} K (Jura 1975), this model must be regarded as quite approximate.

The pressure of the gas is directly proportional to its surface density Σ. For a Gaussian density distribution, we have

$$\Sigma = \int_{-\infty}^{\infty} \rho(z)dz = (2\pi)^{1/2} \rho_0 H. \qquad (6)$$

Since the total pressure in the midplane is given by $P_0 = \rho_0 C^2$, equations (5) and (6) imply that the surface density that can be supported by a given pressure is

$$\Sigma = \frac{3.54}{C_6} \left(\frac{\tilde{P}_0}{6000 \text{ cm}^{-3} \text{ K}} \right) \; M_\odot \text{ pc}^{-2} \qquad (7)$$

The maximum possible surface density of warm gas occurs for $P_{th} = P_{th,max}$. For the purpose of illustration, consider the case in which $P_{th,max} \simeq 2000$ cm^{-3} K, the total pressure is three times the thermal pressure because of magnetic fields and cosmic rays, and the temperature of the warm gas is 6000 K (Spitzer 1978); the resulting maximum possible surface density of warm gas is 3.3 $M_\odot$ pc^{-2}, somewhat less than that observed in the ISM ($\simeq 5.5$ $M_\odot$ pc^{-2} from Table I).

For the parameters we have adopted, the two-phase ISM would have the following structure: Gas in excess of 3.3 $M_\odot$ pc^{-2} would be confined to a cold, thin disk, as in the one-phase model described above; the disk would be about 30 pc thick, and its thermal pressure would exceed $P_{th,max}$. At a height above the plane at which the thermal pressure has dropped below $P_{th,max}$, a two-phase structure would become possible, with cold clouds embedded in a warm intercloud medium. The scale height of the warm gas would be about 130 pc. Clouds could continue to exist up to heights somewhat greater than this, until the thermal pressure dropped below $P_{th,min}$; for $\tilde{P}_{th,min} = 150$ cm^{-3} K, this height is about 300 pc. At greater heights, the gas would be homogeneous and warm. If not supported by turbulent motions or magnetic fields, clouds which formed in the two-phase zone would rain down on the cold disk. A drought would ensue—i.e., the pressure at the surface of the cold disk would drop to $P_{th,min}$, eliminating the two-phase zone— unless disturbances occurred which could drive the gas near the surface of the disk back into the warm phase, replacing the fallen cloud material.

This model is a substantial improvement over the one-phase model, but it is a far cry from reality. The possibility that clouds could exist in the two-phase zone above the cold disk could increase the scale height of the cold gas somewhat above 15 pc, but it would still be much less than the observed 135 pc. Warm gas is observed to have an effective scale height $\Sigma/2\rho_0$ of 430 pc (see Table I), several times greater than this idealized model predicts. This discrepancy is due to the fact that the model has no turbulence; it is quiescent, in contrast to the observed ISM. It is possible to add turbulence to a two-phase model of the ISM, but since the dominant source of interstellar turbulence is supernovae (Spitzer 1978), which produce large volumes of hot gas, we consider turbulence in the context of a three-phase model.

c) Three Phases: The Violent ISM

The actual ISM is rent by stellar winds and supernova explosions, leading McCray and Snow (1979) to term it the "violent ISM". Supernovae are the dominant source of energy; they heat large volumes of the ISM and agitate it, inflating the ISM to its observed height. Cold clouds have a scale height much greater than 15 pc because of their turbulent motions. The fact that WNM is observed near the plane shows that, in some parts of the plane at least, the thermal pressure is less than $P_{th,max}$. The total pressure in the plane is much greater: McKee (1990) estimates $\bar{P}_0 \simeq 2.2 \times 10^4$ cm^{-3} K. The scale height for the hot gas at a temperature of 5×10^5 K, the temperature calculated by MO, is about 3 kpc (this is higher than implied by equation [5] because the hot gas extends well above the stellar disk, so that $g \ll g'_0 z$). If magnetic fields or cosmic rays contribute to the support of the hot gas, the scale height would be yet higher. The hot gas can confine clouds far from the plane, as envisaged by Spitzer (1956); absorption line observations indicate that these clouds extend to about 3 kpc from the plane. Since most of the cold clouds are close to the plane, the gas in the lower halo (the "extracloud layer" in Cox's [1989] terminology) forms a *two-phase* medium consisting of the warm medium and the HIM. In the Galactic plane, supernova remnants create a third, hot phase of the ISM. There is general agreement, both theoretical and observational, that the third phase exists, but there is a debate, as yet unresolved by observation, as to its filling factor.

IV. THE SNR FILLING FACTOR

A central issue in the theory of the ISM is the determination of f_h, the filling factor of hot, low density gas generated by supernova remnants. Cox and Smith (1974) were the first to show that this filling factor might be substantial, and MO argued that this conclusion was quite robust. Here we shall go through the arguments leading to an estimate of f_h carefully, highlighting the points of controversy.

Consider that subset of supernovae which are effectively random in space and time; this includes supernovae from low-mass progenitors (Type Ia) and those supernovae from high-mass progenitors which are not in large associations. The rate of these supernovae per unit volume is denoted by S. Let $V(t)$ be the volume of an isolated SNR as a function of its age t. For a one-phase or two-phase ISM, this volume first increases as the remnant expands, and then decreases as the hot gas cools off. In a three-phase ISM, the remnants cannot be approximated as isolated; the SNR expansion is well defined, but thereafter the remnant merges into the hot medium. Let $dQ(t)$ be the probability that a given point is in a remnant of age between t and $t+dt$; in terms of the SN rate, we have $dQ(t) = SV(t)dt$. The expected number of SNRs younger than t encompassing a given point is then

$$Q(t) = S \int_0^t V(t')dt'. \tag{8}$$

The expected number of SNRs of *any* age which encompass a point is $Q(t \to \infty)$; we denote this by Q, without an argument. Q is sometimes termed the "porosity" of the medium (Cox and Smith 1974). Since $V(t)$ is not well-defined for an SNR in a three–phase medium after its expansion stops, Q is not precisely defined in such a medium, but it is large: any point may be regarded as being in an SNR of some age. In the absence of interactions among the SNRs, the probability of not being in any remnant is $\exp(-Q)$, and the probability of being in at least one SNR is $1 - \exp(-Q)$. In a multi–phase medium, the filling factor of hot gas, f_h, is that fraction of the volume of SNRs not in clouds, either cold or warm:

$$f_h = [1 - \exp(-Q)](1 - f_{c+w}). \tag{9}$$

The filling factor of the clouds in the SNR, f_{c+w}, itself varies with the age of the SNR, so the appropriate mean must be used in equation (9).

The meaning of Q may be clarified by considering a related quantity, the expected number of SNRs of volume $\leq V$ which engulf a given point in a time interval Δt. This expected number is simply $SV\Delta t$; for any t such that the SNRs are still expanding, $SV(t)t > Q(t)$. An example which emphasizes the distinction between $SV\Delta t$, the expected number of SNRs to overtake a point in Δt, and $Q(t)$, the expected number of SNRs engulfing a point at a given time, is the case of hypothetical SNRs that expand essentially instantaneously beyond $V(t)$ and then disappear. In this case $SV\Delta t$ is finite, but $Q(t)$ vanishes. We see that Q is directly proportional to the *volume* in space–time occupied by SNRs, $\mathcal{V} \equiv \int V dt$.

a) The SNR Four-Volume

We can estimate the volume in space–time (the four–volume) occupied by an SNR with a simple model in which the SNR expands as a power–law in time, $R \propto t^\eta$, until it reaches a radius R_m at time t_m; at this point, its expansion velocity v_m is a factor β times the ambient isothermal sound speed C_0,

$$v_m = \frac{\eta R_m}{t_m} \equiv \beta C_0. \tag{10}$$

Recall that $C_0 \equiv (P_0/\rho_0)^{1/2}$, where P_0 is the total pressure, including turbulent pressure; in a multiphase ISM, P_0 and ρ_0 are measured in the intercloud medium. The numerical factor β is uncertain, and in principle depends on the morphology of the magnetic field and on the details of the interaction of the SNR with cosmic rays. We shall generally set $\beta \simeq 1$, except that for a strongly magnetized ISM we shall take $\beta = \sqrt{2}$, so that βC_0 is the Alfven velocity. At late times an SNR in a one or two phase medium consists of a bubble of hot gas surrounded by a cold shell of swept–up gas; if the shell is thick, as in the case of a highly magnetized ISM (Cox and Slavin 1989), we take R_m to be the radius of the hot bubble, corresponding to the inner radius of the shell.

After the expansion has ceased, the SNR will subsequently contract provided the filling factor of the hot gas is not too large (the importance of this contraction was emphasized by Heiles [1987] and by B.–C. Koo, private communication). Under the assumption that this contraction occurs at a constant velocity over a time $t_{\rm con}$, the total four-volume of the SNR is

$$\mathcal{V} = \left(\frac{V_m t_m}{3\eta + 1} + \frac{V_m t_{\rm con}}{4} \right) \equiv q V_m t_m, \tag{11}$$

where $V_m = (4\pi R_m^3/3)$ and q is a numerical coefficient of order unity. If the contraction occurs at the velocity v_m, we have

$$q = \frac{1}{3\eta + 1} + \frac{1}{4\eta}. \tag{12}$$

This model is certainly over-simplified: it ignores the transition from expansion to contraction, which is accompanied by an overshoot in which the internal pressure drops below the ambient value (e.g., Ostriker and McKee 1988), and which would tend to increase Q; and it does not include the effects of the embedded clouds, which tend to destroy the remnant from within toward the end of its evolution, and which would tend to decrease Q.

Cioffi, McKee, and Bertschinger (1988) have studied the late evolution of spherically symmetric SNRs in a uniform medium. They found that the expansion can be approximated with $\eta = 0.3$ at late times. The maximum radius of the remnant is

$$R_m = 77.3 \left(\frac{E_{51}^{0.316}}{n_0^{0.153} \zeta_m^{0.051} \beta^{0.429} \tilde{P}_{04}^{0.214}} \right) \text{ pc}, \tag{13}$$

which is reached at a time

$$t_m = 2.97 \times 10^6 \left(\frac{E_{51}^{0.316} n_0^{0.348}}{\zeta_m^{0.051} \beta^{1.429} \tilde{P}_{04}^{0.714}} \right) \text{ yr}, \tag{14}$$

where n_0 is the ambient density of hydrogen nuclei, $\tilde{P}_{04} = \tilde{P}_0/(10^4 \text{ cm}^{-3} \text{ K})$, and ζ_m allows cooling rate to depend on the metallicity of the gas; we shall take $\zeta_m = 1$ in our numerical estimates. With $q = 1.36$ from equation (12), the SNR four-volume is

$$\mathcal{V} = 7.82 \times 10^{12} \left(\frac{E_{51}^{1.26}}{n_0^{0.110} \zeta_m^{0.204} \beta^{2.72} \tilde{P}_{04}^{1.36}} \right) \text{ pc}^3 \text{ yr}. \tag{15}$$

This result is quite close to the value found by MO over a decade ago, aside from the factor $q \simeq 1.36$. Note in particular that $\mathcal{V}$ is insensitive to the ambient density n_0. On the other hand, the four-volume is sensitive to the parameter β; for the case in which the interstellar magnetic field does not dominate the pressure, we assume that the expansion of the hot SNR bubble stops when the velocity drops to the ambient isothermal sound speed, so that $\beta \simeq 1$. For a *quiescent* (i.e., non-turbulent) two-phase ISM in which the mean intercloud density is 0.3 cm^{-3} and the ambient pressure is $\tilde{P}_{04} = 0.62$ (comprised of a thermal pressure $\tilde{P}_{th} = 3600$ cm^{-3} K and a magnetic pressure corresponding to a 3 μG field, but omitting the pressure of the cosmic rays, which have an uncertain effect), the four-volume of an SNR is $\mathcal{V} = 1.7 \times 10^{13} E_{51}^{1.27}$ pc^3 yr.

Recently, Cox and Slavin (1989) have studied the evolution of SNRs in a strongly magnetized ISM in which the magnetic pressure is 7200 cm^{-3} K (corresponding to $B = 5$ μG) and the thermal pressure only about 700 cm^{-3} K. For an assumed SNR energy of 5×10^{50} erg, they found that the hot interior of the SNR expanded to a radius of 65 pc at $t \simeq 8 \times 10^5$ yr, and then stalled; the bubble eventually collapsed at a velocity of about 20 km s^{-1}. Altogether, they estimated $\mathcal{V} \simeq 3 \times 10^{12}$ pc^3 yr. Using equation (11), and even allowing the bubble to stall at R_m for 5×10^5 yr, we estimate a somewhat smaller value, $\mathcal{V} \simeq 2 \times 10^{12}$ pc^3 yr. The analysis of Cioffi et al. (1988) did not explicitly include a magnetic field, so its application to this problem must necessarily be approximate. The analysis did include a detailed accounting for the energy losses experienced by the hot interior due both to adiabatic expansion and radiative losses, however, so we are encouraged to try it. Since the ISM in the Cox and Slavin model is strongly magnetized, we set βC_0 equal to the Alfven velocity ($\beta \simeq \sqrt{2}$). With $\tilde{P}_{04} = 0.8$ and $n_0 = 0.1$ cm^{-3}, equation (15) gives $\mathcal{V} = 2.2 \times 10^{12}$ pc^3 yr, for the four-volume of the hot interior of the SNR, remarkably close to Cox and Slavin's result. Equation (13) gives $R_m = 80$ pc (somewhat too large), equation (14) gives $t_m = 7.7 \times 10^5$ yr (essentially correct), and equation (12) with $\eta = 0.3$ gives $q = 1.36$ (somewhat too small). Had Cox and Slavin used a supernova energy of 10^{51} erg, as is appropriate for SN 1987A at least (Chevalier 1990), they would have found $\mathcal{V} = 5.3 \times 10^{12}$ pc^3 yr. Comparing this with the low-field case above, we see that the higher field decreases the four-volume of an SNR by a factor of about 3. These estimates should be treated with caution, however; Cox and Slavin's calculations (and Cioffi et al.'s) were spherically symmetric and thus could not treat the multi-dimensional effects associated with magnetic fields.

b) The Porosity of the ISM, Q

The porosity of the ISM is simply $Q = S\mathcal{V}$, the product of the SN rate per unit volume and the SNR four-volume. The Galactic SN rate $\nu_{\rm Gal}$ is discussed in the Appendix. Evidence from historical SN and from the observed rate of massive star formation suggests that $\nu_{\rm Gal} \simeq 0.022$ yr^{-1} is a conservative estimate of this rate. This value is consistent with the rate estimated from extragalactic SN by Tammann (1982) and by van den Bergh (1990) provided the Hubble constant is in the range 50–70 km s^{-1} Mpc^{-1} (if the Hubble constant is larger than this, the implied Galactic SN rate increases as h^2). The corresponding SN rate per unit area at the solar circle is $S_{\rm 2D}(R_0) \simeq 2.6 \times 10^{-11}$ pc^{-2} yr^{-1}.

To determine S, we must determine the fraction of the SN that are effectively random in space and time, and their scale height (correlated SN are discussed below). According to Evans, van den Bergh, and McClure (1989), almost 90% of SN are of Types Ib or II, so we focus on them. These stars are believed to have massive progenitors. Humphreys and McElroy (1984) surveyed over 5000 OB stars, and found them to be nearly evenly divided between those in associations and those in the field; we therefore assume that about half of the supernovae are random. Heiles (1990) comes to the same conclusion based on an analysis of the size of an association required to create a superbubble; he estimates that associations containing less than about 20 massive stars have supernovae that are effectively random. The observed scale height of OB stars is about 60 pc (e.g., Allen 1963). The scale height of these stars when they die might be somewhat larger; indeed, the scale height of the youngest pulsars is about 150 pc (Taylor and Manchester 1977). We shall adopt the latter value as a reasonable estimate, which gives

$$S = 0.5 \left(\frac{2.6 \times 10^{-11} \text{ pc}^{-2} \text{ yr}^{-1}}{300 \text{ pc}} \right) = 4.4 \times 10^{-14} \text{ pc}^{-3} \text{ yr}^{-1} \quad (16)$$

for the solar neighborhood. This is about half the nominal rate $S = 10^{-13}$ pc^{-3} yr^{-1} adopted by MO, primarily because we are excluding correlated SN from the rate. To be consistent, we should then exclude the volume occupied by the correlated supernovae as well; since their filling factor appears to be about 0.1–0.2 (see below), this would increase the estimate of S in equation (16) by 10–20 %. Note that the conversion from the SN rate per unit area, $S_{2D}(R_0)$, to the SN rate per unit volume, S, is essentially the same for SN Ia as for the massive SN: for SN Ia, all the SN contribute to the porosity, but their scale height is larger (Heiles [1987] estimates 325 pc). Thus, if SN Ia make a larger contribution to the SN rate than estimated by Evans et al., the value of S in equation (16) would be unaffected.

A completely independent estimate of S can be obtained from observations of radio SNRs. Large, shell-like radio SNRs are interacting with a reasonably dense ISM, and therefore contribute to the porosity. Caswell and Lerche (1979) find that the observed remnants have a scale height of 80 pc and a birthrate of 0.005 yr^{-1}. They argue that remnants above the plane evolve more rapidly and are therefore undercounted; allowing for these SNRs ups the birthrate to 0.0125 yr^{-1} and the scale height to 200 pc, leaving S unchanged at $S = 3.7 \times 10^{-14}$ pc^{-3} yr^{-1}. This is probably a lower bound on the true value of S because regions of lower than average density are likely to underrepresented in radio surveys; it is quite close to the conservative estimate made in the previous paragraph.

For the typical quiescent two-phase medium described below equation (15), the resulting porosity is $Q = SV = 0.75$. For such a large value, the turbulent pressure is significant (see §V below), which will tend to reduce the SNR four-volume V; on the other hand, the filling factor of hot gas, f_h, is larger than the estimate in equation (9), since, as Cox and Smith (1974) indicated, interactions among SNRs help to maintain the hot gas. The gas pressure in the ISM is correlated with the number of SNRs present: Regions with no SNRs have a lower pressure than the typical value, making it easier for SNRs to expand into them, and regions with more than one SNR will expand to a larger size than given by equation (13). In his simulation, Smith (1977) found that f_h is slightly greater than Q, rather than the $1 - \exp(-Q)$ expected in the absence of interactions. The high magnetic field in Cox and Slavin's (1989) model reduces the porosity considerably, to $Q \simeq 0.23$ (assuming $E_{51} = 1$) or 0.12 (for their value, $E_{51} = 0.5$). It appears that additional evidence, both observational and theoretical, must be brought to bear in order to determine the filling factor of hot gas in the ISM; new theoretical evidence will be presented in §V.

c) Correlated Supernovae and the Formation of Superbubbles

One of the major advances in our understanding of the ISM in the past decade is the realization that the clustering of supernovae in stellar associations can have a dramatic effect on the ISM. HI shells far too large to have been created by individual supernovae have been observed throughout the Galaxy, with large shells and supershells occurring primarily outside the solar circle (Heiles 1979) and vertical structures ("worms") occurring inside (Heiles 1984). A number of authors, beginning with Bruhweiler et al. (1980), have attributed these structures to energy injection by associations of massive stars, leading to the creation of superbubbles in the ISM. A discussion of this topic is beyond the scope of this paper; a thorough review has recently been given by Tenorio–Tagle and Bodenheimer (1988). It should be noted that a number of the theoretical models of superbubbles assumed that the gas disk of the Galaxy is thin, so that the energy could easily break out into the halo; however, the observations discussed in §II above show that the gas disk is in fact rather thick, so that most superbubbles should remain confined (Cox 1989). Observationally, superbubbles appear to occupy only about 10–20% of the area of the Galactic plane in the solar neighborhood (Heiles 1980), leaving most of the volume available for a two– or three–phase ISM.

V. INTERSTELLAR TURBULENCE AND THE POROSITY OF THE ISM

The observations of the ISM described in §II show that the scale height of the gas is several times greater than can be explained in the simple one– or two–phase models of the ISM considered in §III above. Observationally, the large scale height is associated with a high degree of turbulence in the ISM. Here we show that the level of turbulence in the ISM is directly associated with the porosity of the ISM (McKee 1990).

Consider supernovae occurring randomly at a rate S per unit volume. Assume the porosity Q is small, so that interactions can be neglected and the SNRs expand into the warm intercloud medium. Furthermore, assume that thermal pressure forces dominate the evolution of the SNRs; the remnants are small compared to a scale height, so that gravity may be neglected, and the magnetic field does not dominate the pressure. After becoming radiative, the SNRs slow down until they merge with the ISM, carrying a radial momentum $(\rho_0 V_m)\beta C_0$ (see §IVa). This momentum is approximately conserved, and is concentrated in a thin shell which expands at about the sound speed (see Landau and Lifschitz 1959), which we take to be βC_0. The shell overlaps another shell at a radius R_{ov} (corresponding to a four-volume $\mathcal{V}_{ov}$) given by

$$S\mathcal{V}_{ov} = \frac{S}{4}\left(\frac{4\pi R_{ov}^3}{3}\right)\left(\frac{R_{ov}}{\beta C_0}\right) \equiv 1, \quad (17a)$$

from equation (11); in this case, $\eta = 1$. Numerically, we have

$$R_{ov} = 93.1 \left(\frac{\beta}{S_{-13}}\right)^{1/4} \left(\frac{\tilde{P}_{04}}{n_0}\right)^{1/8} \text{ pc}, \quad (17b)$$

where $S_{-13} \equiv S/(10^{-13}\text{ pc}^{-3}\text{yr}^{-1})$. We assume that the momentum is annihilated at R_{ov}. We obtain an upper limit on the turbulent pressure by assuming that the momentum flux is carried by the dynamic pressure, ρv^2. Approximating the shell as a delta function, we can express the momentum flux density as

$$\rho v^2 = \left(\frac{\rho_0 V_m \beta C_0}{4\pi r^2}\right) \delta\left(t - \frac{r}{\beta C_0}\right) \quad (18)$$

Averaging this over space and time yields

$$\int \rho v^2 \, dV \, dt = \langle \rho v^2 \rangle V_{ov} = \frac{\langle \rho v^2 \rangle}{S} = R_{ov} \rho_0 V_m \beta C_0. \quad (19a)$$

The turbulent pressure is proportional to the one–dimensional velocity dispersion, $P_{turb} = \langle \rho v^2 \rangle / 3$; numerical evaluation yields

$$P_{turb} = 1.05 \times 10^{-12} \frac{(S_{-13} E_{51}^{1.26})^{3/4}}{n_0^{0.084} \zeta_m^{0.153}} \quad \text{dyne cm}^{-2}, \quad (19b)$$

where the dependence on $\tilde{P}_{04}$ and β is so weak (exponent less than 0.04) that it has been omitted. We see that the level of turbulence in the ISM depends on the rate of energy injection, but is almost independent of everything else.

To compare the turbulent pressure with the total pressure in the ISM, we note that

$$Q = SV = qSV_m t_m = qSV_m \left(\frac{\eta R_m}{\beta C_0}\right) \quad (20)$$

from equations (10) and (11), so that

$$\frac{\langle \rho v^2 \rangle}{\rho_0 C_0^2} = \frac{\beta^2}{q} \left(\frac{Q R_{ov}}{\eta R_m}\right). \quad (21)$$

Eliminating R_{ov}/R_m with the aid of equations (17) and (20), we then find

$$\frac{P_{turb}}{P_0} = \frac{4^{1/4} \beta^2}{3} \left(\frac{Q}{q\eta}\right)^{3/4} \simeq 0.9 Q^{3/4}, \quad (22)$$

where the numerical evaluation is for the values used in §IV. This estimate of the turbulent pressure does not include the contribution of the motions due to SNRs prior to merging with the ISM or during their subsequent collapse (§IVa), but one can show that this contribution is of higher order and does not alter the result in equation (22) when Q is small. We conclude that *the turbulent pressure due to explosive energy injection is significant if and only if the resulting porosity is large.* Thus, provided the interstellar magnetic field is not too strong, the observation that the gas disk of the Galaxy is thick and highly turbulent implies that the porosity of the ISM is large. In their calculation of the turbulent velocity of the clouds, MO found $P_{turb} \simeq P_{th}$, consistent with the large porosity they inferred. More generally, if the turbulent pressure is about half the total pressure, then equation (22) implies that $Q \simeq 0.5$. It is unlikely that the conclusion that Q is large can be avoided by appealing to superbubbles: the fact that the ISM extends to great heights implies that most superbubbles are confined within the ISM (Cox 1989), so the relation between the turbulent pressure and the porosity given above should apply to them as well, at least approximately. Since superbubbles are observed to have a small filling factor (Heiles 1980), they cannot account for the observed level of interstellar turbulence.

What if the interstellar magnetic field is large? In that case, a fraction of the energy from the supernovae can be stored in the oscillations of the field (Cox 1988). Since $\beta^2 = 2$ in this case, equation (22) implies $Q \simeq 0.2$. However, if the interstellar field is large, it must be tangled on scales $\lesssim 100$ pc, since the mean field observed toward pulsars is $\lesssim 2.5$ μG, and perhaps as low as 1.6 μG (Rand and Kulkarni 1989). Insofar as the tangled field is isotropic, it would contribute a pressure of only 1/3 that of an ordered field, so it is difficult to see how such a field could inflate the ISM to the required height.

Salpeter, and Terzian 1983), so that low mass clouds would have a larger radius than in the MO model. The average column density (weighted by the cloud area) is 2.7×10^{20} cm^{-2}, close to Spitzer's (1978) "standard cloud" with 3.6×10^{20} cm^{-2}.

Energy dissipation in the disk. An essential feature of the three–phase model is that the SNR energy is dissipated in the disk, not in the halo as in galactic fountain models. The model is thus consistent with the failure to detect soft X–ray emission from two edge–on spirals by Bregman and Glassgold (1982). Songaila, Bryant, and Cowie (1989) have recently pointed out that their upper limit on the ionization of high velocity clouds severely constrains the energy dissipated in a Galactic fountain. Cowie (1987) has suggested that the gas observed in O VI absorption lines cannot cool in the disk, but must cool in the halo. However, his argument does not apply to the three–phase model. MO explicitly demonstrated that energy balance occurs in the disk: evaporative cooling lowers the temperature below the nominal 10^6 K Cowie assumed (in fact, evaporative cooling dominates radiative cooling throughout the HIM— Begelman and McKee 1990); and the rate of radiative cooling is enhanced by a factor $\beta_{MO} \sim 10$ over the value assumed by Cowie because of density fluctuations and, more important, non–equilibrium cooling in the freshly evaporated gas. A more precise numerical calculation, including an accurate evaluation of β_{MO} and allowing for deviations from spherical symmetry, would be worthwhile in order to verify these points.

Vertical structure of the ISM. The ISM is observed to extend to great heights above the plane and to be highly turbulent (§II); indeed, the turbulent motions are largely responsible for supporting the gas (Kulkarni and Fich 1985, McKee 1990). As discussed in §V, the high degree of turbulence in the ISM demands that the porosity of the ISM be large (McKee 1990), as required for the three–phase model. MO showed that the three–phase model can account for the observed velocity dispersion in the clouds; they did not attempt to calculate the velocity dispersion of the stripped cloud envelopes, most of which will be part of the WIM.

c) Comparison with Observation: Unresolved Issues

There are a number of unresolved issues in the comparison between the three phase model and observation. Some of these issues are fundamental to any theory, such as the topology of the clouds, and the strength and topology of the magnetic field. The issues of particular relevance to the three–phase model are:

O VI absorption lines. The observation of widely distributed absorption by highly ionized ions in the ISM (Jenkins 1978) is one of the strongest pieces of evidence for a widely distributed hot component of the ISM. MO attributed these absorption lines to conductive interfaces of clouds embedded in the hot gas, and the case for this interpretation has been strengthened by the work of Cowie et al. (1979). In his recent review of hot gas in the Galaxy, Spitzer (1990) does not regard the case as conclusive, however. Böhringer and Hartquist (1987) and Borkowski, Balbus, and Fristrom (1990) have found that the column density of O VI in a typical conductive interface is about 10^{13} cm^{-2}, just as inferred by Jenkins (1978). This all appears consistent with the three–phase model; the difficulty is that MO predicted about 10 times as many interfaces, each with one tenth the column density. In the MO model, the conductive interfaces surround the WIM envelopes of the clouds; since the properties of the WIM and the number density of clouds appear to be in good agreement with observation (see above), this appears to be a fundamental problem. On the other hand, the global rate of evaporation inferred from observation is not that different from the model: Cowie (1987) infers a volume–averaged evaporation rate of $\dot{n} \simeq 2 \times 10^{-9}$ cm^{-3} yr^{-1}, whereas the MO model has $\dot{n} = 7.5 \times 10^{-9} S_{-13}$ cm^{-3} yr^{-1}; if, in contrast to our normal practice in this section, we adopt the estimated SN rate in equation (16), we obtain $\dot{n} \simeq 3 \times 10^{-9}$ cm^{-3} yr^{-1}, in reasonable agreement with Cowie's estimate.

There are several possible resolutions of the discrepancy between the predicted and observed number of O VI features. One possibility is that each feature seen by Jenkins is actually the set of features in an individual SNR; however, it is then difficult to account for the small velocity dispersion in each oberved feature. Another possibility is that the number of features is substantially smaller than estimated by MO because the warm clouds are much larger than in the MO model (Cowie 1987). In fact, McKee and Ostriker have realized that a proper calculation of the conductive interfaces in the MO model would indeed increase their effective size: since the filling factor of the warm cloud envelopes is typically 0.23, the isotherms in the hot gas would percolate and form structures with radii of curvature large compared to that of an individual cloud. A quantitative evaluation of this idea remains to be carried out, however.

An alternative model has been suggested by Cox and Slavin (1989): the observed lines occur in cooling, magnetically confined SNR bubbles. However, there appear to be too few of these to account for the observations: The probability per unit length of intersecting an SNR between the ages t and $t + dt$ is $SA(t)dt$, where $A(t) = \pi R(t)^2$ is the area of the SNR. Following the approach laid out in §IVa, we find that the number of SNRs of any age per unit length is

$$\frac{1}{\lambda} = S \int_0^\infty A(t')dt' = S \left(\frac{A_m t_m}{2\eta + 1} + \frac{A_m t_{\text{con}}}{3} \right), \quad (23)$$

where $A_m = \pi R_m^2$. With the value of S from equation (16) (which is essentially the same as that needed by Cox and Slavin to account for the mean density of O VI observed by Jenkins [1978]), and the bubble dynamics described in §IVa, the Cox and Slavin model predicts that the nearest OVI feature along a typical line of sight is at a distance $\lambda \simeq 800$ pc, far greater than the 170 pc inferred by Jenkins. Cox and Slavin suggest that this discrepancy would be reduced if the SNRs are highly distorted, but this remains to be demonstrated. Spitzer (1990) has pointed out another difficulty with this model: it predicts that several broad lines should be seen in the Jenkins data, but none are observed.

The filling factor of hot gas, f_h. The most critical comparison between the three–phase model and observation is the value of the filling factor of the hot gas. Locally, f_h is large, but the mean density $\bar{n}_H$ is substantially lower than average, so this cannot be regarded as typical (Cox and Reynolds 1987). Two estimates of f_h have been made under the assumption that the hot phase displaces the cooler phases over large volumes (Jenkins 1978, Heiles 1980); these estimates (both of which give $f_h \sim 0.1 - 0.2$) do not apply to the three–phase model because the HIM is well mixed with the other phases. Heiles's estimate, which is based on the absence of H I, does provide an estimate of the filling factor of superbubbles in the solar neighborhood, however. Conductive interface models for the O VI absorption lines have been interpreted as implying $f_h \sim 0.1 - 0.2$ (Cox and Slavin

1989, Spitzer 1990), but until the origin of these lines is understood this argument must be regarded as inconclusive. From an analysis of Reynolds' observations of the WIM (e.g., Reynolds 1989b), Kulkarni and Heiles (1987) have estimated the filling factor of the hot gas well above the Galactic plane as about 0.5. Finally, Van Buren (1989) has developed a clever method for inferring the filling factor of the infrared cirrus based on the heating of the dust by nearby stars, and finds that it occupies only about 20% of the volume; the remainder of the volume has little dust, and is consistent with being HIM. This technique has the promise of providing the most accurate measure of the filling factors of the various phases of the ISM, particularly if it can be extended to the gas phase (i.e., if it is possible to measure the mean density of gas in the vicinity of an appropriate random sample of main sequence stars by observing the effect of the local stellar radiation field on the gas). Determining these filling factors is one of the most important problems in interstellar astrophysics today.

The WNM. Studies of the heating and cooling of the ISM show that warm ($T \gtrsim 10^{3.5}$ K) neutral gas can exist in equilibrium only below a maximum thermal pressure $P_{th,max}$, and this plays a crucial role in two–phase models of the ISM (see §III). The actual value of $P_{th,max}$ depends on the heating mechanism and on the depletion of the coolants, and is uncertain. MO assumed that soft X–rays were the dominant heating agent, and found that the pressure was low enough for WNM to exist in about half the volume of the ISM; in this half, the average filling factor of the WNM is 0.36. Because the pressure is lower than average in this half of the ISM, the density of the WNM is low, $n_H(WNM)=0.16$, and consequently the mean density averaged over the entire volume of the ISM is low as well, $\bar{n}_H(WNM)=0.03$ cm^{-3}. Including the neutral hydrogen in the WIM brings the total density of warm H I to about 0.04 cm^{-3}. Reference to Table I shows that this is lower than observed by a factor of about 4, as recognized by MO. This problem has been reduced, and perhaps eliminated, by the recent work of Wang and Cowie (1989): using more modern estimates of the heating rate of the WNM and making a more detailed model for the effect of time dependent pressure fluctuations, they found a significantly larger filling factor for the WNM. MO predicted that the WNM should exist in layers surrounding cold cloud cores; Payne et al. (1983) have found that such a model is indeed consistent with 21 cm observations, but substantially more WNM was required than in the MO model (as expected). In addition, they found that there is a substantial component of WNM not associated with cloud cores; much of this is presumably the high–z H I described in §II. All in all, with the Wang and Cowie modification, the three–phase model may account for the WNM reasonably well.

Evaporation and SNR Evolution. One of the novel aspects of the MO model was a new theory of SNR evolution: because the ambient density is so low, the gas inside the SNR is evaporated cloud material rather than swept up ambient gas. This model predicts that the mean density inside the SNR should drop with time; in practice, this sets in for radii greater than about 20 pc (Cowie et al. 1981). Unfortunately, there is little direct evidence for the importance of evaporation in SNRs (McKee 1982, 1988). A major reason for this is that observed SNRs from massive progenitors, which comprise the bulk of the SNRs (Evans et al. 1989), are interacting primarily with gas processed by the progenitor star rather than with the pristine ISM (McKee 1988). One of the few examples in which evaporating gas may have been identified in an SNR is in Puppis A (Teske and Petre 1987). Most models for the O VI absorption lines involve evaporation of clouds (MO) or a shell (Cox and Slavin 1989), but in view of the uncertainties described above, this does not allow us to infer the effects of evaporation on SNR evolution. On the other hand, if isolated SNRs evolve in a two–phase ISM (e.g., Cox 1988), then they should be visible in emission (Cowie and York 1978a, McKee 1982): when the shock velocity is 100 km s^{-1}, the emission measure averaged over the shell is about $75n_0$ cm^{-3} pc, which is readily detectable for $n_0 \gtrsim 0.1$; with 1.3×10^{-11} uncorrelated SNRs pc^{-2} yr^{-1} (Appendix A), there should be about 15 such remnants within 1 kpc for $n_0 \sim 0.1$ and $E_{51} \sim 1$. The fact that such remnants have not been detected weighs against a two–phase model of SNR evolution. Evaporation plays a key role in making the hot gas stable (MO, Cowie et al. 1981, Begelman and McKee 1990), so direct observations of evaporation in the ISM are of critical importance.

VII. CONCLUDING REMARKS

This review has been somewhat like a trip through time, seeing how our knowledge and understanding of the diffuse interstellar medium has evolved over the past decade. Theoretical models, such as the three–phase model, are useful insofar as they organize existing knowledge and suggest new observations and experiments; based on the latter criterion, the model has been highly successful. It is gratifying that some of the properties of ISM that have become apparent only recently, such as the thickness of the gas disk, have a natural explanation in terms of the three–phase model: In order to lift the gas to heights well above its natural scale height (even allowing for magnetic and cosmic ray pressure), an enormous injection of kinetic energy is required; since supernovae are the dominant source of kinetic energy in the ISM, their remnants must have a large filling factor f_h if they are to supply the required turbulence. Since f_h is large, the ISM in the plane is in three phases, cold, warm, and hot; in the lower halo, above the cold clouds, this argument suggests that the ISM should be in two phases, warm and hot. On the other hand, it is disconcerting that one of the observational underpinnings of the three–phase model, the O VI absorption lines, has yet to receive a comprehensive, quantitative explanation. Furthermore, uncertainties in the magnitude and structure of the magnetic field in the diffuse ISM lead to a significant uncertainty in the theoretically expected value of f_h. The observations anticipated over the next decade should shed new light on the structure and evolution of the ISM, providing fresh challenges for the three–phase model and its successors.

APPENDIX A: THE GALACTIC SUPERNOVA RATE

An accurate determination of the Galactic supernova rate is essential if we are to understand the effect of supernova energy injection on the interstellar medium. Indeed, as Heiles (1987) has emphasized, it is important to know not only the global rate $\nu_{\rm Gal}$, but also the rate per unit volume, S, and per unit area, $S_{\rm 2D}$, throughout the Galaxy. Fortunately, recent work on this problem, particularly by van den Bergh and his collaborators, enables us to make a somewhat more accurate estimate than in the past.

Following Ratnatunga and van den Bergh (1989), we assume that the SN rate scales with the density of the older disk stars, $S_{2D}(R) \propto \exp(-R/H_R)$, where $H_R \simeq 4$ kpc (Freeman 1987) is the disk scale length. The Galactic SN rate is then related to the local rate per unit area, $S_{2D}(R_0)$, by an effective area A:

$$\nu_{\text{Gal}} = \int_0^\infty S_{2D}(R) 2\pi R dR \equiv A S_{2D}(R_0), \qquad (A.1)$$

with

$$A = 2\pi H_R^2 \exp(R_0/H_R). \qquad (A.2)$$

For a Galactocentric distance $R_0 = 8.5$ kpc (Kerr and Lynden–Bell 1986), this effective area is $A = 840$ kpc^2. We shall use equation (A.1) to transform back and forth between the local SN rate per unit area and the Galactic rate.

A1) Galactic Observations

The most direct determination of the local supernova rate is from historical observations. Tammann (1982) infers $S_{2D}(R_0) = 6.4 \times 10^{-11}$ pc^{-2} yr^{-1} from 5 historical SNRs (1006, 1054, 1572, 1604, and 1667=Cas A) which have occurred within 5 kpc of the Sun within the last millenium, corresponding to $\nu_{\text{Gal}} = 0.054$ yr^{-1}. Estimating rates in this fashion is equivalent to estimating probabilities *a posteriori*, and is fraught with pitfalls. Tammann's estimate might be viewed as optimistic because two of the SN barely slip into the 1000 year time frame; for example, had we measured a millenium using numbers with a base 12 instead of base 10, the rate would have been lower by almost a factor 2. Clark and Stephenson (1982) consider the supernovae that have been seen in the last two millenia, and assume that SN are visible out to 6 kpc. Including SN185 (=RCW 86) and SN1181 (=3C58), even though the latter is at a nominal distance of 8 kpc (but see below), implies an SN rate $S_{2D}(R_0) = 3.1 \times 10^{-11}$ pc^{-2} yr^{-1}, corresponding to $\nu_{\text{Gal}} = 0.026$ yr^{-1}. This estimate must be regarded as conservative: The coverage of supernova searches may not have been as complete in the first millenium as in the second, since only Chinese records are available for that period. Furthermore, 6 kpc appears generous as a radius for SN visibility, since only 1 or 2 of the supernovae are more distant than $6/\sqrt{2}$ kpc, whereas 3.5 are expected. (Fesen [1983] has argued that 3C58 is only about 3 kpc distant, which would make Kepler's SN the only historical SN more than 4 kpc from the Sun.)

A second method of estimating the SN rate is to infer it from the birthrate of massive stars. Ratnatunga and van den Bergh (1989) and McKee (1989) each find that the Scalo (1986) IMF implies a massive star SN rate of about 0.01 yr^{-1} if all stars above 8 $M_\odot$ become supernovae. McKee (1989) then suggested two modifications to this IMF: First, its amplitude should be increased by a factor 1.36 to bring it into agreement with the observed mean stellar surface density (Kuijken and Gilmore 1989a; had Bahcall's [1984] surface density been used, the increase would have been yet larger). Second, the fraction of stars that are more massive than 8 $M_\odot$ should be increased by a factor $C_F \lesssim 2$ in order to bring the predicted massive star birthrate closer to that measured by Garmany, Conti, and Chiosi (1982) and by Van Buren (1985). For $C_F = 1.4$, an intermediate value, the local high–mass SN rate is then 1.9×10^{-11} pc^{-2} yr^{-1}, and the global rate is 0.016 yr^{-1}. If 11% of SN are from low-mass progenitors, as found by Evans, van den Bergh, and McClure (1989) for Sab-Sd galaxies, then the total SN rate is 0.018 yr^{-1}. McKee (1989) also estimated the global high–mass SN rate implied by the luminosity of the Galaxy in thermal radio emission (Güsten and Mezger 1982), and obtained a high–mass SN rate of 0.013 yr^{-1}; this estimate is sensitive to the IMF of the most massive stars, which dominate the production of ionizing photons, and is therefore somewhat uncertain.

Another estimate of the high–mass SN rate comes from the pulsar birthrate; since not all massive stars leave neutron star remnants, and not all neutron stars need to be observable pulsars, this is a lower limit to the true high–mass SN rate. Narayan (1987) finds a rate 0.018 yr^{-1} for a Galactocentric radius $R_0 = 10$ kpc; assuming that this rate scales as R_0^2, we find a total birthrate of 0.013 yr^{-1} for $R_0 = 8.5$ kpc. This estimate of the pulsar birthrate is quite uncertain, however; in a subsequent analysis, Narayan and Ostriker (1990) estimate that the birthrate is in the range 0.004–0.014 yr^{-1}(after rescaling to $R_0 = 8.5$ kpc), depending on the assumptions made for the pulsar beaming angle, the pulsar spatial distribution, etc. Their "standard model" gives a rescaled birthrate of 0.0062 yr^{-1}, significantly less than other estimates of the high–mass SN rate in the Galaxy; this estimate could be reconciled with the other estimates if a significant fraction of the massive stars do not produce observable pulsars when they explode as supernovae.

Finally, van den Bergh (1983) has given a fourth estimate of the Galactic SN rate by scaling the ratio of bright radio SNRs to historical SN observed in the Galactic anticenter, where the historical coverage should have been relatively complete, to the Galaxy as a whole. He finds $\nu_{\rm Gal} = 0.022 \pm 0.013$ yr^{-1}; the large uncertainty is due to the fact that the estimate is based on only 3 historical SN.

Of the various estimates of the Galactic SN rate, the argument from the direct observation of historical SN seems best. With only 7 SN to base it on, this estimate is necessarily uncertain. As remarked above, however, Clark and Stephenson's (1982) estimate is conservative in that it is almost certain that other SN occurred in the volume of space–time used for their estimate. In order to obtain a lower limit for the effect of supernovae on the ISM, we shall be more conservative yet and adopt van den Bergh's value, $\nu_{\rm Gal} = 0.022$ yr^{-1}, corresponding to $S_{\rm 2D}(R_0) = 2.6 \times 10^{-11}$ pc^{-2} yr^{-1}. It is difficult to assign a meaningful error bar to this estimate, but in view of the other estimates above, it is unlikely that the true rate is lower than this by more than 20%.

A2) Extragalactic Observations

By observing SN in other galaxies, it is possible to get around the small number statistics that plague internal estimates of the Galactic SN rate. This advantage comes with a number of problems of its own, however: in addition to the problem of obtaining a well-defined sample from which to measure the extragalactic rate, the resulting estimate depends on the square of the Hubble constant, on the luminosity of the Galaxy, and on its Hubble type.

Van den Bergh (1990) has reviewed the analyses of the data on extragalactic SN obtained by Evans (van den Bergh, McClure, and Evans 1987; Evans et al. 1989). We assume that the Galaxy is intermediate in type between Sab, Sb and Sbc-Sd; the results of Evans et al. (1989) imply that such a galaxy has an SN rate per 10^{10} $L_B(\odot)$ of $\nu_L = 0.023h^2$ yr^{-1}, where h is the Hubble constant normalized to 100 km s^{-1} Mpc^{-1}. Following van den Bergh (1990), we adopt a blue luminosity of $2 \times 10^{10} L_B(\odot)$ for the Galaxy. The inferred Galactic SN rate is then $\nu_{\rm Gal} = 0.046 h^2$ yr^{-1}. This is consistent with the rate derived from Galactic observations for $h \simeq 0.7$. (Interestingly enough, van den Bergh [1989] has recently determined the Hubble constant to be 0.67±0.08.) Van den Bergh (1990) derived a somewhat lower Galactic SN rate, $\nu_{\rm Gal} = 0.032h^2$ yr^{-1}, by treating the Galaxy as a typical Shapley–Ames galaxy, which includes early–type galaxies with low SN rates.

The selection effects affecting Evans' survey are difficult to quantify. Van den Bergh (1990) remarks that the use of a more powerful telescope did not result in the discovery of fainter SN, which opens the possibility that the survey is incomplete. For this reason, it is worthwhile to compare the results of Evans' survey with those of other surveys. Tammann (1982) has analyzed a survey of Shapley–Ames galaxies that discovered 77 supernovae, several times more than in Evans' survey. Tammann's SN rates for spirals have been criticized (Maza and van den Bergh 1976) as being too high because he increased the observed rates by a factor of 2.8 to allow for the increased obscuration due to the inclination of the galaxies; such an effect was not found in the Evans data (van den Bergh *et al.*, 1987). However, as Tammann emphasized, his rates included only observed SN, and (in the absence of the inclination factor) are lower limits to the true rates. For example, the average of his rates for Sab, Sb galaxies and Sbc–Sd galaxies gives $\nu_L = 0.041h^2$ yr^{-1}; if there were no inclination factor, this would be reduced by a factor 2.8 to $0.015h^2$ yr^{-1}. However, his sample has more Type I's than Type II's, whereas Evans *et al.* (1989) find that the intrinsic ratio of II's to I's is about 2. This discrepancy is presumably due to the fact that SN of Types Ib and II are more difficult to detect than those of Type Ia because of their lower luminosity, which reduces both the distance out to which they can be detected and the time for which they are observable at a given distance. If Evans *et al.* are correct, it follows that Tammann's sample is indeed incomplete. If the ratio of SN Ia to SN I in the Tammann sample is the same as in the Evans *et al.* sample, then Tammann's SN Ia rate is $\nu_L({\rm Ia})=0.0046h^2$ yr^{-1} (after removal of the inclination effect). Since this represents only 11% of the total SN rate according to Evans *et al.*, the revised Tammann total rate becomes $\nu_L = 0.0046h^2/0.11 = 0.042h^2$ yr^{-1}. After removal of the inclination effect and allowance for incompleteness, the rate is just about the same as he found originally! This rate is almost twice the Evans *et al.* rate; because of the heterogeneous nature of the data on which Tammann's rate is based, it is difficult to pinpoint the origin of the difference. We note that the SN Ia fraction taken from Evans *et al.* is quite uncertain, being based on only 4 SN Ia's; if the true fraction were higher than 11%, the discrepancy would be decreased.

In sum, the inferred Galactic SN rate from extragalactic SN is $\nu_{\rm Gal} \simeq (0.046 - 0.084)h^2$ under the assumptions that the Galaxy is intermediate between Sab, Sb galaxies and Sbc–Sd galaxies, and that it has a blue luminosity of $2 \times 10^{10}\ L_B(\odot)$. These rates are consistent with that estimated from Galactic data if the Hubble constant is in the range 0.7–0.5 (in accord with the values favored by van den Bergh and Tammann, respectively [!]).

APPENDIX B: ERRATA IN MO

Several minor typographical and numerical errors have been found in MO. Although only of historical interest at this point, the correct expressions are given here. In equation (9), $t_c \propto \alpha^{-0.58}$; $t_c = 10^{5.75}$ yr; and $v_{bc} \propto \alpha^{0.78}$; $v_{bc} = 10^{2.28}$ km s^{-1}. Above equation (49), $\ln(a_{0u}/a_{0l})$ is to be evaluated from equation (37).

In equation (51), the numerical coefficient is $10^{-1.42}$. In Table 1, the column density of the smallest clouds is 6.5×10^{19} cm. In equation (D3), Σ should be replaced by $10^{4.44}\Sigma^{-1}$. (The parameter Σ defined in MO should not be confused with the surface density in Table I.)

ACKNOWLEDGMENTS

I wish to thank my collaborators, Jerry Ostriker and Len Cowie, for their contributions to the development of the three-phase model and for their comments on this manuscript; Don Cox for stimulating remarks over the years and for suggestions which improved this paper—in particular, for emphasizing the importance of the turbulent pressure in the porosity argument in §V; and Alex Filippenko, Bon-Chul Koo, Shri Kulkarni, and Lyman Spitzer for very helpful remarks. I wish to express my appreciation to the organizers of the Centennial Meeting of the Astronomical Society of the Pacific for the opportunity to speak, and I thank Leo Blitz for being an understanding editor. My research is supported in part by NSF grant AST86-15177.

REFERENCES

Allen, C.W. 1963, *Astrophysical Quantities* (2d ed.;London: Athlone Press), p. 241.
Anderson, S., Kulkarni, S.R., Prince, T., and Wolszczan, A. 1989a, *IAU Circular 4819*.
———. 1989b, *IAU Circular 4853*.
Bahcall, J.N. 1984, *Ap. J.*, **276**, 169.
Begelman, M.C., and McKee, C.F. 1990, *Ap. J.*, in press.
Bloemen, J.B.G.M. 1987, *Ap. J.*, **322**, 694.
Böhringer, H., and Hartquist, T.W. 1987, *M.N.R.A.S.*, **228**, 915.
Borkowski, K.J., Balbus, S.A., and Fristrom, C.C. 1989, *Ap. J.*, submitted.
Bregman, J.N. 1980, *Ap. J.*, **236**, 577.
Bregman, J.N., and Glassgold, A.E. 1982, *Ap. J.*, **263**, 564.
Bronfman, L, Cohen, R.S., Alvarez, H., May, J., and Thaddeus, P. 1988, *Ap. J.*, **324**, 248.
Bruhweiler, F.C., Gull, T.R., Kafatos, M., and Sofia, S. 1980, *Ap. J. (Letters)*, **238**, L27.
Caswell, J.L., and Lerche, I. 1979, *M.N.R.A.S.*, **187**, 201.
Chevalier, R.A. 1990, this volume.
Chevalier, R.A., and Fransson, C. 1984, *Ap. J. (Letters)*, **274**, L43.
Chevalier, R.A., and Oegerle, W.R. 1979, *Ap. J.*, **227**, 398.
Chieze, J.P., and Lazareff, B. 1980, *Astr. Ap.*, **91**, 290.
———. 1981, *Astr. Ap.*, **95**, 194.
Cioffi, D.F., McKee, C.F., and Bertschinger, E. 1988, *Ap. J.*, **334**, 252.
Clark, D.H., and Stephenson, F.R. 1982, in *Supernovae: A Survey of Current Research*, eds.M. Rees and R. Stoneham (Dordrecht: Reidel), p.355.
Corbelli, E., and Salpeter, E.E. 1988, *Ap. J.*, **326**, 551.
Cowie, L.L. 1987, in *Interstellar Processes*, ed.D. Hollenbach and H. Thronson (Dordrecht: Reidel), p 245.

Cowie, L.L., Jenkins, E.B., Songaila, A., and York, D.G. 1979, *Ap. J.*, **232**, 467.
Cowie, L.L., McKee, C.F., and Ostriker, J.P. 1981, *Ap. J.*, **247**, 908.
Cowie, L.L., and Songaila, A. 1986, *Ann. Rev. Astr. Ap.*, **24**, 499.
Cowie, L.L., and York, D.G. 1978a, *Ap. J.*, **220**, 129.
——. 1978b, *Ap. J.*, **223**, 876.
Cox, D.P. 1981, *Ap. J.*, **245**, 534.
——. 1988, in *Supernova Remnants and the Interstellar Medium*, ed.R.S. Roger and T.L. Landecker (Cambridge: Cambridge University Press), p. 73.
——. 1989, in *Structure and Dynamics of the Interstellar Medium*, eds. G. Tenorio-Tagle, M. Moles, and J. Melnick (New York: Springer-Verlag), in press.
Cox, D.P., and Reynolds, R. 1987, *Ann Rev. Astr. Ap.*, **25**, 303.
Cox, D.P., and Slavin, J.D. 1989, in *EUV Astronomy*, eds.R.F. Malina and S. Bowyer (New York: Pergamon), in press.
Cox, D.P., and Smith, B.W. 1974, *Ap. J. (Letters)*, **189**, L105.
Evans, R., van den Bergh, S., and McClure, R.D. 1989, *Ap. J.*, **345**, 752.
Falgarone, E., and Lequeux, J. 1973, *Astr. Ap.*, **25**, 253.
Ferriere, K.M., Zweibel, E.G., and Shull, J.M. 1988, *Ap. J.*, **332**, 984.
Fesen, R.A. 1983, *Ap. J. (Letters)*, **270**, L53.
Field, G.B., Goldsmith, D.W., and Habing, H.J. 1969, *Ap. J. (Letters)*, **155**, L149 (FGH).
Freeman, K.C. 1987, *Ann. Rev. Astr. Ap.*, **25**, 603.
Garmany, C.D., Conti, P., and Chiosi, C. 1982, *Ap. J.*, **263**, 777.
Güsten, R., and Mezger, P.G. 1982, *Vistas in Astronomy*, **26**, 159.
Habe, A., Ikeuchi, S., and Tanaka, Y.D. 1981, *Pub. Astr. Soc. Japan*, **33**, 23.
Hartquist, T.W., and Morfill, G.E. 1986, *Ap. J.*, **311**, 518.
Heiles, C. 1976, *Ap. J.*, **204**, 379.
——. 1979, *Ap. J.*, **229**, 533.
——. 1980, *Ap. J*, **235**, 833.
——. 1984, *Ap. J. Suppl.*, **55**, 585.
——. 1987, *Ap. J.*, **315**, 555.
——. 1990, *Ap. J.*, in press.
Hobbs, L.M. 1974, *Ap. J.*, **191**, 395.
Humphreys, R.M., and McElroy, D.B. 1984, *Ap. J.*, **284**, 565.
Ikeuchi, S. 1987, in *Starbursts and Galaxy Evolution*, ed.T.X. Thuan and T. Montmerle (Gif sur Yvette: Editions Frontieres), p. 27.
Ikeuchi, S., Habe, A., and Tanaka, Y.D. 1984, *M.N.R.A.S.*, **207**, 909.
Jenkins, E.B. 1978, *Ap. J.*, **220**, 107.
Jenkins, E.B., Jura, M., and Loewenstein, M. 1983, *Ap. J.*, **270**, 88.
Jura, M. 1975, *Ap. J.*, **197**, 581.
Kerr, F.J., and Lynden-Bell, D. 1986, *M.N.R.A.S.*, **221**, 1023.
Klein, R.I., McKee, C.F., and Colella, P. 1990, this volume.
Kuijken, K., and Gilmore, G. 1989a, *M.N.R.A.S.*, **239**, 605.
——. 1989b, *M.N.R.A.S.*, **239**, 651.
Kulkarni, S.R., and Fich, M. 1985, *Ap. J*, **289**, 792.
Kulkarni, S.R., and Heiles, C. 1987, in *Interstellar Processes*, ed.D. Hollenbach and H. Thronson (Dordrecht: Reidel), p. 87.
Landau, L.D., and Lifschitz, E.M. 1959, *Fluid Mechanics* (Reading:Addison-Wesley).

Larson, R.B. 1981, *M.N.R.A.S.*, **194**, 809.
Lockman, F.J. 1984, *Ap. J.*, **283**, 90.
Lockman, F.J., Hobbs, L.M., and Shull, J.M. 1986, *Ap. J.*, **301**, 380.
Mathis, J.S. 1986, *Ap. J.*, **301**, 423.
Maza, J., and van den Bergh, S. 1976, *Ap. J.*, **204**, 519.
McCray, R., and Snow, T.P. 1979, *Ann. Rev. Astr. Ap.*, **17**, 213.
McKee, C.F. 1982, in *Supernovae: A Survey of Current Research*, ed. M.J.Rees and R.J. Stoneham (Dordrecht: Reidel), p. 433.
———. 1988, in *Supernova Remnants and the Interstellar Medium*, ed. R.S.Roger and T.L. Landecker (Cambridge: Cambridge University Press), p. 205.
———. 1989, *Ap. J.*, **345**, 782.
———. 1990, *Ap. J.*, to be submitted.
McKee, C.F., Hollenbach, D.J., Seab, C.G., and Tielens, A.G.G.M. 1987, *Ap. J.*, **318**, 674.
McKee, C.F., and Ostriker, J.P. 1977, *Ap. J.*, **218**, 148 (MO).
Narayan, R. 1987, *Ap. J.*, **319**, 162.
Narayan, R., and Ostriker, J.P. 1990, *Ap. J.*, submitted.
Norman, C., and Ikeuchi, S. 1989, *Ap. J.*, **345**, 372.
Ostriker, J.P., and McKee, C.F. 1988, *Rev. Mod. Phys.*, **60**, 1.
Parker, E.N. 1969, *Space Sci. Rev.*, **9**, 651.
Payne, H.E., Salpeter, E.E., and Terzian, Y. 1983, *Ap. J.*, **272**, 540.
Rand, R.J., and Kulkarni, S.R. 1989, *Ap. J.*, **343**, 760.
Rand, R.J., Kulkarni, S.R., and Hester, J. 1990, *Ap. J.*, in press.
Ratnatunga, K.U., and van den Bergh, S. 1989, *Ap. J.*, **343**, 713.
Reynolds, R.J. 1985, *Ap. J.*, **294**, 256.
———. 1989a, *Ap. J. (Letters)*, **339**, L29.
———. 1989b, in *Galactic and Extragalactic Background Radiation*, ed.S. Bowyer and C. Leinert, in press.
———. 1989c, *Ap. J.*, **345**, 811.
Savage, B.D. 1987, in *Interstellar Processes*, ed.D.J. Hollenbach and H. Thronson (Dordrecht: Reidel), p. 123.
Scalo, J.S. 1986, *Fund. Cos. Phys.*, **11**, 1.
Scoville, N.Z., and Sanders, D.B. 1987, in *Interstellar Processes*, ed.D. Hollenbach and H. Thronson (Dordrecht: Reidel), p. 21.
Shapiro, P.R., and Field, G.B. 1976, *Ap. J.*, **205**, 762.
Songaila, A., Bryant, W., and Cowie, L.L. 1989, *Ap. J. (Letters)*, **345**, L71.
Smith, B.W. 1977, *Ap. J.*, **211**, 404.
Spitzer, L. 1956, *Ap.J.*, **124**, 20.
———. 1978, *Physical Processes in the Interstellar Medium* (NewYork: Wiley).
———. 1990, *Ann. Rev. Astr. Ap.*, **28**, in press.
Tammann, G. 1982, in *Supernovae: A Survey of Current Research*, ed. M.J. Rees and R. Stoneham (Reidel: Dordrecht), p 371.
Taylor, J.H., and Manchester, R.N. 1977, *Ap. J.*, **215**, 885.
Tenorio-Tagle, G., and Bodenheimer, P. 1988, *Ann. Rev. Astr. Ap.*, **26**, 145.
Teske, R.G., and Petre, R. 1987, *Ap. J.*, **314**, 673.
van den Bergh, S. 1983, *Pub. A.S.P.*, **95**, 388.
———. 1989, *Astr. Ap. Rev.*, **1**, 111.
———. 1990, in *Supernovae*, ed.S. Woosley (Berlin: Springer-Verlag), in press.
van den Bergh, S., McClure, R.D., and Evans, R. 1987, *Ap. J.*, **323**, 44.
Van Buren, D. 1985, *Ap. J.*, **294**, 567.

———. 1989, *Ap. J.*, **338**, 147.
Wang, Z., and Cowie, L.L. 1988, *Ap. J.*, **335**, 168.
Weaver, H. 1974, in *Highlights of Astronomy, Vol. 3*, ed.G. Contopoulos (Dordrecht: Reidel), p. 423.
Wolff, M.T., and Durisen, R.H. 1987, *M.N.R.A.S.*, **224**, 701.

Section II

The Contents of the Interstellar Medium

PROPERTIES OF THE ISM: GAS IN THE HALO

BLAIR D. SAVAGE
Department of Astronomy, University of Wisconsin, 475 N. Charter St.,
Madison, Wisconsin 53706

ABSTRACT The properties of interstellar gas in the galactic halo are reviewed. Halo gas is found to have a wide range of physical conditions with temperatures ranging from less than 170 K to more than 200,000 K. The gas extending away from the plane of the Milky Way has density scale heights ranging from less than 300 pc for certain species in the neutral medium to approximately 3000 pc for the most highly ionized gas. The complex kinematical characteristics of the gas provides important clues about its origin. The gas phase elemental abundances in the neutral halo gas are closer to solar than is found for the highly depleted gas of the Milky Way disk. The possible origin of gas at large distances away from the galactic plane is discussed.

INTRODUCTION

The gaseous galactic halo is now recognized as an important region of interstellar space. The halo represents a place for heated and overpressurized disk gas to vent its energy. A study of matter in the halo can provide information about significant galactic phenomena such as correlated supernova explosions in OB associations. The circulation of gaseous matter in the Milky Way may involve a flow of hot gas away from the galactic plane followed by the cooling of that gas and the subsequent raining of cooler clouds of material back down upon the plane. The structural control of the interstellar medium provided by magnetic fields and the pressure of cosmic rays might be clarified from a better understanding of the properties of the thermal gas in the halo. The extent to which matter in the halo is supported by the combined effects of magnetic fields and the pressure of cosmic rays versus the dynamic phenomena and thermal pressures associated with a fountain is currently a subject of considerable discussion. As the interface region between the galactic disk and intergalactic space, the halo might eventually provide clues about the nature of the intergalactic medium and the ionizing extragalactic radiation field.

Although the possible existence of a gaseous halo or corona was first postulated by Lyman Spitzer Jr. (1956), progress in studying the nature of the matter was relatively slow until the decade of the 1980's. As recently as 1979, very few astronomers dared to advocate the existence of much gaseous matter beyond distances away from the plane, z, of about 500 pc. In fact the common view of the distribution of the neutral gas phase was that it occupied a very thin layer with a z distribution approximately described by a gaussian function with a scale height of 120 pc. It is now known that such a simple view for the description of the neutral gas is very wrong and that other components of the gas (e.g. the highly ionized atoms) have scale heights as large as 3000 pc.

In the following sections we first review the various diagnostic techniques which have been used to study gas at large distances from the galactic plane and then summarize the known properties of the gas. We conclude with discussions of the competing views regarding the origin of the gas in the Milky Way halo. Other recent reviews of this subject are those of Jenkins (1987a) and Savage (1987a and 1987b). The publication of the NRAO 1986 Conference on the Gaseous Halos of Galaxies (eds.J. N Bregman and F.J. Lockman; NRAO Press) is particularly useful. Also, the theory of the hot phase of the ISM including gas in the halo has been recently reviewed by Spitzer (1990).

DIAGNOSTIC TECHNIQUES

Table 1 provides a list of some of the many techniques that have yielded some information about gas and other particles in the galactic halo. The literature is extensive and the listed references are for either recent review papers or for relatively recent research papers which provide a suitable introduction to the literature. The emphasis of this review will be on the properties of the thermal gas. However, it is important to recognize that an ultimate understanding of the phenomena occurring in the halo will require a determination of how the thermal gas couples to cosmic rays and the magnetic field (e.g. see references 17,18, and 19).

TABLE I Techniques for the Study of Milky Way Halo Gas

Technique	Species or Gas Phase Sampled	References*
21 cm emission	H I	1,2,3
optical absorption	Ca II, Ti II, Na I, K I	4,5,6,7
ultraviolet absorption	many ions	8,9,10
optical emission	Hα, O III	11
ultraviolet emission	C IV, O III	12
pulsar dispersion	e's	13
free-free absorption and emission	e's	14
X-ray emission	hot gas (10^6 K)	15,16
non-thermal radio emission	high energy e's and B field	17
Faraday rotation	e's and B field	18
γ Ray emission	high energy e's & protons	19

* References (This list identifies recent review papers and/or research papers which provide a survey to the astronomical literature).

1. Lockman (1984)
2. Lockman (1986)
3. Giovanelli (1986)
4. Albert (1983)
5. Jenkins (1986)
6. Morton and Blades (1986)
7. Edgar and Savage (1989)
8. Savage and deBoer (1979,1981)
9. Pettini and West (1982)
10. Savage and Massa (1987)
11. Reynolds (1986)
12. Martin and Bowyer (1989)
13. Reynolds (1989)
14. Spitzer (1978 and references therein)
15. Marshall and Clark (1984)
16. Nousek et al. (1982)
17. Beuermann, Kanbach and Berkhuijsen (1985)
18. Sofue, Fujimoto and Wielebinski (1986)
19. Stecker (1979)

PROPERTIES OF HALO GAS

In this section we will draw upon many of the results in the literature as introduced in Table I in order to summarize the known properties of halo gas. The summary will include information about scale heights, densities, temperatures, pressures, elemental abundances, kinematics, and structures in the gas.

Scale Heights and Densities

A number of techniques have been used to determine the stratification of interstellar gas away from the plane of the Milky Way. The results are summarized in Table II.

Lockman(1984) studied the z distribution of H I in the inner Galaxy through subcentral (e.g. tangent) point analysis of 21 cm emission line profiles and discovered that in addition to the traditional confined components of HI with scale heights of 110 to 250 pc the neutral gas has a distribution with an exponential tail with a scale height of nearly 480 pc. This extended distribution of H I is known to exist locally from H I Lyman α absorption line studies (Bohlin, Savage and Drake 1978; Hobbs et al. 1982). It was also partially described more than two decades ago by Oort (1962) and Shane (1971) but was largely ignored until recently.

From optical and ultraviolet absorption line spectroscopy it is possible to obtain estimates of the stratification of the interstellar gas by measuring column densities, $N(X)$, of various species toward a large number of stars whose line of sight distances can be estimated. Plots of log $[N(X)\sin|b|]$ versus log $|z|$, where z is distance of each star away from the galactic plane, can be studied to infer the scale height of the gas. Results of such studies for a number of interstellar species are found in Table II. Care must be exercised in evaluating the results of such studies because the selection criteria for the sample of stars used in the analysis can have a significant effect on the result. For example, if the low $|z|$ objects included in the sample favor high density regions of interstellar space, the inferred value of the midplane density for the particular species being studied will be higher than the true average value and the inferred scale height will be smaller than the true value. Because of this problem of sample bias it is important to evaluate how the inferred scale heights of various species compare to that of a well studied species such as that of H I *for the same sample of stars*. For example, the measures of H I, Fe II, Ca II and Ti II scale heights for a sample of about 65 stars studied by Edgar and Savage (1989), reveal that while the H I has a relatively small scale height of 240 pc for the particular sample of stars, the refractory elements Fe II, Ca II and Ti II have much larger apparent scale heights. A similar result is found in van Steenberg and Shull's (1988,1989) study which revealed that H I, Mn II, S II , Zn II and Si II have have scale heights of 250 to 350 pc while Fe II was found to have a significantly larger scale height of 1110 pc. The different scale height estimates for the same species from one data set to another in part illustrate the effects of sample biases. Another effect influencing the results is that some observers have reported exponential scale heights while others have reported gaussian scale heights (see Table II).

The most significant result emerging from the studies of species normally found in the neutral gas (e.g. Mn II, S II, Zn II, Si II, Fe II and Ti II) is that for some species the scale heights are similar to that found for H I, while for others

TABLE II Scale Heights for Milky Way Gas

Ion	$n(0)$ (cm^{-3})	H (pc)	Type[a]	$N_\perp$[b] (atoms cm^{-2})	Reference
H I	0.16	110	G		Lockman (1984)
"	0.09	250	G		"
"	0.053	480	E	7.85×10^{19}	"
H I	0.46	330	G		van Steenberg&Shull(1989)
Mn II	7.0×10^{-9}	250	G		van Steenberg&Shull(1988)
S II	8.4×10^{-7}	320	G		"
Zn II	4.0×10^{-9}	350	G		"
Si II	8.0×10^{-7}	350	G		"
Fe II	8.0×10^{-8}	1110	G		"
H I	0.37	240	E		Edgar and Savage (1989)
Fe II	8.0×10^{-8}	500	E		"
Ca II	7.5×10^{-10}	1160	E		"
Ti II	2.5×10^{-10}	>2000	E		"
e's	0.015	70	E		Reynolds (1989)
"	0.025	1500	E	1.16×10^{20}	"
H I	0.10	800	E	2.45×10^{20}	Savage and Massa (1987)
Si IV	2×10^{-9}	~3000	E	1.85×10^{13}	"
C IV	7×10^{-9}	~3000	E	6.48×10^{13}	"
N V	3×10^{-9}	~2000	E	1.85×10^{13}	"
O VI	2.0×10^{-8}	300[c]	E		Jenkins (1978)

[a] The G indicates Gaussian scale heights while the E indicates exponential scale heights. Some of the differences between scale height estimates from one investigation to the next can be explained as being due to the differences in the function or functions fitted to the data. Other differences are undoubtedly due to object sample biases as discussed in the text.

[b] Total column densities perpendicular to the galactic plane are listed for the more extended components of the gas. Note that the mid-plane density and scale height estimates from Savage and Massa (1987) for HI are both nearly a factor of 2 larger than the corresponding estimates from Lockman(1984). Therefore, the value of $N_\perp$ is a factor of 3 larger. The total column density of highly ionized gas in the halo can be inferred from the values of $N_\perp$ for N V by allowing for the H to N abundance and for the fact that at most only 25% of the N will be in the form of N V for gas near 200,000 K. Applying these two factors, to 1.85×10^{13} yields ~1×10^{18} for the column density of ionized gas in the temperature range which favors the production of N V.

[c] This value of the O VI scale height should be considered as a lower limit. The Copernicus satellite was unable to observe very many stars at large distances from the galactic plane.

(particularly Fe II and Ti II) the scale heights are considerably larger. In fact for Ti II, the results of Edgar and Savage (1989) provide only a lower limit to the scale height. It is noteworthy that the heavy elements exhibiting the largest scale heights are those elements that exhibit strong depletion onto dust in neutral gas regions of the galactic disk. Undoubtedly, the processing of interstellar matter from the dust phase to the gas phase is playing an important role in influencing the scale heights of the various heavy elements listed in Table II (see Edgar and Savage 1989). However, the possibility for anomalous abundances in halo gas due to enrichment from such galactic phenomena as supernova explosions should not be ignored (see Jenkins 1986). Also, if the Galaxy is accreting unprocessed matter or matter with anomalous abundances it might eventually be possible to recognize the existence of such processes through accurate abundance studies.

Recent measures of pulsar dispersion toward several pulsars in globular clusters has permitted Reynolds (1989) to infer the scale height of free electrons in the Galaxy. The result of fitting the data toward 38 pulsars with a two component exponential distribution of free electrons reveals the existence of an extended component of ionized gas with an exponential scale height of 1500 pc.

The gas component with the largest scale height so far found is the highly ionized gas as recorded by the IUE satellite. In a sample of about 40 stars including objects in the disk and halo, Savage and Massa (1987) found that while the H I scale height was 800 pc, the highly ionized gas as traced by Si IV and C IV had exponential scale heights of about 3000 pc. The large scale height for H I nicely illustrates the effects of star selection biases. The disk stars in the Savage and Massa sample were selected to avoid dense interstellar regions of nebulosity. In fact, the disk sight lines mostly measure intercloud and interarm gas. Thus a large scale height for H I was expected. The fact that the Si IV and C IV scale heights are a factor of four larger than that of H I for the same sample of stars conclusively reveals the greatly extended distribution of the most highly ionized gas.

Unfortunately, the Copernicus satellite was not sensitive enough to probe O VI absorption toward distant halo stars. Therefore, the scale height estimate for O VI from Jenkins(1978) should only be considered as a lower limit.

In summary we see from Table II that the stratification of the interstellar gas away from the galactic plane is complex. In the best studied cases (e.g.H I), it is found that there exist two or more components to the stratification. There are substantial differences in the stratification of the highly refractory elements compared to H I and compared to other various weakly depleted heavy elements. The ionized gas as traced by electrons has a large scale height. The very highly ionized gas as traced by Si IV and C IV has the largest scale height measured.

Temperatures

Halo gas is highly inhomogeneous in its properties. It contains cool clouds, and warm and hot gas. The range of temperatures so far determined is from less than 170 K to more than 200,000 K.

The evidence for cool neutral halo gas comes from the H I 21 cm emission line widths for clouds at $|z| \sim 500$ pc in the inner Galaxy studied by Lockman (1984). Some of the cooler clouds detected have line widths implying $T < 350$ K. Additional evidence is from a very high resolution measurement of the Na I absorption line width in the 60 km s^{-1} cloud toward SN 1987A (Pettini 1988).

The observed line is very narrow and implies T < 170 K.

The evidence for warm neutral gas in the halo comes from the extensive measures of neutral H I gas temperatures through comparison of H I emission and absorption (Payne, Salpeter and Terzian 1983). Many of the attempts to measure the temperature of the warm neutral medium have resulted in lower limits for the temperature. Many of the lower limits are near 3000 K. The highest measured temperature for the warm neutral gas is 6000 K. The temperature of the warm neutral medium is probably in the range from about 3000 K to 6000 K. See Kulkarni and Heiles (1987) for a review.

An analysis of the line widths of Hα and [S II] provides an upper limit of 20,000 K and a most probable value of 8000 K to the temperature of the the ionized gas in the low halo (Reynolds 1985).

Evidence for hot gas in the halo comes from several sources. IUE ultraviolet absorption line measurements toward halo stars have revealed the presence of interstellar N V absorption by a gas phase estimated to have a scale height of about 2000 pc (Savage and Massa 1987; Savage, Massa, and Sembach 1990). Among the ions accessible to the IUE spacecraft, N V is the most difficult to produce by photoionization because 77 eV is required to convert N IV to N V. Since most hot stars have strong He II absorption edges in their spectra at 54 eV, the production of N V by starlight photoionization would appear difficult. Thus the presence of N V is often taken as direct evidence for the existence of hot gas in the halo. Under conditions of collisional equilibrium, the abundance of N V peaks at about 200,000 K. However, the absorption lines are more likely formed in cooling gas under conditions of non-equilibrium ionization as it cools from temperatures exceeding 200,000 K (Edgar and Chevalier 1986).

The far-UV measures of diffuse interstellar emission lines of C IV and O III] by Martin and Bowyer (1989) from high galactic latitude directions are also evidence for hot halo gas. The observations are most easily understood as being produced by gas cooling through the temperature regime where C IV peaks in abundance (e.g. ~100,000 K).

It is difficult to determine if the halo is a source of diffuse X-rays indicative of gas with T ~ 1 to 3×10^6 K. This is because the distant halo gas must be viewed against the bright X-ray emission from the local hot ISM. Also the soft X-ray emission from a distant hot medium would be attenuated by photoelectric absorption occurring in cool foreground gas. Because of these complications, the interpretation of the X-ray emission measurements has been ambiguous. Data which have been interpreted as containing a halo gas contribution to the diffuse X-ray background are the M1 band (0.5 to 1.0 KeV) data of Nousek et al. (1982) and the SAS-3 carbon band (0.16 to 0.284 keV) data of Marshall and Clark (1984). Although the data are consistent with hot halo gas having an emission measure of 0.004 $cm^{-6}pc$ and a temperature of about 2 to 3×10^6 K, a more local origin is also possible. However, Burrows et al. (1984) argue against a significant fraction of the C-band emission coming from a galactic halo.

Pressure

Observations of Hα emission from a $10° \times 12°$ region centered near the direction l =144° and b = - 21° reveal structures that extend in some cases to approximately 1000 pc below the galactic plane (Reynolds 1980). The inferred electron density in these structures is ~ 0.2 cm^{-3}. Assuming a temperature of

~8000 K, this implies a pressure P/k of ~ 3200 for the ionized gas is appropriate at $|z| \sim 1000$ pc.

A comparison of the strength of C IV absorption (Savage and Massa 1987) and C IV emission (Martin and Bowyer 1989) permits an estimate of the intrinsic density, pressure and filling factor, f, of the gas producing the absorption and emission. With the assumptions that the absorbing and emitting gas concide in space and that the temperature of the gas is 10^5 K the result is $n_e \sim 0.01$ cm^{-3}, p/k ~ 1000 and f ~ 0.01 (Martin and Bowyer 1989).

Elemental Abundances

In the galactic disk it is found that many of the heavy elements are selectively removed from the gas phase because of their incorporation into interstellar dust. A study of elemental abundances in halo gas must therefore consider the consequences of the cycling of matter from the solid phase to the gas phase. For a review of abundances of the elements in the interstellar gas see Jenkins(1986, 1987b). The different scale heights found for H I and various heavy elements (see Table II) convincingly reveal that there are substantial variations in the gas phase abundances of certain elements with distance from the galactic plane. Elements such as Fe and Ti ,which in the disk are depleted by factors of 50 to 100, are sometimes found with nearly solar gas phase abundances in clouds known to be located at large distances away from the galactic plane (Albert 1983; Edgar and Savage 1989; van Steenberg and Shull 1988).

While the survey data referenced in Table II are valuable in obtaining a picture about the phenomena influencing the gas phase composition of heavy elements in halo gas in a global sense, it is also important to obtain detailed measures of many elements for several clouds known to be situated in the halo. There are only a few cases where detailed abundance estimates have been made for such gas. These include studies with the IUE satellite of the intermediate velocity clouds toward the LMC (Savage and de Boer 1981; Blades et al. 1988) and a study with the Copernicus satellite of the intermediate velocity gas toward HD 93521 (Caldwell 1979).

In their study of Milky Way halo gas toward the LMC, Savage and deBoer (1981) found that the gas toward the LMC star HD 36402 in the velocity range from about 50 to 140 km s^{-1} has gas phase abundances of Fe, Mg, Si, and Al much closer to solar than for normally depleted disk gas. An analysis of higher quality IUE data for the sightline to SN 1987A by Blades et al. (1988) revealed that in the 70 km s^{-1} cloud, the elements Si, Cr, Mn, Fe, and Ni all show relative gas phase abundances close to the solar value. Considering that Cr, Ni and Fe are generally depleted by factors of about 100 in disk gas, this result implies that the 70 km s^{-1} cloud toward SN 1987A is probably nearly devoid of dust.

The most accurate abundance analysis for gas in matter known to be situated at a large distance from the galactic plane is for the intermediate velocity gas toward HD 93521 a star at about 1000 pc in the direction l = 183° and b = +62°. Copernicus satellite data for HD 93521, have been analyzed by Caldwell (1979) who finds that the clouds with $v_{lsr} < -30$ km s^{-1} have gas phase abundances of Si, P, S, Ar, and Fe within a factor of 3 of the solar values. In the case of Ti, the measures of Albert (1983) reveal a gas phase abundance about four times less than in the sun in the same velocity range. That particular Ti abundance is among the largest ever measured in the interstellar gas.

In summary, the abundance measurements so far obtained for neutral halo matter reveal gas phase abundances much closer to solar than for the depleted gas of the galactic disk. Apparently the process injecting gas into the halo destroys solid matter and produces gas with nearly population I gas phase heavy element abundances. However, ths suggestion (Jenkins 1986) that heavy elements may be injected into the halo directly by type 1 supernovae should not be ignored.

Kinematics

The kinematics of the gas in the halo is important since the state of motion of the gas can provide important clues about processes that inject gas into the halo. Although the literature is enormous, the understanding of the meaning of the data is quite limited. An extensive literature exists for 21 cm emission line studies of H I, and the high velocity hydrogen ($|v_{LSR}| > 70$ km s^{-1}) has received most of the attention. A recent review of high velocity H I is that of Giovanelli (1986). While the high velocity clouds appear to relate to a number of different phenomena, it is noteworthy that many of the high velocity clouds have motions consistent with a z component of motion, v_z, of up to -100 km s^{-1} , a galactic radial component of motion, v_R , of up to -100 km s^{-1} and a galactic rotational component of motion which implies slower rotation than matter in the underlying disk (Kaelble, deBoer and Grewing 1985). Such motions have been interpreted as being consistent with the expectations of a galactic fountain flow (Bregman 1980; Houck 1988; Houck and Bregman 1989). The maximum velocities expected for falling condensations in such a flow depend on the distance above the plane to which the flow extends. That height depends on the base temperature of the hot gas driving the fountain. The true high velocity clouds with $v \sim 150$ km s^{-1} require flows to $z \sim 5$ kpc and base temperatures of 1 to 2×10^6 K . More modest speeds ($v \sim 100$ km s^{-1}) are predicted for lower temperature fountains (e.g. $T \sim 3 \times 10^5$ K) with flow heights of about 1 to 2 kpc.

Intermediate velocity hydrogen ($30 < |v_{LSR}| < 70$ km s^{-1}) has received much less attention than its higher velocity counterpart . However, Kulkarni and Fitch (1985) have amply demonstrated that the intermediate velocity H I is an important component of the interstellar gas. In particular the high velocity dispersion H I they studied with a mean dispersion of 35 km s^{-1} is probably the large scale height component of neutral hydrogen studied by Lockman (1984), and most of the kinetic energy of the ISM appears to reside in the intermediate velocity gas.

Our view of the sky is strongly influenced by local galactic structure and the recent history of our immediate environment. For example, the event that produced the local hot bubble in which the Sun is located may have produced local kinematical disturbances. Toward the north galactic pole there is a substantial amount of H I falling toward the galactic plane with velocities of up to about -80 km s^{-1} (Weaver 1974; Wesselius and Fejes 1973; deBoer and Savage 1984; Danly 1989). Toward the south galactic pole the neutral medium is relatively quiescent but still exhibits inflow. The two average H I profiles shown by Kulkarni and Fich (1985) for the north and south galactic pole region nicely illustrate the pronounced inflow of H I toward the plane in the local region of the Milky Way and the kinematical differences between the north and south polar regions (also see Weaver 1973; Fig.1).

Ultraviolet absorption studies of the strong resonance lines of heavy elements toward distant stars provide a sensitive probe of the interstellar gas. Danly (1989) has studied the kinematics of the neutral gas toward 70 stars through IUE absorption line measurements of such species as C II, Si II and O I. The emphasis was on gas at |b| > 40° in order to study the z motions of the gas. The kinematic differences between the north and south polar regions seen in the H I emission data are also revealed in the absorption line data. By combining the emission line and absorption line data it is possible to infer approximate distances to various high velocity phenomena seen in the sky. In particular, the measures of Danly (1989) reveal that the extensive region of gas over the north galactic pole moving with v of about -60 km s^{-1} lies somewhere between z = 340 and 700 pc, that the high velocity gas associated with complex M II is beyond z ~ 700pc and that the high velocity gas of Complex C I and C III is beyond z ~ 1700 pc. The detection of Complex C toward the RR Lyrae star BT Draconis at z ~ 1.7 kpc and not toward other nearer stars by Songaila, Cowie, and Weaver (1988) appears to place at least part of the complex at a large distance away from the galactic plane.

The high ionization absorption lines toward halo stars studied by Danly (1987) also exhibit a north-south galactic polar asymmetry. The lines toward distant stars at the north galactic pole are wide (FWHM up to 110 km s^{-1}) and centered at velocities of about -30 to -50 km s^{-1}. Toward the south pole the line widths are relatively narrow (FWHM ~ 50 to 70 km s^{-1}) and the average velocities are close to +10 km s^{-1}. The high ionization data provide a similar kinematic picture as that revealed by the low ionization data; a highly disturbed infalling region toward the north galactic pole and a relatively quiescent region toward the south galactic pole.

The effect of differential galactic rotation on the appearance of interstellar absorption lines of the neutral and highly ionized gas in the halo was first studied by Savage and de Boer (1979, 1981) and more recently by Savage and Massa (1987) and Savage, Massa, and Sembach (1989). A preliminary result from the 1987 study of 40 stars suggested that substantial deviations from corotation occur in halo gas and that rotation may cease completely at about |z| ~ 3 kpc. However, much additional work will be needed to confirm that result. In a new observing program Savage, Massa and Sembach (1990) are obtaining high quality IUE line profiles for selected halo stars for which galactic rotation effects are expected to be large. In a detailed analysis of the profiles toward HD 163522, a B1 Ib halo star at a line of sight distance of 9 kpc and a z distance of -1.5 kpc in the direction l = 350° and b = -9°, it was found that the profiles of interstellar species known to have large scale heights (e.g. Si IV, C IV and N V) are significantly more affected by galactic rotation than the profiles of species having smaller scale heights (e.g. H I, Fe II, Mg II, S II, etc). Simple model calculations were performed to understand this result. In the case of gas with a small scale height, the sight line simply runs out of gas before the effects of galactic rotation become appreciable. Such studies demonstrate that once the galactic rotation curve is known for matter away from the galactic plane, it will be possible to infer from the observed line profiles the actual run of density with distance away from the galactic plane for a large number of interstellar species.

Interstellar Structures

Radio emission line studies of the H I 21 cm line and diffuse X-ray emission studies have revealed the existence of structures in the interstellar medium that may have a direct bearing on gas found at large distances from the galactic plane (see the review of Tenorio-Tagle and Bodenheimer 1988). The neutral hydrogen structures studied by Heiles (1979,1984) may represent the dynamic products of multiple supernova explosions. The largest structures with sizes up to 3 kpc, known as supershells, likely represent regions of space where explosive phenomena have actually caused matter to burst out of the galactic disk of gas. From an analysis of narrow velocity cuts of 21 cm emission filtered to enhance the detection of small angular scale structure, Heiles found structures resembling worms crawling out of the galactic plane. Such structures suggest that the dynamic phenomena that create the shells and supershells observed locally, may be rather common on the galactic scale.

The Cygnus superbubble discovered by Cash et al. (1980) is an example of an extended (~500pc) soft X-ray emitting region indicative of hot gas with T near 2×10^6 K. Other related nearby regions of disturbed gas seen in X-rays, Hα emission or optical and ultraviolet absorption are the Orion-Eridanus region, the Gum nebula and Loop I/ North Polar Spur (Reynolds and Ogden 1979; Cowie et al. 1979; McCammon et al. 1983).

ORIGIN OF HALO GAS

The theories for the origin of gas at large distances from the galactic plane have moved in two quite separate directions in recent years. In one class of theories known as galactic fountain models, it is proposed that the gas in the halo is present because of dynamic phenomena occurring in the galactic disk which result in the ejection of hot gas into the halo where it eventually cools and then rains back down upon the disk as condensations in a flow resembling a fountain. In another class of theories referred to as the magnetic and cosmic ray supported galactic halo models, it is proposed that the pressure of cosmic rays interacting with the galactic magnetic field is adequate to support the gas found at large distances from the galactic plane. In the following, we examine how these two classes of theories are able to explain some of the existing observations of halo gas.

Galactic Fountain Models

The term "Galactic Fountain" was first introduced by Shapiro and Field (1976) to describe a process in which hot gas rises above the galactic plane before cooling and condensing to form clouds which then fall to the plane. McKee and Ostriker (1977) and Cox (1981) incorporated the basic idea of a fountain into their comprehensive descriptions of the overall structure of the ISM and the important processes controlling that structure. Shapiro and Field suggested that a fountain may provide an interpretation for the high velocity neutral hydrogen clouds seen in 21 cm emission. The height to which hot gas will rise and the expected velocities of the condensations in the cooling gas depend on the temperature of the gas at the base of the fountain, the rate of cooling of the upflowing gas, and whether or not there are heating processes occurring at large z. Estimates of the expected velocities show that fountains driven by hot gas (1 to 2×10^6 K) can roughly reproduce the pattern of motions in the high velocity cloud phenomena seen in the neutral hydrogen 21 cm line (see Bregman 1980) while fountains driven by cooler gas (2 to 3×10^5 K) are expected to have velocities which are more compatible with the existing optical and ultraviolet data for gas in the low halo (see Houck and Bregman 1989).

The filling factor and morphology of the hot gas in the galactic disk are currently very uncertain. If the hot gas filling factor were large, the flow of hot bouyant gas into the halo would occur quite freely. Gas with a temperature of 10^6 K has a thermal scale height of 6 kpc in the galactic gravitational field and it will attempt to assume a distribution in |z| compatible with that scale height. If the filling factor of hot gas in the galactic disk is small, individual bubbles of hot gas may experience difficulty in pushing the cooler matter away. With the recognition that the cooler gas of the Milky Way also has an extended component with a scale height of perhaps 500 pc for the neutral phase and 1500 pc for the ionized phase, it has become clear that the flow of hot gas into the halo may not occur as freely as previously imagined. Recent models for the flow of gas into the halo therefore have generally considered the phenomena occurring in regions of multiple supernovae which will create superbubbles of hot gas that may have a chance of breaking through the cooler matter of the galactic disk (e.g. see MacLow and McCray 1988 ; Norman and Ikeuchi 1989). In this new type of model referred to by Norman and Ikeuchi (1989) as "chimney models," it is proposed that the connection between gas in the disk and halo is through "chimneys" which are the consequence of superbubbles bursting out of the galactic disk, forming collimated structures through which hot gas flows into the halo. The evidence cited for such a model include: 1. The direct observations of local examples of hot superbubbles and their cooler counterparts the H I supershells. 2. The known existence of OB associations throughout the Galaxy, some of which are currently exhibiting the phenomena which should ultimately create supperbubbles. 3. Observations of hot gas in the halo as traced by N V absorption and C IV emission. and 4. The kinematical information provided by the high and intermediate velocity cloud phenomena.

The required circulation rate of gas from the disk into the halo and back can be estimated from the measurements of N V absorption and C IV emission associated with halo gas. Cooling gas of a galactic fountain can explain the IUE observations of N V absorption, provided the fountain flow rate is about $4 M_\odot yr^{-1}$ to each side of the galactic plane (Edgar and Chevalier 1986) which corresponds to a local mass flux of $\sim 6\times 10^{-9}$ $M_\odot$ yr^{-1} pc^{-2}. A somewhat larger flow rate can explain the C IV emission observations of Martin and Bowyer (1989). The observed flux of high velocity clouds is estimated to be 10^{-8} to 10^{-9} $M_\odot$ yr^{-1} pc^{-2} (Kaelble, deBoer and Grewing 1985). Although these numbers are comparable, care must be exercised in drawing conclusions about the implications of the similarity. The high velocity cloud estimate only allows for clouds with $|v| > 70$ km s^{-1} and the calculated mass flux depends on uncertain assumptions regarding the distances to the clouds. Motions in the solar region of the Galaxy also involve substantial motions of intermediate velocity gas which may be associated with a low fountain (Houck and Bregman 1989). The mass flux of intermediate velocity gas would add significantly to the total local mass flux.

Magnetic and Cosmic Ray Supported Galactic Halo Models

An alternate expanation for the support of gas at large distances away from the galactic plane is found in those models which involve the pressure support from the galactic magnetic field and cosmic rays. These models, which build on the work of Ipavich (1975) and Jokipii (1976), have been advanced by Hartquist, Pettini and Tallant (1984), Chevalier and Fransson (1984), Hartquist and Morfill (1986) and Bloemen (1987). In one class of models the B field is parallel to the galactic plane and the support is via magnetic pressure which is affected by the cosmic ray pressure (see Bloemen 1987). For the other class of models the B field is perpendicular to the galactic plane and the pressure support is from the streaming motions of cosmic rays along the B field (e.g. see Hartquist and Morfill 1986).

In some of the models it is proposed that the ionization of the gas and in particular the production of the highly ionized species is by photoionization from radiation produced by hot galactic stars (Bregman and Harringtion 1986) or from the extragalactic EUV background (York 1982; Hartquist, Pettini and Tallant 1984; Fransson and Chevalier 1985). Photoionized halo models have been successful at providing a possible explanation for the observed amounts of Si IV and C IV in gas at large distances from the galactic plane. However, the photoionization models have not been successful in explaining the observed amount of N V which appears to require the existence of collisionally ionized gas near 200,000 K.

A strong motivation for theorists to consider photoionization for the production of Si IV and C IV found in the halo was the observation by Pettini and West (1982) that Si IV and C IV only become abundant at heights of about 1 kpc above the galactic disk. However, the work of Savage and Massa (1987) revealed that while the z distribution of Si IV and C IV may exhibit an enhancement near 1 kpc, substantial densities of these species are also found in the low density medium of the galactic disk. Also, the very patchy distribution of interstellar Si IV and C IV absorption causes the N(ion)sin |b| versus |z| plots to have a considerable width, thus making it difficult to see jumps in N(ion)sin|b| near 1 kpc. While a jump of a factor of 2 or 3 in N(ion)sin |b| may exist near near 1 kpc (see Savage and Massa 1987) its reality is not certain.

In a recent extension of the photoionization models by Ito and Ikeuchi (1988), it was found necessary to include three gas phases: 1. A neutral gas phase with $T < 10^4$ K. 2. A photoionized gas phase containing CIV and Si IV with T near 10^4 K and 3. A hot collisionally ionized gas phase with $T \sim 2 \times 10^5$ K to explain the existence of N V. The support of gas in this model was proposed to be a galactic fountain driven by superbubble phenomena. However, another possibility is that the gas is supported by diffusing cosmic rays and that the high temperatures required to produce N V arise from cosmic ray heating of the gas (Hartquist and Morfill 1986).

The Pettini and West (1982) data also indicated that the bulk motions of C IV and Si IV toward or away from the galactic plane were small (e.g. < 20 km s^{-1}). Hartquist and Morfill took this as evidence favoring a relatively quiesent cosmic ray supported halo. With additional data it is now clear that the velocity structure of the highly ionized gas is more complex (see Danly 1987). In particular, toward the north galactic pole substantial negative velocity shifts are seen. The high ion profiles are broad with FWHM approaching 110 km s^{-1} for the most distant stars in the sample. Also from the theoretical point of view, it is not at all obvious that large velocities would be expected for these ions even if they are participating in a fountain flow. The velocities expected will depend sensitively on where in the flow a particular species becomes visible. If the lines of C IV and N V are produced in cooling condensations as advocated by

Edgar and Chevalier (1986), we might sometimes expect the species to occur near the peak height of the flow where the velocities are at a minimum. Houck and Bregman's (1989) kinematical calculations have demonstrated that a low halo driven by gas near 3×10^5 K can produce cool falling gas with a velocity range comparable with that observed. Toward the south galactic pole the neutral and highly ionized gas are relatively quiescent even to z distances of -3 kpc. Unless we just happen to be looking through a quiescent region of a more generally disturbed medium, it would appear the motions of a fountain have difficulty explaining the behavior of the gas toward the south galactic pole.

Composite Models

It is clear from the discussions above that although substantial progress has been made in studying halo gas from the observational and theoretical perspectives, much work still remains. It may be true that the eventual understanding of the support and ionization of gas in the halo will need to involve aspects of many of the different phenomena discussed above. It seems that all of the following statements are probably true:

1. The effects of supernovae in creating bubbles and superbubbles must be included in any theory of halo gas.
2. The kinematics of the gas provides evidence for highly disturbed regions (e.g. toward the north galactic pole) and relatively quiescent regions (e.g. toward the south galactic pole) . In the disturbed regions various types of flows (e.g. fountains and other disturbances) are probably influencing the z distribution of the gas. In the relatively quiesent regions other phenomena are probably important (e.g magnetic fields and cosmic rays).
3. Hot (~200,000 K) collisionally ionized gas exists and processes must be advanced to explain the heating of that gas.
4. Photoionization plays an important role in creating the weakly ionized gas in the halo and may also strongly affect the distribution of the highly ionized gas with the probable exception of N V and O VI.
5. It is reasonably likely that the relative importance of processes 1 to 4 above varies with position and time in the galaxy. There may be regions where the gas distribution is controlled by fountain activity and other regions where the magnetic field and cosmic rays control the thermal gas. The dynamical activity near the galactic center may be so great that the central region of the Galaxy exhibits a wind rather than a bound circulating fountain.
6. Given the relative youth of this subject we should be wary that many of our current prejudices may be completely wrong.

Acknowledgements

Support by NASA grant NAG5-186 is gratefully acknowledged. I have benefited from discussions with D.Cox, R. Edgar, J. Mathis , and R. Reynolds about the diffuse and the hot ISM. T. Hartquist communicated valuable comments about the support of halo gas by cosmic rays. R. Edgar and K. Sembach were helpful in providing comments about the manuscript.

REFERENCES

Albert,C.E. 1983, *Ap.J.*, **272**,509.
Beuermann,K.,Kanbach,G.,and Berkhuijsen, E.M. 1985, *Astr. Ap.*, **153**,17.

Blades, J.C., Wheatley, J.M., Panagia, N, Grewing M. Pettini, M., and Wamsteker, W. 1988, *Ap.J.(Letters)*, **332**, L75.
Bloemen, J.B.G.M. 1987, *Ap.J.*, **322**,694.
Bohlin, R.C., Savage, B.D., and Drake, J.F. 1978, *Ap.J.* , **224**,132.
Bregman,J.N. 1980, *Ap.J.*, **236**,577.
Bregman, J.N., and Harrington, P.J. 1986, *Ap.J.*, **309**,833.
Burrows,D.N., McCammon, D., Sanders, W.T., and Kraushaar, W.L. 1984, *Ap.J.*, **287**, 208.
Caldwell, J.A.R. 1979, Ph.D.Thesis, Princeton University.
Cash, W., Charles, P.,Bowyer, S., Walter, F. Garmire,G. and Riegler, G. 1980, *Ap.J.(Letters)*, **238,** L71.
Chevalier, R. A., and Fransson, C. 1984, *Ap.J. (Letters)*, **279**,L43.
Cowie, L.Songaila, A., and York, D.G. 1979, *Ap.J.*, **232**, 467.
Cox, D.P. 18-981, *Ap.J.*, **245**,534.
Danly, L. 1987, Ph.D. Thesis, University of Wisconsin-Madison.
_____. 1989, *Ap.J.*, (in press)
de Boer, K.S. and Savage, B.D.1984, *Astr. Ap.*, **136**, L7.
Edgar, R.J., and Chevalier, R.A. 1986, *Ap.J. (Letters)*, **310**, L27.
Edgar, R.J., and Savage, B.D. 1989, *Ap.J.*, **340**,762.
Fransson, C., and Chevalier, R.A. 1985, *Ap.J.*, **296**, 35.
Giovanelli, R. 1986, in *Proceedings of the NRAO Conference on Gaseous Galactic Halos*, eds. J.N.Bregman and F.J. Lockman , (Greenbank:NRAO SP), p.63.
Hartquist, T.W., and Morfill, G.E.1986, *Ap.J.*, **311**, 518.
Hartquist, T.W., Pettini, M., and Tallant, A. 1984, *Ap.J.*, **276**, 519.
Heiles, C. 1979, *Ap.J.*, **229**, 533.
_____. 1984, *Ap.J. Suppl.*, **55**, 585.
Houck, J.C. 1988 Masters Thesis, University of Virginia.
Houck, J.C. and Bregman, J.N. 1989, (preprint).
Hobbs,L.M., Morgan, W.W., Albert, C.E., and Lockman, F.J. 1982, *Ap.J.*, **263**, 690.
Ipavich, F.M. 1975, *Ap.J.*, **196**,107.
Ito,M., and Ikeuchi, S. 1988,*Publ. Astron. Soc. Japan*, **40**, 403.
Jenkins, E. B. 1978, *Ap.J.* , **220**, 107.
_____. 1986, in *Proceedings of the NRAO Conference on Gaseous Galactic Halos*, eds. J.N.Bregman and F.J. Lockman , (Greenbank:NRAO SP),p.1.
_____.1987a,in *Exploring the Universewith the IUE Satellite*,(Dordrecht:D.Reidel.Pub.Co.),p.531.
_____.1987b,in *Interstellar Processes*, eds. D.J.Hollenbach and H.A.Thronson,Jr., (Dordrecht:D.Reidel Pub.Co.), p.533 .
Jokipii,J.R. 1976, *Ap.J.*, **208**, 900.
Kaelble,A., de Boer, K.S., and Grewing, M. 1985, *Astr. Ap.*, **143**, 408.
Kulkarni, S.R., and Heiles, C. 1987, in *Interstellar Processes*, eds. D.J.Hollenbach and   H.A.Thronson,Jr.,   (Dordrecht:D.Reidel Pub.Co.), p.87 .
Kulkarni,S.R., and Fitch,M. 1985, *Ap.J.*, **289**, 792.
Lockman, F. J. 1984, *Ap.J.*, **283**, 90.
_____.1986, in *Proceedings of the NRAO Conference on Gaseous Galactic Halos*, eds. J.N.Bregman and F.J. Lockman, (Greenbank:NRAO SP), p.63.

Lockman, F.J., Hobbs, L.M., and Shull,M. 1986, Ap.J., **301**, 380.
Mac Low,M.-M., and McCray, R.C. 1988, Ap.J., **324**,776.
Martin, C., and Bowyer, S. 1989, Ap.J., (in press).
Marshall,F.J., and Clark,G.W. 1984, Ap.J., **287**, 633.
McCammon, D., Burrows, D.N., Sanders, W.T., and Kraushaar, W.L. 1983, Ap.J., **269**, 107.
McKee, C.F., and Ostriker, J.P. 1977, Ap.J., **218**, 148.
Morton,D.C., and Blades,J.C. 1986, M.N.R.A.S., **220**, 927.
Norman, C.A., and Ikeuchi,S. 1989, Ap.J. (in press).
Nousek,J.A.,Fried,P.M.,Sanders,W.T.,and Kraushaar,W.L. 1982, Ap.J., **258**, 83.
Oort, J.H. 1962, in *The Distribution and Motion of Interstellar Matter in Galaxies*, ed. L. Woltjer (New York:Benjamin), p.71.
Payne, H.E., Salpeter, E.E., Terzian, Y. 1983, Ap.J., **272**, 540.
Pettini, M., and West, K.A. 1982, Ap.J., **260**, 561.
Pettini, M. 1988, *Proc. Astro. Soc. Australia*, **7**, 527.
Reynolds, R.J. 1980, Ap.J., **236**, 153.
_____.1985, Ap.J., **285**, 256.
_____. 1986, in *Proceedings of the NRAO Conference on Gaseous Galactic Halos*, eds. J.N.Bregman and F.J. Lockman (Greenbank:NRAO SP),p.53.
_____. 1989, Ap.J.(Letters), **339**, L29
Reynolds, R.J., and Odgen, P.M. 1979,Ap.J., **229**, 942.
Savage, B.D. 1987a ,in *Interstellar Processes*, eds. D.J.Hollenbach and H.A.Thronson,Jr., (Dordrecht:D.Reidel Pub.Co.), p.123 .
_____. 1987b, in *QSO Absorption Lines:Probing the Universe*, eds. J.C.Blades, D.Turnshek, and C.A. Norman,(Cambridge:Cambridge Univ.Press), p.195.
Savage, B.D., and deBoer, K.S. 1979, Ap.J. (Letters) , **230**, L77.
_____. 1981, Ap.J., **243**,460.
Savage, B.D., and Massa, D. 1987, Ap.J., **314**, 380.
Savage, B.D., Massa, D., and Sembach, K. 1990, Ap.J. (submitted).
Shane, W.W.1971, Astr. Ap. Suppl., **4**,315.
Shapiro, P.R., and Field, G.B. 1976, Ap.J., **205**, 762.
Sofue, Y., Fujimoto,M., and Wielebinski, R. 1986, Ann. Rev. Astr. Ap., **24**, 459.
Songaila, A., Cowie,L.L., and Weaver, H. 1988, Ap.J., **329**,580.
Spitzer, L. 1956, Ap.J., **124**, 20.
_____. 1978, in *Physical Processes in the Interstellar Medium*, (New York: John Wiley and Sons).
_____. 1990, Ann. Rev. Astr. Ap. (in press).
Stecker, F.W. 1979, in *The Large-Scale Characteristics of the Galaxy*, ed.W.B. Burton, (Dordrecht:Reidel), p. 475.
Tenorio-Tagle, G., and Bodenheimer, P. 1988, Ap.J., **26**, 145.
van Steenberg,M., and Shull,J.M. 1988, Ap.J., **330**, 942.
_____.1989, Ap.J.,
Weaver, H. 1973, in *Highlights of Astronomy*, **13**, ed. G.Contopoulos, (Dordrecht:Reidel), p.423.
Wesselius, P.R., and Fejes,I. 1973, Astr.Ap., **24**,15.
York, D.G. 1982, Ann. Rev. Astr. Ap., **20**, 221.

DENSE GAS IN THE GALAXY

N.Z. SCOVILLE Astronomy Department, Mail Code 105-24, California Institute of Technology, Pasadena, California 91125

I. INTRODUCTION

Extensive observations of carbon monoxide emission in the disk of our galaxy have shown that molecular (H_2) rather than atomic (HI) hydrogen is the major active component of the interstellar medium. In the Galaxy, virtually all known regions of star formation activity are associated with molecular clouds. The relationship of these molecular clouds to other phases of the interstellar medium is now rather different than what was imagined two decades ago. The giant molecular clouds (GMCs) in which most of the molecules reside are self-gravitating with an effective internal pressure due to supersonic, bulk internal motions approximately an order of magnitude higher than the pressure of the other phases of the interstellar medium. Thus, these clouds are not in pressure equilibrium with the hotter, more diffuse phases; yet, the young stars which form within the GMCs are ultimately responsible for energizing the rest of the ISM through high mass OB stars and their subsequent supernovae explosions.

Surveys of CO emission in the disk of the Milky Way have shown that the abundance of molecular gas in the interior of the galaxy exceeds that of atomic hydrogen, and that there exists a much stronger radial gradient than is seen in the 21 cm HI emission. High resolution CO surveys also clearly demonstrate that the molecular gas is mostly contained in clouds of mass 10^5-10^6 $M_\odot$. These clouds may have associated atomic gas in a dissociative envelope but typically, except in regions of high mass star formation, the atomic gas constitutes <10% of the overall cloud mass. Although the giant molecular clouds are believed to be self-gravitating, their shapes are often elongated and irregular, indicating either that large scale magnetic fields are significant or, more likely, that there is significant energy input from associated star formation and external sources on time scales comparable with the dissipation times ($\sim 10^7$ years).

One of the major developments over the last decade has been the growing recognition that the chemistry and physics of the dense, molecular clouds is much more complex than for the other phases of the ISM. Observations at millimeter and sub-millimeter wavelengths have revealed over 2,000 spectroscopic transitions, which are identified with 65 molecules and their isotopic variants, with complexity ranging up to 11 atoms (eg. HC_9N). With the exception of the H_2 molecule, which probably forms on the surface of dust grains, most of the molecular abundances can be explained adequately by low temperature gas phase ion-molecule reactions. In most of the molecular gas, the temperatures are ~10 K which is the thermal equilibrium temperature expected for a balance between heating by low energy cosmic rays and cooling in the CO rotational transitions. In the immediate vicinity of active star forming regions, the cosmic ray heating is supplemented by H_2-grain heating, since the grain temperatures are elevated in the presence of the strong radiation field of the young stars. Additional mechanical energy input in the form of shocks may also be provided by the supersonic winds associated with the young stars. Thus, the molecular gas in the vicinity of active star forming complexes may have temperatures ranging between 20-3000 K depending on the length scale (1 pc to ~ 10 AU).

This review summarizes the most recent results concerning the large scale distribution of the molecular gas in the galaxy and the statistical properties of the GMCs as derived from CO line surveys. In §II, we review the basis for CO as a tracer of the galactic H_2. The large scale distribution in radius and with respect to spiral arms is discussed in §III, and the properties of the GMCs are described in §IV. §V presents quantitative estimates for the rate of star formation in the GMCs as derived from the IRAS far infrared data and §VI presents a brief summary of the detailed characteristics and phenomena seen in the closest, best studied molecular cloud: the Orion molecular cloud.

II. CO AS A TRACER OF H_2

Since molecular hydrogen itself has a small moment of inertia and no permanent dipole moment (due to the fact that it is made up of two identical nuclei and the center of mass coincides with the center of the charge distribution), it has no permitted, radio, or infrared transitions. Electronic transitions in the Lyman and Werner bands have been detected using the Copernicus satellite in low opacity clouds along the line of sight to nearby ultraviolet stars (cf. Spitzer and Jenkins 1975), but these transitions do not provide a general probe of H_2 in more distant and highly obscured regions. The rotation-vibration transitions in the infrared have been seen in 2000 K, shock-excited gas but are not detectable in lower temperature regions.

On the other hand, the theoretical and observational foundation for the use of the CO line as a tracer of H_2 gas is now well established. The J=1-0 CO transition at 115.271 GHz (λ=2.6 mm), corresponding to $h\nu/k$=5.5 K, is excited readily by collisions with H_2, even in clouds of very low kinetic temperature. The critical density for significant excitation of the J=1 state is approximately $n_{H_2} \geq 3000$ cm^{-3}. This value is appropriate if the excitation is determined solely by a balance between H_2-CO collisions and spontaneous decay. A lower critical density, reduced by a factor of $1/\tau$, is appropriate in cases where the lines are optically thick as is undoubtedly the case in essentially all GMCs. Thus, the low lying levels will be approximately thermalized ($T_x = T_k$) for $n_{H_2} \gtrsim 300$ cm^{-3} since $\tau \geq 10$ in most clouds. Measurements of the rare ^{13}CO isotope, for which the solar abundance ratio is 1/89, show intensities 0.5-0.1 of the CO intensity, providing strong evidence that the CO line is optically thick. In summary, it is generally expected that the J=1-0 CO transition will be thermalized at densities of a few hundred per cc and since the transition is optically thick, the observed brightness temperature will equal the gas kinetic temperature (once corrections are made for the cosmic background radition, radio telescope inefficiencies and beam dilution).

For quantitative analysis of the emission from giant molecular clouds, it has become common to refer to a quantity called the CO luminosity defined by,

$$L_{CO} \equiv T_{CO} \Delta V \pi R^2 \qquad (1)$$

where T_{CO} is the peak brightness temperature in the CO transition, ΔV is the line width, and R is the cloud radius. For clouds in virial equilibrium, $\Delta V = (GM/R)^{\frac{1}{2}}$ and

$$L_{CO} = (\frac{3\pi G}{4\rho})^{\frac{1}{2}} T_{CO} M \qquad (2)$$

In general, the cloud volume (rather than the density) is the dominant factor varying between GMCs of different mass. Thus, to the extent that most of the clouds have similar density (ρ) and temperature (T_{CO}, see below), the CO luminosity will be a proportional measure of the total mass of molecular gas contained in the telescope beam. The reason that this proportionality holds for optically thick lines is that *for a collection of clouds with varying mass but fairly constant density, the virial line-width increases linearly with cloud radius (ie. linearly with the column density of gas). Thus, an increasing mass of gas along the line of sight is registered in a linear increase in the width, ΔV, of the emission lines.* [For extragalactic observations where many unresolved clouds are contained in the antenna beam, Equations (1) and (2) remain valid with the change that R is the beam radius at the distance of the galaxy and T_{CO} is the apparent, beam diluted, brightness temperature.]

There have been several observational attempts to evaluate the constant of proportionality between the CO emission and H_2 column density and to test the linearity between the cloud mass and CO luminosity suggested by Equation 2. Four general techniques, all pseudo-empirical have been applied: measurement of the ^{13}CO emission which is presumably less saturated than CO; correlation of visual extinction in nearby dark clouds with the CO intensity; estimation of virial masses using the CO and ^{13}CO line widths and cloud sizes; and estimates of the total column density of nucleons as derived from gamma ray observations. The range inferred for the constant of proportionality between the H_2 column density and the integrated line of sight CO emission is 1.8-4.8×10^{20} cm^{-2} (K km s^{-1})$^{-1}$ (cf. Scoville and Sanders 1987). The relatively good agreement of these totally independent methods suggests that global mass estimates for molecular hydrogen are probably correct to a factor of 2, although there will undoubtedly be larger uncertainties when the relations are applied to individual clouds. Figure 1 shows the data obtained for a sample of 41 clouds containing giant radio HII regions and seven clouds without HII regions, illustrating a single

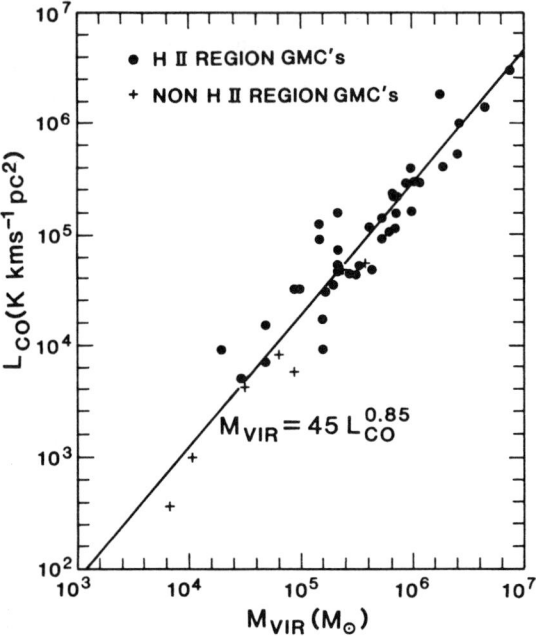

Figure 1. The variation of CO luminosity, L_{CO}, integrated over the projected area of the clouds is shown as a function of the virial mass as derived from the measured cloud diameter and velocity dispersion. The power law, $M_{VIR} = 45\, L_{CO}^{0.85}$ adequately fits the observational data for both cloud samples over a range of more than three orders of magnitude in cloud mass (Scoville and Good 1989).

constant of proportionality between the measured CO luminosity and the derived virial mass over a range of 3 orders of magnitude in cloud mass (Scoville and Good 1989).

Although most observational determinations have indicated an approximately linear relation between the integrated CO emission and the H_2 column density, Equation 1 does suggest that in some instances, the dependence on the gas temperature (T_K) and density $\rho^{-\frac{1}{2}}$ should show up. Thus, in galaxies with abundant high-mass star formation where the gas may be hotter, the H_2 mass may be over estimated using the same constant of proportionality determined for galactic GMCs. On the other hand, in these same regions, the increased cloud temperature may be somewhat offset by a corresponding increase in the mean density of the clouds.

Table 1

Recent Studies of H_2 in the Galaxy

MAJOR RECENT SURVEYS

	Coverage	Resolution	Completion
CO: Massachusetts/Stony Brook	North	0:8	1986 (1)
Goddard	North & South	8'	1987 (2)
^{13}CO: NRAO	North	1:1	1981 (3)
Bordeaux	North	4'	1983 (4)
Bell Labs	North and Gal. Center	2'	1987+ (5)

LARGE SCALE H_2 DISTRIBUTION

Axisymmetric: Sanders, Solomon and Scoville (1984)
              Bronfman et al (1987)

Non-Axisymmetric: Clemens, Sanders, and Scoville (1986)

CLOUD PROPERTIES

Stark (1979)
Liszt, Xiang, and Burton (1981)
Sanders, Scoville, and Solomon (1985)
Scoville et al (1986)

NOTES

(1) Sanders et al (1986), (2) Bronfman et al (1988), (3) Liszt, Xiang, and Burton (1981), (4) Despois and Baudry (1983), (5) Stark et al (1987)

III. THE LARGE SCALE GALACTIC DISTRIBUTION OF H_2

The first large-scale surveys of carbon monoxide in the inner galaxy were carried out by Scoville and Solomon (1975) and shortly thereafter by Burton *et al.* (1975). The principal result of both surveys was the discovery that the molecular emission is concentrated in a ring at 3-7 kpc radius. Scoville and Solomon (1975) also noted that most of the molecular material is confined in discrete clouds. Both of these characteristics are radically different from those of HI, for which the distribution is rather flat in radius with little tendency to be clumped (cf. Kulkarni 1987). Table 1 provides a summary of the characteristics of later much more extensive surveys carred out in both CO and ^{13}CO.

In Figure 2, the radial distribution of H_2 at the midpoint of the Z distribution is compared with the radial distributions for atomic hydrogen and giant radio HII regions. The dominant features in the molecular distribution are a strong concentration within 500 pc of the galactic center and the molecular cloud ring at 3-7 kpc radius. The ring feature is also prominent in the HII region distribution. In contrast, HI is seen to be relatively constant in radius and exhibits a depression in the central region of the galaxy. In Table 2, parameters for the galactic H_2 distribution are summarized (assuming $R_o=8.5$ kpc) and compared with the properties of the HI. Over all the

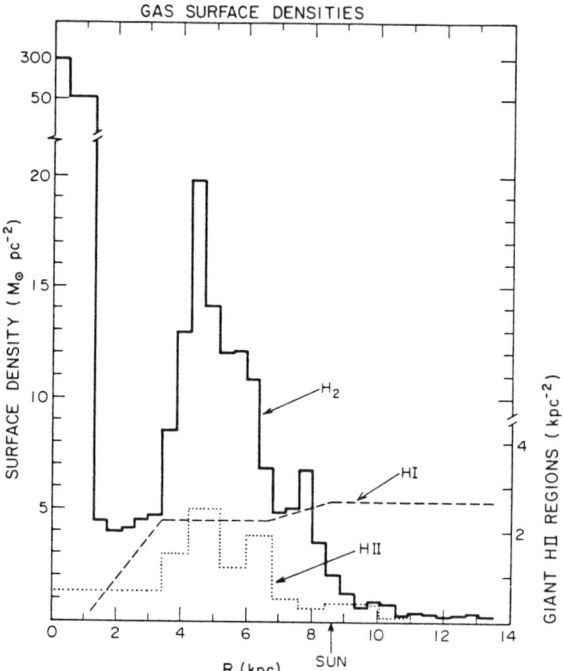

Figure 2. Comparison of gas surface densities in the Milky Way disk. Value of H_2 (Clemens, Sanders, and Scoville 1988) and HI (Burton and Gordon 1978) include a 1.36 correction factor for He and they have been scaled to $R_o=8.5$ kpc.

Galaxy, the total HI and H_2 masses are approximately equal within the uncertainities of their respective determinations; on the other hand, the radial distributions of mass in the two phases are entirely different. Approximately 90% of the H_2 mass lies inside the solar circle whereas only 30% of the HI mass is in the same region.

Whether or not all of the observed CO emission in the galactic disk is confined to a regular pattern of a few spiral arms or is more uniformly distributed into the interarm regions is critical to understanding both the origin and evolution of the molecular clouds. If the molecular clouds are short-lived, forming out of diffuse HI in the spiral arm shocks, and are then destroyed upon leaving the arms, their spatial distribution should follow the spiral potential, and a high contrast between the arm peaks and interarm troughs should be seen. On the other hand, if the clouds form by another mechanism, unrelated to the spiral wave, and survive through several spiral arm passages, they should be widely distributed throughout the disk and exhibit a low contrast between arm and interarm regions. Observational determinations of the CO concentration in spiral arms has been confused due to the uncertain location of the spiral arms and due to blending in the observed longitude-velocity plane as a result of the cloud velocity dispersion. Stark (1979) first pointed out that the classic arm and interarm zones in the inner galaxy contain equal numbers of small clouds (<30pc), but that the largest clouds appear to exist only in the arm reigons. A later analysis by Solomon, Sanders, and Revolo (1985) and Scoville et al. (1987) showed that the distribution of hotter molecular clouds (T_K >11 K) closely follows the l/v distribution of giant HII regions while the cooler clouds representing nearly 60% of the total Galactic CO emission are more uniformily distributed throughout the disk. In summary, there is good consensus that the hotter clouds are associated with the spiral arms and perhaps the larger more massive clouds tend to be found there also. On the other hand, there also appears to be a substantial fraction of the CO emission arising from interarm areas.

A definitive understanding of the distribution and properties of clouds relative to the spiral arms has awaited the advent of *high resolution* observations of other galaxies such as M51 for which the spiral pattern is well defined. Recently, the Owens Valley millimeter-wave interferometer has been used by Vogel, Kulkarni, and Scoville (1988) and Rand and Kulkarni (1989) to map CO in the spiral arms at a resolution of 7". These data show striking concentrations of the CO emission displaced by approximately 300 pc upstream from the arms seen in Hα. The locations of the CO emission maxima coincide precisely with regions of enhanced dust obscuration seen in R band images. Comparison of the interferometer fluxes with those obtained in single dish measurements indicates that approximately 25% of the total CO is in the arm structures, implying an arm-interarm contrast of 3:1 averaged over scales of 500 pc (Vogel *et al.* 1988). Even in the downstream areas where strong concentrations of CO emission are not apparent in the interferometer maps, the single dish data incidate that the surface density of molecular gas still exceeds that of atomic gas.

The M51 results suggest a picture in which pre-existing GMCs become concentrated in the spiral arms as a result of orbit crowding in the potential well of the spiral arm. The CO emission seen in the interferometer maps probably represents massive cloud *associations* rather than gravitationally bound structures. Within such associations, it is plausible that the rate of cloud-cloud collisions increases quadratically with the cloud number density; thus, a 3:1 increase in cloud density leads to an order of magnitude increase in the rate of OB star formation if high mass star formation is linked to cloud-cloud collisions (Scoville, Sanders, and Clemens 1987). The observed displacement between Hα and H$_2$ arms suggests a gestation period of approximately 3×10^7 years for the HII regions to form fully. In addition, the absence of cloud associations at the locations of the HII regions suggests that the associations disperse as they come out of the potential well. The coincidence of the HI ridge line with the optical HII regions suggests rather strongly that the HI is a dissocation product of high mass star formation within the molecular clouds, rather than a precursor. The extent to which this picture applies to the Galaxy is presently unclear since the H_2/HI mass ratio is somewhat larger in M51.

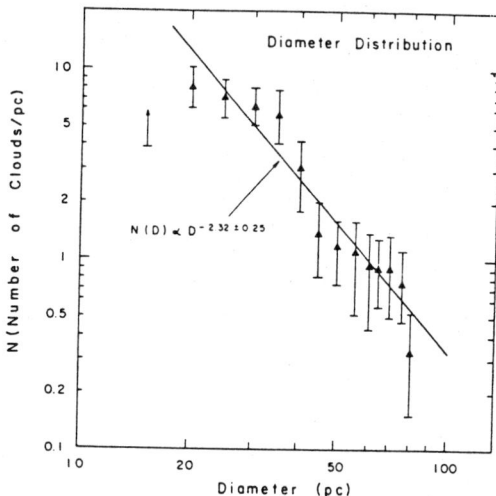

Figure 3. The number distribution of cloud diameters for clouds with D>10 pc within the sample volume studied by Sanders, Scoville and Solomon (1985). The distribution is adequately fit by a power law: $N(D) \propto D^{-2.32}$.

IV. PROPERTIES OF GMCs

The most complete sampling of the molecular cloud properties is provdied by the U. Mass-Stony Brook high resolution Galactic CO survey (Sanders *et al.* 1986). Over 1400 discrete emission regions have been cataloged from this survey (Scoville *et al.* 1987). This cloud sample and the earlier but more limited observations obtained by Sanders, Scoville, and Solomon (1985) provide a basis for the study of the cloud properties: their diameters, velocity dispersions, temperatures and masses. In Figure 3, we show the distribution of cloud diameters obtained by Sanders, Scoville, and Solomon (1985). This distribution may be fit by a power law $N(D) \propto D^{-2.32}$. Combining this cloud size spectrum with the observed relationship between size and velocity dispersion yields a virial mass spectrum

$$N(M) \propto M^{-1.61} \qquad (3)$$

This mass spectrum is similar to that expected if the larger GMCs are built up by aglommeration of smaller clouds (Scoville and Hersh 1979; Kwan 1979; Kwan and Valdes 1982). For the mass spectrum given in Equation (3), approximately 50% of the total galactic H_2 is contained in 1,000 clouds with

Table 3

Properties of a "Typical" GMC

Diameter	40 pc
M_{H_2+He}	4×10^5 $M_\odot$
$\overline{n}_{H_2}$	180 cm^{-3}
T_K	10 K
P_{TH}/k	2000 cm^{-3} K
σ_v	3.8 km s^{-1}

D>40 pc (M>4×10^5 $M_\odot$). The properties of a "typical" GMC of diameter 40 pc are given in Table 3. The time required for the GMCs in the molecular ring to double their mass by sweeping up smaller GMCs is approximately 2×10^8 years.

A similar mass spectrum is derived for clouds in the central disk of M33 by Wilson and Scoville (1989) from high resolution aperture synthesis CO mapping. One notable difference found in M33 is an upper mass cutoff in the spectrum at 4×10^5 $M_\odot$, above which no GMCs are seen despite the large number of smaller clouds detected. The opposite situation is seen in the central disk of the Galaxy. Here there exists a large number of clouds with mass exceeding 10^6 $M_\odot$ (eg. Sgr B2; Scoville *et al.* 1975, Bally *et al.* 1988) with velocity dispersions and internal densities a factor of 5-10 greater than the typical Galactic ring GMCs. The observed variations in the upper mass cutoff between the Galactic ring, M33, and the Galactic center may be due to variations in the mean ISM density or the rate of star formation per unit mass in the different regions (see Wilson and Scoville 1989).

Although the *thermal* pressure of the molecules ($P_{TH}/k \sim 2000$ cm^{-3}K) is similar to that of the diffuse interstellar medium, the observed velocity dispersion within the GMCs (Table 3) is far in excess of that expected for 10 K gas. Thus, the *effective* internal pressure is at least an order of magnitude higher than that of the intercloud medium. Since the virial masses estimated from the observed velocity dispersions are consistent with mass estimates based on molecular column density measurements, we now view the GMCs *as self-gravitating rather than in pressure equilibrium with other phases of the ISM.*

V. STAR FORMATION RATES IN GMCs

Inasmuch as the total column density of a typical GMC is $N_{H_2} = 2 \times 10^{22}$ cm^{-2} (Table 3), corresponding to a visual distinction of 20 magnitudes, most of the luminosity from recently formed stars is absorbed by the dust and reradiated in the far infrared. The IRAS survey has provided us with the first complete accounting of the luminosity from young star forming regions within GMCs. Over the past few years, there have been several studies (Sodroski et al. 1987; Solomon et al. 1988; Scoville and Good 1989) which have attempted to assess the infrared properties of GMCs in the inner disk of the galaxy. Sodroski et al. (1987) analyzed the large scale distribution of CO, H_2, HI, and HII and ascribed approximately equal percentages of the total far infrared emission to each phase. Solomon et al. (1988) and Scoville and Good (1989) analyzed in detail the infrared properties of individual GMCs. Scoville and Good (1989) find a mean luminosity-to-mass ratio $L_{IR}/M_{VIR}=12\pm2$ $L_\odot/M_\odot$ for clouds containing radio HII regions. This luminosity-to-mass ratio is a factor of 10 higher than that obtained for a comparison sample of clouds not associated with HII regions and a factor of 4 greater than the galactic average for molecular clouds. Only a weak correlation was found between the luminosity-to-mass ratio and the cloud virial masses, suggestive that the star formation efficiency is fairly independent of the mass or size of an individual cloud and depends more critically on other properties such as galactic location (Scoville and Good 1989; Solomon et al. 1988).

For the range of longitudes associated with the molecular cloud ring, the mean ratio of L_{IR}/M_{H_2} is 2.8 $L_\odot/M_\odot$ (Scoville and Good 1989). The infrared (1-500 μm) luminosity of the galaxy including both the galactic disk and the galactic center region is estimated to be approximately 7×10^9 $L_\odot$. If this luminosity is generated by O, B, and A stars, the required steady state star formation rate for the galaxy is $\dot{M}_{OBA}=0.5$ $M_\odot yr^{-1}$. For a standard Salpeter initial mass function, the total rate of star formation is approximately 5-10 times greater than this estimate. Adopting an overall star formation rate of 3 $M_\odot yr^{-1}$ which is consistent with that derived from the radio continuum luminosity of the galaxy (Mezger 1978), the time required to cycle the entire galactic mass of molecular hydrogen through stars is approximately 10^9 years. Although the cycling time is short compared to the age of the galaxy, it is long compared to the free fall time ($\sim10^7$ years) for the GMCs. Thus, the effective star formation efficiency judged relative to the free fall time is only $\sim1\%$ and the clouds should be viewed as relatively stable objects, not in free fall collapse. That is, the macroscopic bulk motions, evidenced by the internal velocity dispersions, are either not readily damped out on a free fall timescale (equivalently the cloud crossing time scale) or if these motions are dissipated, then the energy must be replenished on a similar time scale.

VI. THE ORION MOLECULAR CLOUD

The Orion molecular cloud at a distance of 450 pc is the nearest and best studied star formation region; it thus, provides an illustrative example of the diverse and complex phenomena presumably occuring in most Galactic GMCs with active OB star formation.

The full extent of the molecular gas in Orion is best judged from the low resolution CO survey of Maddalena et al. (1986) and the infrared emission mapped by IRAS (cf. Robinson 1984). The complex consists of two GMCs, one associated with the HII region Orion B on the north, and the other associated with the Orion nebula (Orion A) to the south. Both clouds are elongated parallel to the galactic plane with long dimension approximately 5°(50 pc) and total gas masses of approximately 10^5 $M_\odot$ each (Maddelena et al. 1986). Within these two molecular clouds are at least six core regions with recent or ongoing high mass star formation. The overall mean density of the Orion A and B clouds is approximately 10^3 cm^{-3} but both clouds have a spine of higher density material (n_{H_2} >10^4 cm^{-3}) running along their long dimension. Bally et al. (1987) have mapped the Orion A ridge in ^{13}CO , finding a S-shaped filament in the central 10 pc and more than 100 individual density condensations of size typically 1 pc. The masses of these condensations are in the range 10-10^2 $M_\odot$. The center of the S-shaped structure coincides with the two high mass star forming regions: Orion A and the BN-KL infrared cluster. At high resolution Mundy et al. (1988) resolved the center of this ridge into at least five dense molecular gas clumps of size 0.02 pc and mass ~10 $M_\odot$. Two of these condensations appear to be rotating about an axis perpendicular to the molecular cloud ridge.

The magnetic field structure associated with the Orion A molecular cloud has been studied through optical, near infrared, and far infrared polarization measurements. On large scales, the optical polarization indicates a magnetic field direction roughly aligned with the long axis of the cloud (Appenzeller 1966, Vorba, Strom, and Strom 1988). Similarily, for the core region of OMC1, the near infrared (Dyck and Lonsdale 1979) and far infrared measurements (Hildebrand et al. 1984) suggest magnetic fields parallel to the molecular ridge. The field strengths derived from 21 cm HI and 18 cm OH Zeeman measurements are in the range 50-100 μG (Heiles and Stevens 1986; Heiles and Troland 1982; Kazes and Crutcher 1986). Higher field strengths (5-40 mG) are deduced for the OH and H_2O masers in the BN-KL complex (Chaisson and Beichman 1975; Hansen et al. 1977; and Norris 1984). Magnetic fields with these strengths could be significant to the overall structure of the cloud; nevertheless, the orientation of the field parallel to the long axis of the cloud implies that the overall cloud collapse occured perpendicular to the magnetic fields, thus, suggesting that the magnetic forces were not a significant determinant for the present cloud structure.

Aside from the five regions of high mass star formation distributed within the two Orion clouds, recent near infrared studies have shown much more extensive population of lower mass stars (Nakajima et al. 1986; McCaughrean 1988). Within the selected area surrounding the Trapezium cluster, the near infrared survey reveals a sufficiently large number of lower mass stars that one infers a high efficiency of star formation (>10% of the initial gas mass) in these core regions (cf. Genzel and Stutzki 1988). At present it is unclear whether the high and low mass stars form contemporaneously or whether the low mass star formation occurs for a much longer time and is more uniformly distributed throughout the cloud.

In addition to the phenomena described above which relate mostly to the pre-star formation structure of the cloud, there exists a wealth of observations for the core of OMC1 which pertain to the after effects of star formation on the surrounding gas and dust. Within the central 0.2 pc surrounding the BN-KL Trapezium clusters highly supersonic gas motions (up to 150 km s^{-1} with respect to the cloud rest velocity) are observed in the molecular gas and strong near infrared and far infrared emission associated with shock front cooling has been observed in the vibrational transitions of H$_2$ and high rotational transitions of CO. These phenomena are described comprehensively in the review article by Genzel and Stutzky (1988).

ACKNOWLEDGEMENTS

It is a pleasure to acknowledge Leona Kershaw's preparation of this manuscript and research support under the IRAS Extended Mission and General Investigator programs and NSF Grant AST 87-14405.

REFERENCES

Appenzellar, I. 1966, *Z. Astrophys.*, **64** 296.
Bally, J., Langer, W.D., Stark, A.A., and Wilson, R.W. 1987, *Ap.J. (Letters)*, **312**, L45.
Bally, J., Stark, A.A., Wilson, R.W., and Henkel, C.J. 1987, *Ap.J. Suppl.*, **65**, 13.
Bronfman, L., Cohen, R.S., Alvarez, H., May, J., and Thaddeus, P. 1988, *Ap.J.*, **324**, 248.
Burton, W.B. and Gordon, M.A. 1978, *Astr. Ap.*, **63**, 7.
Burton, W.B., Gordon, M.A., Bania, T.M., and Lockman, F.J. 1975, *Ap.J.*, **202**, 30.
Chaisson, E.J. and Beichman, C.A. 1975, *Ap.J. (Letters)*, **199**, L39.
Clemens, D.P., Sanders, D.B., and Scoville, N.Z. 1988, *Ap.J.*, **327**, 139.
Cohen, R.S. and Thaddeus, P. 1977, *Ap.J. (Letters)*, **217**, L155.
Dame, T.M., Elmegreen, B.G., Cohen, R.S., and Thaddeus, P. 1986, *Ap.J.*, **305**, 892.
Dame, T.M. *et al.* 1987, *Ap.J.*, **322**, in press.
Despois, D. and Baudry, A. 1983, in *Surveys of the Southern Galaxy*, ed. W.B. Burton and F.P. Israel (Dordrecht: Reidel), p. 173.
Dyck, H.M. and Lonsdale, C.J. 1979, *Astron. J.*, **84**, 1339.
Genzel, R. and Stutski, 1988, *Ann. Rev. Astr. and Astrophys.*, **27**.
Gordon, C.P. 1970, *Astr. J.*, **75, No. 8**, 914.
Haslan, C.G.T. and Osborne, J.L. 1987, *Nature* **327**, 211.
Hauser, M.G. *et al.* 1984, *Ap.J.*, **285**, 74.
Heiles, C. and Troland, T.H. 1982, *Bull. AAS*, **17**, 570.
Heiles, C. and Steven, M. 1986, *Ap.J.*, **301**, 331.
Hildebrand, R.H., Dragovan, M., and Novak, G. 1984, *Ap.J. (Letters)*, **284**, L51.
Kazes, I. and Crutcher, R.M. 1986, *Astr. Ap.*, **164**, 328.
Kulkarni, S. and Heiles, C. 1987, in *Interstellar Processes*, eds. D.J. Hollehbach and H.A. Thronson (Dordrecht: Reidel) p. 87.
Kwan, J. 1979, *Ap.J.*, **229**, 567.
Kwan, J. and Valdes, F. 1983, *Ap.J.*, **271**, 604.
Liszt, H.S., Xiang, D., and Burton, W.B. 1981, *Ap.J.*, **249**, 532.

Maddalena, R.J., Morris, M., Moscowitz, J., and Thaddeus, P. 1986, *Ap.J.*, **303**, 375.
McCaughrean, M.J. 1988, Ph.D. Thesis, University of Edinburgh.
Mundy, L.G., Cornwell, T.J., Masson, C.R., Scoville, N.Z., Baath, L.B., and Johansson, L.E.B. 1988, *Ap.J.*, **325**, 382.
Rand, R. and Kulkarni, S. 1989, *Ap.J. Letters*, (in press).
Robinson, L.J. 1984, *Sky and Telescope*, **67**, 4.
Robinson, B.J., Manchester, R.N., Whiteoak, J.B., Sanders, D.B., Scoville, N.Z., Clemens, D.P., and Solomon, P.M. 1984, *Ap.J. (Letters)*, **283**, L31.
Sanders, D.B., Clemens, D.P., Scoville, N.Z., and Solomon, P.M. 1986, *Ap.J. Suppl.*, **60**, 1.
Sanders, D.B. and Scoville, N.Z. 1987 in *Evolution of Galaxies*, Proc. 10th Europ. Reg. Meeting IAU, ed. J. Palous, p. 451.
Sanders, D.B., Scoville, N.Z., and Solomon, P.M. 1985, *Ap.J.*, **289**, 373.
Sanders, D.B., Solomon, P.M., and Scoville, N.Z. 1984, *Ap.J.*, **276**, 182.
Scoville, N.Z. and Good, J.C. 1989, *Ap.J.*, **339**, 149.
Scoville, N.Z. and Hersh, K. 1979, *Ap.J.*, **229**, 578.
Scoville, N.Z. and Sanders, D.B. 1986, in *Interstellar Processes*, ed. H. Thronson and D. Hollenbach (Dordrecht: Reidel).
Scoville, N.Z., Yun, M.S., Clemens, D.P., Sanders, D.B., and Waller, W.H. 1987, *Ap.J. Suppl.*, **63**, 821.
Sodroski, T.J., Dwek, E., Hauser, M.G., and Kerr, F.J. 1987, *Ap.J.*, **322**, 101.
Solomon, P.M., Rivolo, A.R., Barret, J., and Yahil, A. 1987, *Ap.J.*, **319**, 730.
Solomon, P.M., Sanders, D.B., and Rivolo, A.R. 1985, *Ap.J. (Letters)*, **292**, L19.
Spitzer, L. and Jenkins, E.B. 1975, *Ann. Rev. Astr. Ap.*, **13**, 133.
Stark, A.A. 1979, Ph.D. Thesis, Princeton University.
Stark, A.A., Bally, J., Knapp, G.R., Krahnert, A., Penzias, A.A., and Wilson, R.W. 1987, AT& T Bell Labs CO Survey, in progress.
Strom, K.M., Strom, S.E., Wolff, S.C., Morgan, J., and Jaffe, D.T. 1988, *Ap.J. Suppl.*, **62**, 39.
Vogel, S.N., Kulkarni, S.R., and Scoville, N.Z. 1988, *Nature*, **334**, 402.
Vrba, F., Strom, S., and Strom, K.M. 1988, *Astron. J.*, **96**, 680.

INTERSTELLAR DUST AND EXTINCTION

JOHN S. MATHIS
Washburn Observatory, University of Wisconsin,
475 N. Charter St., Madison, WI 53706

ABSTRACT There are substantial differences in the optical and ultraviolet extinction properties of interstellar dust in various directions, but a meaningful mean extinction law can be characterized surprisingly well by just one parameter, which can be taken to be $R_V = A(V)/E(B-V)$. The "standard" extinction laws of (Seaton 1979; Savage and Mathis 1979) are appropriate to $R_V = 3.2$, close to the value for the diffuse ISM. Many "peculiar" extinction laws, such as in the outer regions of dense clouds, are simply other values of R_V. To within present observational errors, near-infrared extinction is the same for all lines of sight. Grain diagnostics other than extinction show that silicates are present. Properties of the $\lambda 2175$ "bump" are discussed. Theories are listed briefly; almost all explain the bump with small graphite, but differ considerably regarding the optical extinction.

1. INTRODUCTION

Recently interstellar dust has been the subject of intensive study, partly because of careful observations of interstellar extinction throughout the entire optical/ultraviolet (UV) wavelength region. The discovery of spectroscopic features in the near-infrared (NIR), 3 - 13 µm, which provide detailed information as to the nature of some interstellar material (hydrocarbons), have also stimulated interest and prompted several laboratory studies of candidate materials.

It has become clear that interstellar grains along various lines of sight have widely varying properties. In this paper, dust observed only through low-density regions will be called "diffuse dust", while dust observable *in the UV* ($\lambda \geq 0.12$ µm) along lines of sight to stars in molecular clouds will be "outer-cloud dust". Lines of sight to sources deep within molecular clouds, where only optical or even NIR observations are possible, will sample "inner-cloud dust". Of course, there is really a continuous gradation between these three types.

Recent general references regarding interstellar dust are: (a) the proceedings of a workshop held at Wye, Maryland, U.S.A.,

February 1985 (Nuth and Stencel 1986); (b) the conference "Dust in the Universe", December, 1987, in Manchester, UK, (Bailey and Williams 1988), and (c) IAU Symposium 135, "Interstellar Dust", held at Santa Clara University, U.S.A., July, 1988 (Allamandola and Tielens 1989).

In this article, I will concentrate upon some recent developments regarding the interstellar extinction law in diffuse dust and outer-cloud dust, since most astronomers want to be able to correct observations for the effects of dust rather than study the nature of the grains. There are excellent reviews of inner-cloud dust in Whittet (1987), Tielens and Allamandola (1987), and Tielens (1989). After discussing extinction in §2, I will mention other diagnostics and what information they provide (§3) and summarize present grain theories (§4).

2. INTERSTELLAR EXTINCTION

The most important property of interstellar grains is that they cause extinction (scattering and absorption) of radiation from the near-infrared (NIR; $\lambda < 5$ µm) through the observable ultraviolet (UV), ionizing UV, and X-ray spectral regions.

2.1. Continuous extinction in the Optical and Observable UV

Let $A(\lambda)$ be the extinction along the line of sight to a particular object. Each line of sight has its own extinction law, or wavelength variation of extinction. The simplest way to express the extinction law is $A(\lambda)/A(V)$, where $A(\lambda)$ is the extinction (in magnitudes) at wavelength λ, and V denotes 0.55 µm. There are alternative ways of expressing the extinction law, such as the ratio of two colors, $E(\lambda-V)/E(B-V)$, where $E(\lambda-V) = A(\lambda) - A(V)$. The information content of expressing the extinction law as $A(\lambda)/A(V)$ and $E(\lambda-V)/E(B-V)$ is the same, provided that R_V, the ratio of total-to-selective extinction [$R_V = A(V)/E(B-V)$], is known. However, the use of $A(\lambda)/A(V)$ seems more physical, since it is directly proportional to the extinction itself.

The extinction law throughout the optical region has been studied for many years (e.g., Whitford 1958); see also Ardeberg and Virdefors 1982), and UV extinction has been discussed in countless papers. Fitzpatrick and Massa (1986, 1988; hereafter FM 86, FM88) have provided a particularly convenient form of expressing UV extinctions determined from low-dispersion spectra of *IUE* (3.3 µm$^{-1} \leq \lambda^{-1} \leq$ 8 µm^{-1}). They considered 45 early-type galactic stars distributed over the sky, including several "peculiar" directions into molecular clouds and H II regions. They used $E(\lambda-V)/E(B-V)$ to describe the UV extinction law. Remarkably, all extinctions in their sample can be represented by the sum of only three mathematical functions: (1) a linear component increasing with λ^{-1}; (2) the "bump", the feature centered on about 4.60 μm^{-1} (= 0.2175 μm) in excess over the linear background; and (3) for $\lambda^{-1} > 5.9\ \mu m^{-1}$ (the "far-UV rise"), a specific cubic polynomial in $(\lambda^{-1}$-$5.9\ \mu m^{-1})$ in excess of the linear component.

The shape of the bump, $A_{bump}(\lambda^{-1})$, is fitted by an asymmetric Drude function (Bohren and Huffman 1983) somewhat better than a symmetric Lorentzian. The Drude function is the amplitude of a damped harmonic oscillator when the width of the resonance is not small with respect to the central frequency. The bump has three parameters (the amplitude, the central wavenumber, λ_0^{-1}, and the full-width-half-maximum, γ). The linear portion has two parameters, and the far-UV rise has one, for a total of only six. These six parameters serve to describe every extinction law in FM86 and FM88 to within the very small observational errors. Note, though, that there is no physical significance to the fitting functions used except for the Drude profile of the bump.

There are considerable differences in both the optical and the UV spectral regions along various lines of sight. Cardelli, Clayton, and Mathis [1988; 1989 (CCM)] found a rather tight relationship between the optical/NIR and UV spectral regions, and one-parameter mean extinction law which holds for both diffuse dust and outer-cloud dust. The optical extinction law can be characterized by R_V. Diffuse dust has $R_V \approx 3.1$, while lines of sight through outer-cloud dust have $R_V \approx 4 - 6$. Figure 1 (from CCM) shows that there are meaningful linear relationships between $A(\lambda)/A(V)$ and R_V^{-1} at various wavelengths, displaying the existence of a relation between the optical extinction law and the UV. The figure shows that there is a continuous change between the "diffuse dust" (small R_V) and "outer-cloud dust" (large R_V) extinction laws.

CCM fitted the slopes of the various $A(\lambda)/A(V)$ - R_V^{-1} relations, examples of which are shown in Figure 1, by an analytic formula which represents the mean extinction law as a function of R_V. The expression will not be reproduced here. Figure 2 shows the mean extinction law for three values of R_V as calculated from the formula. Also shown are actual observations with the same values of R_V. The central panel is about as discrepant as actual observations are from the mean relationship. The dispersion of individual extinction laws around that mean law is shown in Figure 1 (from the spread in the individual observed points) and in Figure 2 (as error bars). The lowest set of curves plotted in Figure 2 are for Herschel 36, an exciting star of M8 and considered to have very "peculiar" extinction. Rather, it has a rather peculiar value of R_V (≈ 5.3), but a "normal" extinction law for that value of R_V. Note, however, that there are real deviations from the mean extinction law for any particular values of R_V. These deviations are especially large at 1200 Å, where the standard deviation of $A(\lambda)/A(V)$ from the mean extinction law is about 0.3. The error bars in the panel in Figure 2 shows the standard deviations from the mean for the stars in the CCM sample at several wavenumbers.

Two "mean" extinction laws (Savage and Mathis 1979, Seaton 1979) are commonly used to correct observations for the presence of dust. Both laws are reproduced closely if $R_V = 3.2$ is substituted into the R_V-dependent mean extinction law given in CCM. The panel in Figure 2 shows the deviations of the CCM mean extinction law, with $R_V = 3.2$, from Seaton (1979). It is not surprising that a mean extinction law mainly based on lines of sight through diffuse dust would correspond to an R_V which is close to 3.1 but slightly higher because some lines of sight penetrate somewhat denser regions than average.

There are real deviations of individual stars from the mean extinction laws. Fig. 3 shows all the extinctions of the stars in the FM88 sample which have values of R_V between 3.3 and 3.5, and also 5.2 and 5.6. The difference in the respective mean extinction laws is clear. The dispersions among the extinction laws within each group, especially at $\lambda \approx 0.12$ µm, is also evident, and is much larger than the photometric errors. Perhaps extreme examples of deviations from the mean extinction law will be found in the future. These deviations will provide valuable information regarding the processes which modify the grains.

The differences of the mean extinction law between diffuse dust ($R_V \approx 3$) and outer-cloud dust ($R_V > 4$ in many cases)

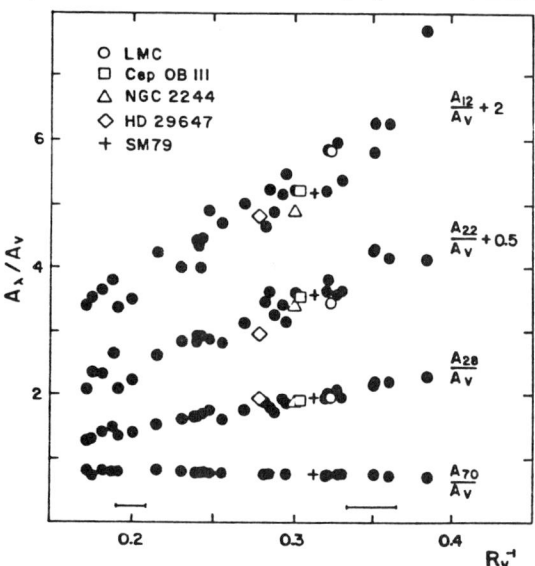

Fig. 1. — Values of $A(\lambda)/A(V)$, plotted against $1/R_V$, where $R_V = A(V)/E(B-V)$ (from Cardelli, Clayton, and Mathis 1989). A12 refers to 1200 Å, A22 to 2175 Å, A28 to 2800 Å, and A70 to 7000 Å (the standard R filter).

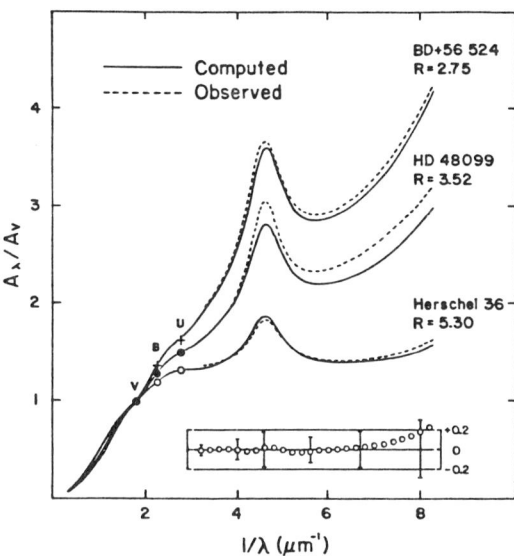

Fig. 2. — Three cases of a mean extinction law, obtained by fitting the slopes of the $A(\lambda)/A(V)$ - R_V^{-1} relationship (above) by an analytic formula, and actual extinction laws of stars with the appropriate values of R_V.

strongly affect any predictions of physical conditions in clouds. Figure 2 shows that the extinction laws for dust with large values of R_V rise much less steeply at shorter wavelengths than does diffuse-dust extinction. Consequently, interstellar radiation incident upon a cloud can penetrate the cloud much more easily than would be predicted from the Seaton or Savage-Mathis extinction laws. Figure 4 shows the mean radiation intensity at the center of a cloud with a radial extinction of $A(V) = 5$ magnitudes, a typical value, computed using $R_V = 3.1$, and also with a typical outer-cloud dust value of $R_V = 5$. The differences in the predicted mean intensities are impressive, and have large implications for the physical processes (molecular dissociation, gaseous chemistry, etc.) in the cloud.

The dependence of the mean extinction law on R_V has considerable implications regarding the nature and evolution of grains. The lines of sight plotted in Figure 1 represent directions all over the sky. In order to produce such regular extinction laws for diverse conditions, the processes which modify the sizes and/or the compositions of grains must operate simultaneously and quite efficiently over a wide range of sizes. It is very conceivable that for various lines of sight the small and large grains would be modified independently, but such is not the case.

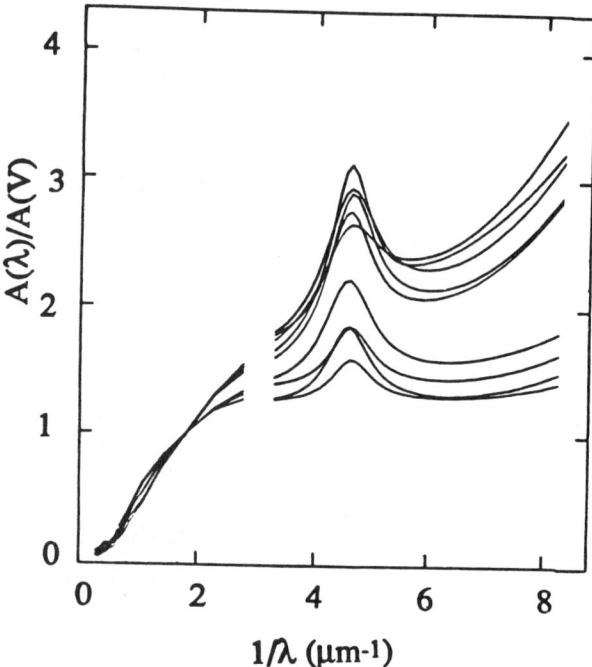

Fig. 3. — The extinction laws for all stars in the FM88 sample which have $3.3 < R_V < 3.5$ and also $5.2 < R_V < 5.6$, showing typical deviations from the mean extinction law. No ground-based ultraviolet is shown because actual observations were not used.

2.3. The λ2175 bump

The bump is the strongest spectral feature in the entire spectrum in terms of equivalent width in frequency units (over 20 times stronger than Lyman-α). Its properties, recently discussed by FM86, Draine (1989), and Cardelli and Savage (1988), are:

1. The bump is so strong that almost surely it is caused by carbon in some form (although Duley, Jones, and Williams 1989 disagree). The equivalent width for the bump in wavenumber units per E(B-V), W, is given by the relation with an oscillator strength, f_{bump}, and column density of absorbers, N_{bump}:

$$W = \int A_{bump}(\lambda^{-1})/E(B-V) \, d\lambda^{-1}$$
$$= (\pi e^2/1.086 m_e c^2) \, f_{bump} \, N_{bump}/E(B-V).$$

FM86 provide values of $A_{bump}(\lambda^{-1})/E(B-V)$. Bohlin, Savage, and Drake (1979) give values of $N(H)/E(B-V)$ for the diffuse ISM. Eliminating E(B-V), we find for the diffuse ISM that N_{bump}

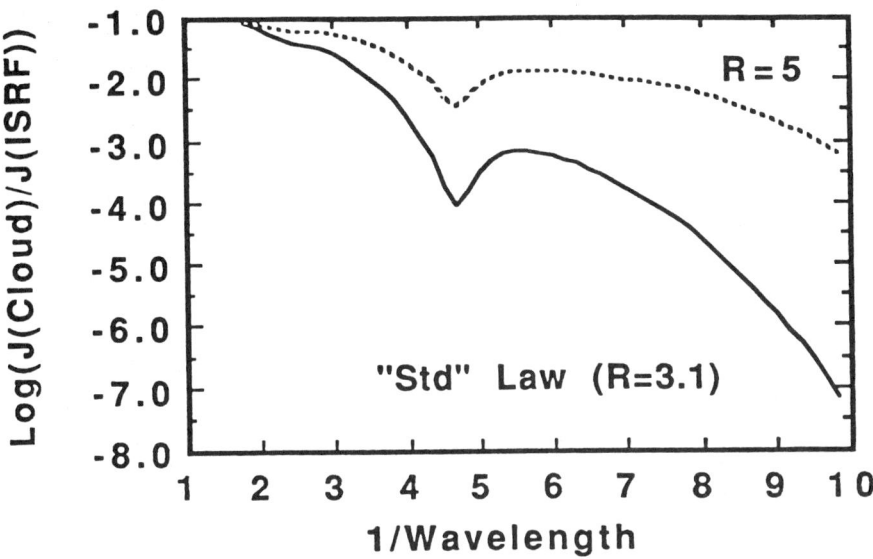

Fig. 4. — The mean intensity of radiation at the center of a cloud with a radial extinction of A(V) = 5 magnitudes and no internal sources, expressed in terms of the mean intensity of the incident interstellar radiation field. Two values of R_V are shown: one corresponding to the mean value for diffuse dust (R_V = 3.1), and one for a typical observed value for lines of sight penetrating clouds (R_V = 5).

f_{bump} = 9.3 x 10^{-5} N(H). Except for noble gases, only C, N, O, Mg, Si, and Fe can possibly provide the requisite atoms. If we assume f_{bump} = 1 as an upper limit (over twice as strong a transition as Lyman-α!), we need >30% of any of the elements Mg, Si, Fe, 10% of N, but <3% of C and O. Polarization studies show that aligned, *large* (radius > 0.1 µm) silicate grains are present, and much of the Si, Fe, and Mg must be contained in those. Material in the interior of large grains cannot provide the bump because of shielding by the outer portions, so virtually every atom of Si, Fe, or Mg outside of the large grains must be assumed to provide the bump, even if the bump transition has an oscillator strength of unity. This scenario is possible, but I think must be considered unlikely.

    2. The central wavenumber of the bump, λ_0^{-1}, is almost invariant among diverse extinction laws. In the FM86 sample, $<\lambda_0^{-1}>$ = (4.599 ± 0.022) µm^{-1}, or λ_0 = (2174.4 ± 10.3) Å, a

standard deviation of only 0.5%. The extreme values of λ_0^{-1} are 4.633 μm^{-1} (2158 Å) and 4.559 μm^{-1} (2193 Å).

Even within the FM86 sample, there are real deviations of λ_0. The intrinsic measurement error of λ_0 was estimated to be about ± 0.01 μm^{-1}, or about half of the observed dispersion. There is no relationship between the values of λ_0 and other parameters, including the width.

Since FM86, two stars with values of λ_0 of 2128 Å (HD 29647) and 2110 Å (HD 62542) have been discovered (Cardelli and Savage 1988), and another with a similar λ_0 will follow (Cardelli, private communication). These bumps are all broad but with a weak central absorption, so that their total equivalent width is not remarkable. They all have notably strong far-UV rises. Their lines of sight are very diverse; HD 62547 is in the Gum Nebula, near ζ Pup, while HD 29647 is in a quiescent region in Taurus.

3. The width of the bump, γ, varies considerably from one line of sight to another, with extreme variations over a factor of two (from $\gamma = 0.768$ μm^{-1} to 1.62 μm^{-1}). There is no relation between γ and λ_0 or any other bump parameter, which is interesting because coatings on grains, if they are responsible for producing the bump, should relate λ_0 and γ. The only significant correlation is that γ increases with the mean gas density along the line of sight, while $A_{bump}(\lambda_0)$ and λ_0^{-1} vary independently of mean gas density.

4. Witt, Bohlin, and Stecher (1986) have reported scattering in the bump in two reflection nebulae, each with a different profile. Many nebulae show no scattering. The scattering indicates that large particles can produce the bump, since small ones produce only absorption. The albedo may drop across the bump (Lillie and Witt 1976), but this result should be taken with caution since it relies on a simple modeling of the UV sky brightness, which is still controversial (compare Tennyson et al. 1988 with Jakobsen et al. 1984).

2.4. NIR Extinction

At present there is little evidence for any variation of the extinction law with R_V or any other parameter for $\lambda > 1$ μm (Jones and Hyland 1980; Whittet 1988 and refs therein). The extinction for $\lambda < 5$ μm is fitted well by a power-law such as $A(\lambda)/A(J) = (\lambda/1.25 \mu m)^{-1.70}$.

2.5. NIR absorptions

The main absorption features in the NIR are the two silicate bands at 9.7 μm and 18 μm, seen in the diffuse ISM towards sufficiently reddened stars and also in circumstellar dust surrounding oxygen-rich stars. The 9.7 μm band is the Si-O stretch; the 18 μm is the Si-O-Si bending mode. The 9.7 μm is seen in emission in both warm interstellar dust, such as in the

Orion Nebula (Gillett et al 1975a) and in circumstellar dust (Pegourie and Papoular 1985; Little-Marenin 1986). The profile of the 9.7 μm feature varies over 40% in width among various objects (Roche and Aitken 1984 and above refs.), but does not show the spectral structure of crystalline cometary or lunar silicates (Sandford and Walker 1985), which have their peak absorption at 10.5 μm. Laboratory measurements of amorphous or radiation-damaged silicates (Day 1979; Krätchmer and Huffman 1979) fit the interstellar profile fairly well. The 18 μm band is 40% as strong as the 9.7 μm at maximum (McCarthy et al 1980; Pegourie and Papoular 1985; Volk and Kwok 1988; Aitken et al 1988).

The optical depth of the silicate band at maximum, relative to A(V), has been measured as $\tau_{9.7}/A(V) = 18.5 \pm 1.0$ (Roche and Aitken 1984) for local diffuse dust and ≈ 9 towards the galactic center (Roche and Aitken 1985). The variations in both the profile and strength of the band suggest that there are real variations in the types and impurities present in interstellar and circumstellar silicates.

The well-known ice band at 3.08 μm, observed within molecular clouds, has a wing extending to 3.6 μm which varies greatly in strength (Smith et al 1989) from source to source. In IRS 7 near the galactic center, the wing is a distinct feature at 3.4 μm, half as strong as the ice band, which is the C-H stretch in aliphatic hydrocarbons (Butchart et al 1986). This 3.4 μm band might signify the presence of organic mantles on grains. A local star, VI Cyg 12, has $\tau_{3.4}/A(V) \leq 0.3$ of IRS 7 (see Gillett et al 1975b). IRS3 in Lynga 8 (Tapia et al 1989) has $\tau_{3.4}/A(V) = 0.4$ of IRS 7, and the ice band 3.4 times stronger than the 3.4 μm. There is no ice band for A(V) < 20 mag in the Ophiuchus molecular cloud (Harris, Woolf, and Rieke 1978).

3. OTHER DIAGNOSTICS OF DUST

3.1. NIR emission

The discovery of "unidentified infrared bands" (UIBs) (Gillett et al. 1973), and the realization that they are ubiquitous and very strong throughout the Galaxy (Giard et al 1989) and elsewhere (see Tielens and Allamandola 1987; Puget and Léger 1989 for reviews) has made it clear that very small grains or large molecules of hydrocarbons are present in space. The emission bands (3.3, 6.2, 7.7, and 11.3 μm for the strongest, plus several weaker ones) correspond to the C-H stretch, C-C bending, and C-H out-of-plane bending in planar aromatic molecules ("polycyclic aromatic hydrocarbons", or PAHs) or slightly less well-ordered, three-dimensional hydrocarbons such as are produced in laboratory conditions (Sakata et al 1984; Borghesi et al 1987) or are found in coal extracts (Papoular et al 1989). The bands indicate which bonds are present but are not specific for individual molecules. The excitation of the bands requires that the carriers be small ($\approx$ 50 C atoms and up), so that the absorption of a single UV photon can produce the observed emission. The carrier must be an important component of the ISM, absorbing 10 -20% of the UV energy from starlight.

The 11.3 μm band, the out-of-plane C-H bending mode, arises when there is a single C-H bond per C atom (a second or third bond on the same atom produces emission at longer wavelengths); the hydrocarbons observed so far in space are highly unsaturated.

There is an associated continuum emission in the NIR underlying the UIBs. It presumably arises from emission from very small grains which are heated by a single photon, but which have such disordered structure that they have virtually a continuum of energy levels.

A red continuum emission is strong in some "reflection" nebulae (Witt, Schild, and Kraiman 1984; Witt and Schild 1988) but varies in strength with respect to the UIBs. Near the star γ Cas, the red emission is associated with a nebula in which UV H_2 fluorescence is strong. This emission in not found in the Merope reflection nebula, or in some areas of others. Possibly special conditions, such as a strong UV radiation field acting upon H_2 coatings on grains, might be required.

3.2. Scattering

True reflection has, of course, been observed, especially in the blue and UV spectral regions. It is found in reflection nebulae, in the diffuse galactic light (DGL), and in high-latitude clouds illuminated from the general galactic radiation field. The results (see Witt 1988 for a review and references) are: (a) grains have a high albedo ($\approx$ 0.7) in the optical and are forward-throwing (g, the mean cosine of the scattering angle, is $\approx$ 0.7); (b) uncertain geometry makes quantitative assessment of the UV properties of grains difficult, except that grains are less forward-throwing at the shortest *IUE* wavelengths (0.12 μm) than at longer.

3.3. Polarization

Aligned elongated dust grains cause more extinction along their long axes than perpendicular to it, producing linear polarization in the transmitted light. The wavelength dependence of the resulting polarization can be expressed as "Serkowski's Law" (Coyne et al 1974), which later observations have slightly modified (Wilking et al 1980). The observed polarization typically rises slowly towards shorter wavelengths through the NIR, peaks at some optical wavelength, λ_{max}, and declines in the ground-based UV. The mean λ_{max} is 0.55 μm, showing that only larger grains participate in polarization. By contrast, extinction keeps rising towards shorter wavelengths (excepting the local maximum at the bump).

Grain alignment is not well understood (see Hildebrand 1988 for a review). Possibly there are superparamagnetic particles of magnetite or metallic iron (Spitzer and Tukey 1951), one or more of which might be included into large grains and dissipate rotational energy (Mathis 1986).

3.4. Depletions

Studies of interstellar absorption lines show that the gas does not have solar abundances of heavy elements (see Jenkins 1987 for a review and references). Some elements are only lightly depleted (O, N, Zn, S), others (P, Mg, and Cl) are lightly depleted in low-density gas but depleted by an order of magnitude at higher densities, and some (Al, Fe, Mn, Ca, Ti, Cr, and Si) are depleted by one to three orders of magnitude as the mean gas density along the line of sight is increased. These depletions show (a) that grains contain almost all of the heavy, highly depleted elements, and (b) that there is a surprisingly good coupling between the *mean* gas density along the line of sight and depletions, and there is probably a better relationship between local density and depletions. The relationship is unexpected because grains supposedly have such small areas, per H atom, that collisions between the heavy atoms and dust should be very infrequent ($< 10^{-9}$ yr^{-1}) in diffuse gas.

3.5. Far-Infrared Emission

The Galaxy emits about 25% of its total luminosity in the far-infrared (FIR; $\lambda > 60$ μm) (see a review by Cox and Mezger 1989). The shorter wavelength range of this radiation shows that grains must be heated by individual photons, much like the process which produces the UIBs, except that the radiation is emitted as a continuum. The spectrum of the FIR emission is not easy to interpret because it depends upon the local radiation field of the grains, but the longest wavelengths are useful for providing information regarding the wavelength dependence of the FIR opacity of grains. The results indicate a λ^{-2} law (Chini et al 1989), as required for long wavelengths and for grains small as compared with the wavelength (see discussion in Draine and Lee 1984, hereafter DL).

4. THEORIES OF GRAINS

Space does not permit a real discussion of the various theories which have been proposed to explain some or all of the observational results described above. They fall into various classes: (a) Core/mantle grains are advocated by J.M. Greenberg and collaborators (refs. in Greenberg 1989). The optical extinction is produced by mantles of organic refractory material, similar to that produced in laboratory conditions, coated onto grains. The bump is produced by small graphite. Very small grains (silicates?) produce the far-UV rise in extinction. (b) <u>Bare silicate-graphite grains</u> were shown by DL to fit the observed extinction law over the huge range 0.1 μm - 1000 μm quite well. This theory is an improvement and extension of the "MRN" theory (Mathis, Rumpl, and Nordsieck 1979). The size distribution of both materials is a power-law in the radii, $n(a) \propto a^{-3.5}$. The distribution was extended to very small particles by Draine and Anderson (1985). Presumably, the smallest "grains" could well be PAHs. (c) <u>Composite grains</u> by Mathis and Whiffen (1989) (c.f. Tielens 1989) also have a power-law distribution in sizes, but each grain contains graphite, amorphous C, and silicates, and is porous as well. The 0.1 μm - 5 μm opacity is good, but the FIR opacity might be too high. (d) <u>Fractal grains</u> (Wright 1987, Hawkins and Wright 1988) are the product of coagulation of similar grains into loose, airy structures whose volumes increase more slowly than a^3 because the fraction of vacuum increases with size. Fractals can have a very large extinction per unit mass, and resemble the very open structures observed in interplanetary dust particles (Brownlee 1989). (e) <u>Silicate core-Carbon-mantle grains</u> (Duley, Jones, and Williams 1989) is the only theory to produce the bump by something other than graphitic carbon: OH^- ions in connection with Si atoms in silicates. The UIBs are produced by tiny islands of amorphous C on the surfaces of small grains. (f) <u>Biological grains</u> have been proposed by Hoyle, Wickramasinghe, and others (Wallis et al 1989 and refs. therein) for both interstellar and cometary grains. Organisms seem to require more phosphorus is available (Duley 1984; Whittet 1984, but see Hoyle and Wickramasinghe 1984), especially since P is almost undepleted in low-density gas. Bacteria, even when dried, have a C-H and O-H stretch absorption near 3.0 - 3.4 μm which is not seen in ordinary interstellar dust. In any case, the band can be explained by ices rather than organisms.

These theories may be summarized by: (a) all have silicates, some with mantles and others not; (b) almost all use small graphite (or at least well-ordered carbon) to produce the bump; (c) all have a large range of sizes. Not many have addressed the difficult question of accounting for the transitions between diffuse- and outer-cloud dust in a simple way.

The future of grain research seems exciting, driven by observations in the infrared and also UV regimes, especially when the full capabilities of the Hubble Space Telescope are realized.

This review has been partly supported by contract 957996 with the jet Propulsion Lab and grant NAGW-1768 with NASA.

REFERENCES

Aitken, D. K., Roche, P. F., Smith, C. H., James, S. D., and Hough, J. H. 1988, *M.N.R.A.S.*, **230**, 629.
Allamandola, L.J., and A.G.G.M. Tielens, eds. 1989, IAU Symposium 135, *Interstellar Dust*. Dordrecht: Reidel Publishing Company. In press.
Ardeberg, A., and Virdefors, B. 1982, *Astr. Ap.*, **115**, 347.
Bohlin, R. C., Savage, B. D., and Drake, J. F. 1978, *Ap. J.*, **224**, 132.
Bohren, C. F., and Huffman, D. R. 1983, *Absorption and Scattering of Light by Small Particles*, John Wiley and Sons (New York).
Borghesi, A., Bussoletti, E., and Colangeli, L. 1987, *Ap. J.*, **314**, 422.
Brownlee, D.E. 1989, *Highlights of Astronomy*, **8**, 281.
Butchart, I., McFazdean, A.D., Whittet, D.C.B., Geballe, T.R., and Greenberg, J.M. 1986, *Astr. Ap.*, **154**, L5.
Cardelli, J.A., and Savage, B.D. 1988, *Ap. J.*, **325**, 864.
Cardelli, J. A., Clayton, G.C., and Mathis, J.S. 1988, *Ap. J. (Lett.)*, **329**, L33.
Cardelli, J.A., Clayton, G.C., and Mathis, J.S. 1989, *Ap. J.*, **345**, in press. (CCM)
Chini, R., Krügel, E., and Kreysa, E. 1989, *Astr. Ap.*, **216**, L5.
Cox, P., and Mezger, P.G. 1989, *The Astr. Ap. Rev.*, **1**, 49.
Coyne, G.V., Gehrels, T., and Serkowski, K. 1974, *A. J.*, **79**, 581.
Day, K.L. 1979, *Ap. J.*, **234**, 158.
Draine, B. T., and Anderson, N. 1985, *Ap. J.*, **292**, 494.
Draine, B.T. 1989, in proceedings of IAU Symposium 135, *Interstellar Dust* (L.J. Allamandola and A.G.G.M. Tielens, eds.), D. Reidel and Co. (Dordrecht), in press.
Draine, B.T., and Lee, H.M. 1984, *Ap. J.*, **285**, 89. (DL)
Duley, W. W. 1984, *Quar. J. R. A. S.*, **25**, 109.
Duley, W.W., Jones, A.P., and Williams, D.A. 1989, *M.N.R.A.S.*, **236**, 709.
Fitzpatrick,E.L., and Massa, D. 1986, *Ap. J.*, **307**, 286 (FM86).
Fitzpatrick,E.L., and Massa, D. 1988, *Ap. J.*, **328**, 734. (FM88)
Giard, M., Pajot, F., Lamarre, J.M., Serra, G., and Caux, E. 1989, *Astr. Ap.*, **215**, 92.
Gillett, F.C., Forrest, W.J., Merrill, K.M. 1973, *Ap. J.*, **183**, 87.
Gillett, F.C., Forrest, W.J., Merrill, K.M., Capps, R.W., and Soifer, B.T. 1975a, *Ap. J.*, **200**, 609.

Gillett, F.C., Jones, T.W., Merrill, K.M., and Stein, W.A. 1975b, *Astr. Ap.*, **45**, 77.
Greenberg, J.M. 1989, *Highlights of Astronomy*, **8**, 241.
Harris, D.H., Woolf, N.J., and Rieke, G.H. 1978, *Ap. J.*, **226**, 829.
Hawkins, I., and Wright, E.L. (1988, *Ap. J.*, **324**, 46.
Hildebrand, R.H. 1988, *Quar. J. R. A. S.*, **29**, 327.
Hoyle, F., and Wickramasinghe, N.C. 1984, Ap. Sp. Sci., **103**, 189.
Jakobsen, P., Bowyer, S., Kimble, R., Jelinsky, P., Grewing, M., Krämer, G., and Wulf-Mathies, C. 1984, *Astr. Ap.*, **139**, 481.
Jenkins, E.L. 1987, in *Interstellar Processes* (D. J. Hollenbach and H. A. Thronson, Jr., eds.), Reidel, Dordrecht, p. 533.
Jones, T. J., and Hyland, H. 1980, *M. N. R. A. S.*, **192**, 354.
Krätchmer, W., and Huffman, D.R. 1979, *Ap. Sp. Sci.*, **61**, 195.
Lillie, C.F., and Witt, A.N. 1976, *Ap. J.*, **208**, 64.
Little-Marenin, I.R. 1986, *Ap. J. (Lett.)*, **307**, L15.
Mathis, J.S., and Whiffen, G. 1989, *Ap. J.*, **341**, 808.
Mathis, J.S., Rumpl, W., and Nordsieck, K.H. 1977, *Ap. J.*, **217**, 425.
Mathis, J.S. 1986, *Ap. J.*, **308**, 281.
McCarthy, J.F., Forrest, W. J., Briotta, D. A., and Houck, J. R. 1980, *Ap. J.*, **242**, 965.
Nuth, J.A. III, and Stencel, R.E, eds. 1986, *Interrelationships Among Circumstellar, Interstellar, and Interplanetary Dust* (N.A.S.A. Conf. Publ. 2403), U.S. Government Printing Office.
Papoular, R., Conard, J., Giuliano, M., Kister, J., and Mille, G. 1989, *Astr. Ap.*, **217**, 204.
Pegourie, B., and Papoular, R. 1985, *Astr. Ap.*, **142**, 451.
Puget, J.L., and Léger, A. 1989, *Ann. Rev. Astr. Ap.*, **27**, 161.
Roche, P.F., and Aitken, D.K. 1984, *M.N.R.A.S.*, **208**, 481.
Roche, P.F., and Aitken, D.K. 1985, *M.N.R.A.S.*, **215**, 425.
Sakata, A., Wada, S., Tanabe, T., and Onaka, T. 1984, *Ap. J. (Lett.)*, **287**, L51.
Sandford, S.A., and Walker, R.M. 1985, *Ap. J.*, **291**, 838.
Savage, B.D., and Mathis, J.S. 1979, *Ann. Rev. Astr. Ap.*, **17**, 73.
Seaton, M. J. 1979, *M. N. R. A. S.*, **187**, 73p.
Smith, R.G., Sellgren, K., and Tokunaga, A.T. 1989, *Ap. J.*, **344**, 413.
Spitzer, L., Jr., and Tukey, J.W. 1951, *Ap. J.*, **114**, 187.
Tapia, M., Persi, P., Roth, M., and Ferrari-Toniolo, M. 1989, *Astr. Ap.*, in press.
Tennyson, P.D., Henry, R.C., Feldman, P.D., and Hartig, G.F. 1988, *Ap. J.*, **330**, 435.

Tielens, A.G.G.M. 1989, in proceedings of IAU Symposium 135, *Interstellar Dust* (L.J. Allamandola and A.G.G.M. Tielens, eds.), D. Reidel and Co. (Dordrecht), in press.
Tielens, A.G.G.M., and Allamandola, L.J. 1987, in *Interstellar Processes* (D.J. Hollenbach and H. A. Thronson, Jr., eds.), Reidel, Dordrecht, p. 397.
Volk, K., and Kwok, S. 1988, *Ap. J.*, **331**, 435.
Wallis, M.K., Wickramasinghe, N.C., Hoyle, F., Rabilizirov, R. 1989, *M.N.R.A.S.*, **238**, 1165.
Whitford, A.E. 1958, *A.J.*, **63**, 201.
Whittet, D.C.B. 1984, *M. N. R. A. S.*, **210**, 479.
Whittet, D.C.B. 1987, *Quar.J. R. A. S.*, **28**, 303.
Whittet, D.C.B. 1988, in *Dust in the Universe* (M. E. Bailey and D. A. Williams, eds.), Cambridge University Press, p. 25.
Wilking, B. A., Lebofsky, M. J., Martin, P. G., Rieke, G. H., and Kemp, J. C. 1980, *Ap. J.*, **235**, 905.
Witt, A.N. 1988, in *Dust in the Universe* (M. E. Bailey and D. A. Williams, eds.), Cambridge University Press, p. 1.
Witt, A.N., and Schild, R.E. 1988, *Ap. J.*, **325**, 837.
Witt, A.N., Bohlin, R.C., and Stecher,T.P.1986, *Ap. J. (Lett.)*, **305**, L23.
Witt, A.N., Schild, R.E., and Kraiman, J.B. 1984, *Ap. J.*, **262**, 708.
Wright, E.L. 1987, *Ap. J.*, **320**, 818.

THE HIGH-ENERGY COMPONENT OF THE ISM:
COSMIC RAYS AND GAMMA RAYS

HANS BLOEMEN
Leiden Observatory & Laboratory for Space Research Leiden,
P.O. Box 9513, 2300 RA Leiden, The Netherlands

ABSTRACT Studies of cosmic-ray particles in the Galaxy are summarized, with emphasis on the insight provided by gamma-ray astronomy.

1. INTRODUCTION

Gamma-ray observations at energies $E_\gamma \lesssim 100$ GeV provide unique possibilities to study the origin and propagation of cosmic rays through the interactions of these particles with interstellar gas and the interstellar radiation field. Direct measurements of cosmic-ray (CR) particles near Earth mainly provide information on CR characteristics in the local region of the Galaxy, without any substantial directional information. Radio-synchrotron studies are limited to the CR electron component, making up only about 1% of the CR particle flux near Earth. After a relatively quiescent period, γ-ray astronomy is soon entering a new era with the launch of several satellites. Observational information is sofar largely restricted to 'high-energy' γ-rays ($E_\gamma \gtrsim 50$ MeV), obtained with NASA's SAS-2 satellite and ESA's COS-B. At these energies, new insight will be given by the Soviet-French satellite experiment GAMMA-1 and the EGRET experiment aboard NASA's Gamma Ray Observatory (GRO), which are both expected to be brought in orbit in 1990. At lower γ-ray energies, where hitherto only very limited information has been obtained, largely with balloon-borne instruments, considerable progress can be expected with the French SIGMA telescope (0.1 – 2 MeV) of the Soviet mission GRANAT (starting at the end of 1989) and the OSSE (0.1 – 10 MeV) and COMPTEL (1 – 30 MeV) telescopes aboard GRO.

Cosmic rays are responsible for four main components of interstellar γ-ray emission:
1. Continuum emission from the decay of π°-mesons, which are produced by nuclear interactions between GeV CR protons (and heavier nuclei) and nuclei of the interstellar gas (π°-decay γ-rays).
2. Continuum emission from the Coulomb scattering of CR electrons with energies $E_e \approx 3E_\gamma$ on the nuclei and electrons of the interstellar gas (bremsstrahlung γ-rays).
3. Continuum emission from the scattering of high-energy ($E_e \gtrsim 10$ GeV) CR electrons on low-energy photons (optical, infrared, and the 3 K universal background) (inverse-Compton γ-rays).

4. Gamma-ray lines from the decay of radioisotopes produced by the interactions of low-energy ($\lesssim 100$ MeV/nucleon) CR nuclei with interstellar matter.

This paper concentrates on the continuum component, which is best studied so far. The line emission occurs at energies of typically ~ 1 MeV, where information is scanty. The continuum emission traces broad parts of the spectra of the CR proton/nuclear component and CR electrons, including those particles that carry the bulk of the CR energy (i.e., 1 - 10 GeV protons), which play a dominant role in high-energy astrophysics of the ISM.

Section 2 reviews basic aspects of CR origin and propagation. Sections 3 and 4 summarize γ-ray continuum studies of cosmic rays, with emphasis on recent results, mainly obtained from the COS-B γ-ray survey. Section 3 concentrates on large-scale aspects of the CR distribution and CR transport, Section 4 on small-scale aspects. The prospects of γ-ray astronomy for CR studies are discussed in Section 5. The relevance of γ-ray astronomy to studies of molecular gas is not addressed in this paper; the state of the art has been reviewed by Wolfendale (1988) and Bloemen (1989).

2. COSMIC RAYS — SOME BACKGROUND

The CR particle flux measured near Earth consists principally of protons, to which helium nuclei contribute about 10% and heavier nuclei and electrons (+ positrons) each add about one percent. A comprehensive review of the CR composition and its energy dependence is given by Simpson (1983). The abundance distribution of cosmic rays is similar to that of the solar system and the local region of the Galaxy, but there are some significant deviations. The most obvious differences are the overabundances in cosmic rays of the spallation products lithium, beryllium, boron, and the sub-iron elements (Z = 21 - 25), which are largely produced by CR interactions with interstellar matter. These secondary particles play an important role for the understanding of CR propagation and confinement (see the review by Cesarsky 1980 and references therein). Several other deviations remain after the fragmentation processes in the ISM have been taken into account. These differences give important clues on the origin of cosmic rays. An comprehensive review is given by Meyer (1985).

Shock acceleration, i.e. first order Fermi acceleration in strong shocks induced mainly by supernova remnants (Axford et al. 1977; Krymsky 1977; Bell 1978; Blandford and Ostriker 1978), is at present the favoured mechanism for the production of cosmic rays with total energies up to $\sim 10^5$ GeV (see, e.g., the review by Axford 1987). It produces quite naturally a power-law CR spectrum with a spectral index of 2 or somewhat larger, which is consistent with the CR source spectrum inferred from observations, and can account for the power required to produce the observed CR energy density. Cosmic rays may get reaccelerated when encountering shocks in the ISM (Blandford and Ostriker 1980) or by second order Fermi acceleration in interstellar turbulence; continuous acceleration, occuring solely during propagation, however, is very hard to reconcile with the observed abundance ratios of secondary and primary particles and with the observed power-law shape of CR spectra (Hayakawa 1969; Eichler 1980; Cowsik 1980, 1986). Cesarsky (1987) has reviewed the state of the art of this topic. Although the non-linear aspects of shock acceleration are not fully understood yet (e.g Völk 1984), the mechanism is very attractive and the problem of the origin of cosmic rays with energies $\lesssim 10^5$ GeV seems to be shifting toward finding the CR injectors, i.e. the sources that provide the CR material and speed it up to moderate suprathermal energies. On the basis of energetics as well as CR composition, stellar flares and winds are considered to be likely candidates (Cassé and Goret 1978; Arnaud and Cassé 1985; Meyer 1985), although others have argued that CR injectors are not needed and that shocks can (or have to) accelerate particles directly out of an interstellar plasma (Eichler 1979, 1980; Axford 1981).

There is no conclusive proof of the confinement region of cosmic rays. At least the CR electrons have a Galactic origin, because they cannot survive the Compton losses in the microwave background. It is in principle possible, however, that (part of) the CR proton-nuclear component fills a much larger volume than that of a galaxy, such as a (super)cluster (Brecher and Burbidge 1972; Burbidge 1974, 1983), although this possibility has frequently been criticized (see, e.g., Ginzburg 1987). As cosmic rays are closely attached to the interstellar magnetic field lines, because of their relatively small gyroradii (for 10 GeV protons in a field of a few μG, for instance, $r_g \approx 10^{-6}$ pc), their propagation is largely determined by the characteristics of the field. Starting with the work of Shklovskii (1952), Pikelner (1953), and Ginzburg (1953), many observational and theoretical studies have indicated that the distributions of the magnetic field and cosmic rays extend far beyond the Galactic disk. Direct evidence for the existence of such halo's or thick disks follows from low-frequency radio-continuum observations of our Galaxy (Baldwin 1976; Brindle 1978; Phillipps et al. 1981ab; Beuermann et al. 1985) and some edge-on galaxies (Ekers and Sancisi 1977; Allen et al. 1978; Beck et al. 1979; Hummel et al. 1984ab; Klein et al. 1983, 1984), with indirect support from the observed composition and spectra of cosmic rays (Berezinskii et al. 1990 and references therein). Several studies of the polarization of starlight and radio synchrotron emission and of the rotation measures of pulsars and extragalactic radio sources [reviewed by Sofue et al. (1986) and Heiles (1987)] have shown that the Galactic magnetic field consists of a systematic component, preferentially aligned parallel to the Galactic plane, and an irregular component of comparable strength. Owing to the presence of this tangled component, the CR particles will not only diffuse along the field lines, but also be spread around to other field lines (Ptuskin 1979). It has been argued that CR propagation can therefore be described in first order approximation as isotropic diffusion. Alternatively, in a so-called dynamical halo model or galactic wind model, the CR particles can be convected away from the disk in a Galactic wind (Jokipii 1976; Owens and Jokipii 1977; Jones 1979; Kóta and Owens 1980; Lerche and Schlickeiser 1982ab). This convective transport may be provided by a large-scale flow of interstellar gas with a frozen-in magnetic field. Also, a directed flux of MHD waves that is not accompanied by a gas flow can lead to convective transport. Cosmic rays might even help to power a wind (Ipavich 1975; Breitschwerdt et al. 1987). In addition to the CR density gradients on a Galactic scale, which can be expected from these processes, several scenarios that lead to small-scale gradients have been proposed, such as trapping of cosmic rays near their sources, in spiral arms, or in tunnels. Cesarsky (1980) has discussed these transport and confinement models in detail.

3. LARGE-SCALE ASPECTS OF CR DISTRIBUTION & TRANSPORT

3.1 Radial Gradient and Halo Size

If cosmic rays are of Galactic origin then a significant radial gradient of the CR intensity can be expected on a large scale in the Galaxy, simply because more potential CR sources are present in the inner regions. Effective mixing of cosmic rays in the Galaxy (as in a halo diffusion model) may lead, however, to a much weaker fall-off for the particle distribution than for the source distribution, so absence of a strong gradient does not necessarily imply an extragalactic origin. Several studies of the radial CR distribution have been performed in the past, using the γ-ray emissivity as a measure of the CR intensity. The basic findings have been reviewed by Bloemen (1989). There has been some confusion due to uncertainties in the H_2 content of the Galaxy and due to differences between recent results and those obtained from early studies of the SAS-2 survey, before the release of the final data base. At present, the principal conclusion can be drawn that radial emissivity gradients are present over the Galaxy as a whole, for the entire γ-ray energy range studied, but the gradients are much weaker than found in early work. The radial emissivity distribution obtained from the latest and most robust analysis (Strong et al. 1988) is shown in Fig. 1. Thus the CR intensity appears to decrease weakly with Galactocentric radius, although the electron and proton-nuclear components of relevance here may behave somewhat differently (see review by Bloemen 1989).

The observed weak CR gradient sets indirect size constraints on the CR halo in a standard diffusion model as described by Ginzburg and Syrovatskii (1964). This method (first used by Stecker and Jones 1977) was recently reconsidered by Bloemen et al. (1990). They assumed a Galactocentric distribution of CR sources that resembles the distribution of supernova remnants (SNR's) and investigated whether the radial CR density distribution that follows from the diffusion model, for different choices of the halo dimensions, is consistent with the weak CR gradient derived from the COS-B data. Their findings are summarized in Fig. 1. It is clear that even a very large CR halo can barely explain the difference between the strong radial gradient of the source distribution and the observed weak CR density gradient. Obviously, a SNR source distribution may not be the right choice. It is hard to predict, however, what an alternative CR source distribution would look like. SNR's probably play an essential role in the acceleration of CR particles (Völk et al. 1989), but the acceleration efficiency per SNR may depend on Galactocentric radius due to large-scale variations in the characteristics of the ambient ISM (such as, variations in the density of the diffuse ISM, the concentration of molecular clouds in the inner Galaxy, etc.). Furthermore, interpretation of the γ-ray data in terms of a 'CR gradient' may not be correct. The argument that can be made is simple. The observed γ-ray intensities trace only CR particles in those regions along the line-of-sight that contribute significantly to the total gas column density. Hence, the radial gradient of the *volume-averaged* CR density may be stronger, or weaker, than derived from γ-ray observations. It has been argued on theoretical

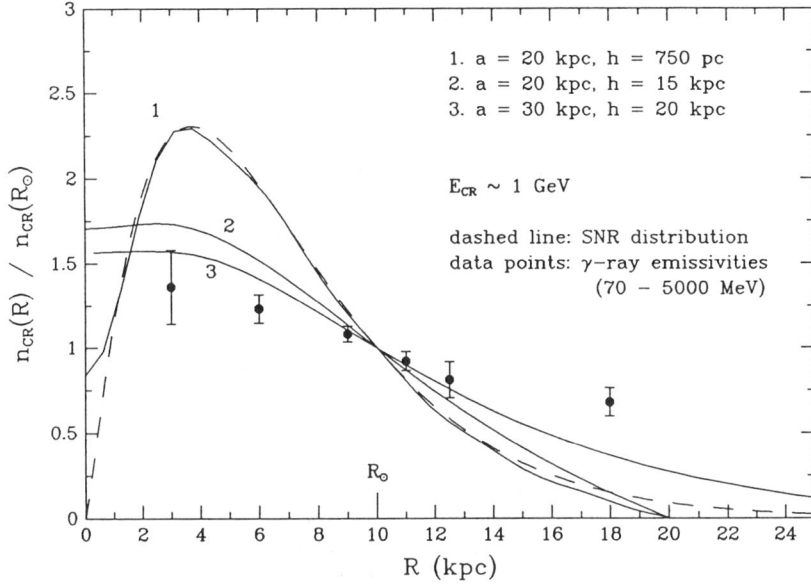

Fig. 1. Examples of the Galactocentric CR density distribution in a diffusion model with a SNR-like CR source distribution and different halo dimensions (a: halo radius in the Galactic plane; h: half thickness). The data points indicate the γ-ray emissivities (70 MeV – 5 GeV) derived by Strong et al. (1988) from the COS-B data. All quantities are normalized at the radius of the solar circle.

grounds, for instance, that the CR density is correlated with matter density. This point is taken up again in Section 4.

3.2 Inverse-Compton Gamma-Ray Halo & 'Medium-Latitude Excess'

Gamma radiation that originates at large distances from the Galactic plane can be expected to be produced predominantly by inverse-Compton (IC) scattering of relativistic electrons on low-energy photons (optical, infrared, and particularly the 2.7 K universal background radiation). The gas densities here are too small to expect significant bremsstrahlung and π^0-decay contributions, which dominate the γ-ray emission from the Galactic disk. The contribution from such an IC halo to the observed γ-ray intensities at medium latitudes, however, was generally estimated to be not very large (Shukla and Paul 1976; Piccinotti and Bignami 1976; Stecker 1977; Kniffen and Fichtel 1981; Bloemen 1985), but may be substantial (Dogiel and Uryson 1988) if the distribution of electrons extends to larger distances from the Galactic plane than assumed in these analyses (a scale height of typically 500 – 1000 pc was generally taken). There is indeed increasing evidence (Bloemen 1989 and references therein) that almost half of the observed γ-ray emission at medium latitudes ($5° \lesssim |b| \lesssim 20°$) in the inner-Galaxy direction cannot be explained by CR-matter interactions. Fig. 2 shows

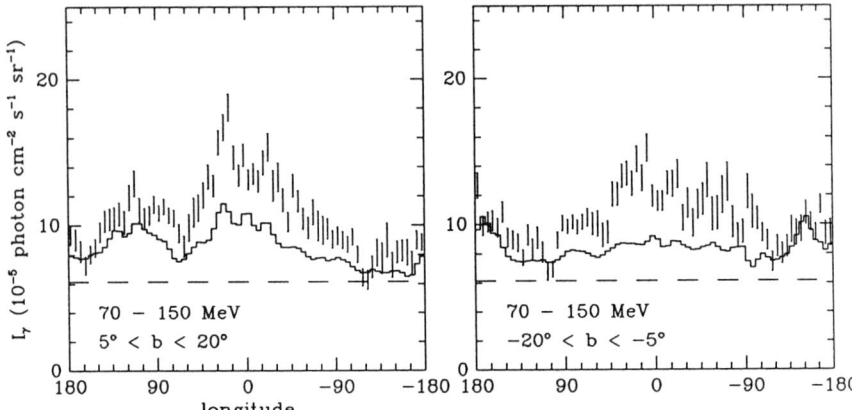

Fig. 2. Longitude distributions of the observed (±1σ error bars, from COS-B) and modelled (histogram) γ-ray intensity at intermediate latitudes. The model predictions include only CR interactions with atomic and molecular gas and were derived from an extension of the model of the Galactic-disk emission presented by Strong *et al.* (1988). The dashed line indicates the isotropic background level, also from Strong *et al.* The excess visible toward the general direction of the inner Galaxy is most pronounced at these energies, but clearly visible at higher energies as well (see Bloemen 1989).

that the excess extends over essentially the entire first and fourth Galactic quadrants.

Bloemen *et al.* (1990) have investigated whether this medium-latitude excess can be attributed to IC emission in the framework of a diffusion model with a large CR halo, as discussed in the previous section. The typical electron energy for IC emission at a γ-ray energy E_γ is given by $E_e \simeq m_e c^2 (E_\gamma/\varepsilon_{ph})^{0.5}$, where ε_{ph} is the average energy of the target photons. Hence, IC scattering of relic photons, which dominate the radiation field at large distances from the plane, into the COS-B energy range requires $E_e \gtrsim 100$ GeV. Electrons with such high energies are subject to severe energy losses and fill no more than ∼ 30% of the halo taken up by the proton-nuclear CR component and low-energy ($\lesssim 1$ GeV) electrons, not suffering these losses. Nevertheless, Bloemen *et al.* (1990) found that the scattering of these electrons on the 2.7 K background leads to intense IC emission if the CR halo at GeV energies is as large as suggested by the observed weak CR gradient. The medium-latitude excess, for instance, can easily be accounted for, although the predicted intensities are in fact too high.

Other interpretations of the medium-latitude excess have been put forward. Bhat *et al.* (1985) and Van der Walt and Wolfendale (1988) considered the possibility of an enhanced CR density inside Loop I, which has a radius of ∼ 60° centered on $\ell \approx 330°$, $b \approx 18°$. Fig. 2 suggests, however, that the angular extent of the excess is larger than that of Loop I. Also, the intensity of the excess is 5 – 10 times larger than the expected value derived by Blandford and Cowie (1982). Another possible reason for the excess is the omission of (warm) ionized gas in the γ-ray modelling. If the scale height of this ionized medium is indeed as large as suggested by pulsar DM data (Reynolds 1989; Clifton et al. 1988; references in these papers), then at least half of the medium-latitude excess can be attributed to CR-matter interactions, with little room for IC emission.

It may be clear from the above that a better understanding of the medium-latitude excess is of great importance for CR astrophysics. Spectral studies over a broad energy range, which will be possible when the GRO observations are available, can be expected to solve this interesting problem.

3.3 Spectral Variations

Specific characteristics of the origin and propagation of cosmic rays may be reflected in the spectral distribution of the γ-ray emission they produce. The COS-B data show evidence for basically two types of large-scale variations of the γ-ray spectrum.

Firstly, the γ-ray intensity spectrum (70 MeV – 5 GeV) is softer toward the inner Galaxy than for the remainder of the Galactic disk (a broad-band spectral variation: 70–150 MeV versus 150 MeV – 5 GeV; Mayer-Hasselwander 1983; Bloemen 1989). Correlation studies with HI and CO observations (Strong et al. 1988 and references therein) have shown that the soft emission can possibly be attributed to phenomena related to molecular clouds (or steep-spectrum γ-ray sources, distributed like the molecular gas). A soft γ-ray spectrum for molecular clouds is most simply explained by an enhanced e/p ratio due to the production of secondary electrons (Brown and Marscher 1977; Marscher and Brown 1978), which may be trapped (Cesarsky and Völk 1978) and even further accelerated (Morfill 1982ab). A useful review is given by Völk (1983). Other interpretations of this broad-band spectral behaviour are reviewed by Bloemen (1989).

Secondly, a peculiar large-scale spectral variation exists within the 300 MeV – 5 GeV range. The COS-B observations show a spectral hardening with increasing latitude in the general direction of the outer Galaxy, systematically over the entire second and third Galactic quadrants (Bloemen et al. 1988; see Fig. 3). Toward the inner Galaxy this spectral flattening with latitude is not seen, which is a strong indication that the findings cannot be attributed to systematic uncertainties in the intensity level and spectrum of the isotropic γ-ray background. The hardening effect suggests that the CR proton spectrum at GeV energies is flatter at medium latitudes, with a change in spectral index of $\sim 0.4 - 0.6$. Toward the inner Galaxy and near the plane in the outer Galaxy, the γ-ray spectra are consistent with direct measurements of the CR proton spectrum near Earth. Bloemen et al. (1990) have considered another possible interpretation, namely whether the presence of an intense IC halo can explain the spectral behaviour — triggered by the facts that a large CR halo may account for the weak CR gradient and that the IC γ-ray spectrum is relatively flat at these γ-ray energies. They found that the presence of an IC halo leads indeed to a hardening of the γ-ray spectrum with latitude, but the observed effect is significantly stronger and they could not explain the fact that it is only seen toward the outer Galaxy.

It is interesting that a recent spectral analysis of radio-continuum surveys of the Galaxy (408 and 1420 MHz) by Reich and Reich (1988) shows a similar spectral flattening with latitude in the outer-Galaxy direction. In this case GeV electrons are traced; the change in the radio spectral index is about 0.2

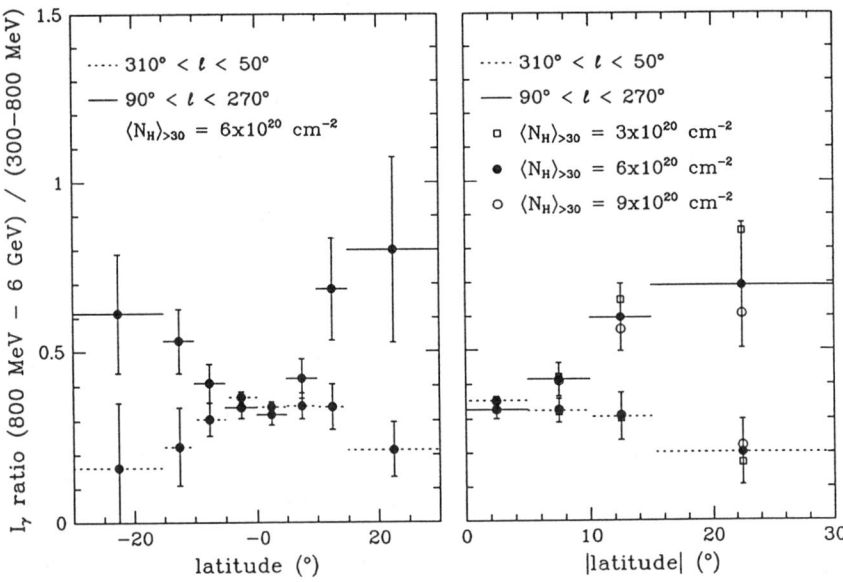

Fig. 3. Latitude distribution of the ratio between the γ-ray intensities for the energy ranges 800 MeV – 6 GeV and 300 – 800 MeV in the general directions of the inner and the outer Galaxy (Bloemen et al. 1988). *(Left)* Regions below and above the Galactic plane are shown separately. *(Right)* Regions on both sides of the plane are combined. The results are shown for different choices of the average gas column density at $|b| > 30°$, which is related to the background correction made — the figure shows that uncertainties in this correction cannot explain the spectral difference seen toward the inner and the outer Galaxy.

– 0.25, so the change in the electron spectral index is $\sim$ 0.4 – 0.5, as for the CR protons. They found some evidence for a similar effect toward the inner Galaxy, but it was not possible to perform a reliable analysis here because only northern-hemisphere data are available at 1420 MHz and a large fraction of the available sky coverage is dominated by emission from Loop I. On the basis of radio-continuum data alone a thermal origin of the flattening could not be excluded, but this interpretation seems unlikely since a similar effect is visible in the γ-ray data.

Although the interpretation of these findings is not straightforward, it is clear that a standard CR halo diffusion model encounters problems: a spectral steepening with distance from the Galactic plane would be expected for the electrons, due to synchrotron and Compton radiation losses, and no spectral changes for the protons. At low radio frequencies a weak spectral steepening with latitude was indeed found in previous work (see, e.g., review by Lawson et al. 1987). A dynamical halo model (Section 2) may provide a viable explanation. The competition of spatial diffusion, convection, adiabatic deceleration, and (electron) radiation losses in such a model may in principle lead to the observed effects, but the problem is very complex and requires robust modelling. Reich and Reich (1988) and Bloemen *et al.* (1988) noted that the findings are to a large extent indeed in agreement with the asymptotic spectral predictions of the Galactic-wind model given by Lerche and Schlickeiser (1982ab). Rogers *et al.* (1988; see also Van der Walt and Wolfendale 1988) have suggested that the spectral changes may have a local origin, related to our location on the inner edge of the Orion spiral arm, and there being acceleration mechanisms at work within the arm such as to give a flatter CR spectrum inside the arm than in the inter-arm region. At the moment there seems certainly room for further speculation on the nature of this peculiar spectral behaviour. Radio studies of other galaxies may shed new light on this topic in the near future.

4. SMALL-SCALE ASPECTS OF CR DISTRIBUTION & TRANSPORT

4.1 Cosmic-Ray – Matter Coupling

Triggered by the first SAS-2 γ-ray observations, Bignami and Fichtel (1974) have proposed a model in which cosmic rays are preferentially located in spiral arms as a result of the weight of the matter tying the magnetic fields and hence the cosmic rays to these regions. Follow-up work (see Fichtel and Kniffen 1984 and references therein) showed that such a model is indeed in agreement with the γ-ray data. Mainly the SAS-2 observations have been analyzed this way, although Fichtel and Kniffen (1984) found consistency for the COS-B data as well. Other analyses of the COS-B survey have shown, however (as discussed above), that a simpler model, in wich the CR distribution has only a weak Galactocentric gradient (hereafter, the *gradient model*), is also in agreement with the data. One can ask whether this implies that the available γ-ray data cannot distinguish between a *coupling model* and a *gradient model*. A detailed comparison of the different studies is not meaningful, because the coupling option was applied only to a model of the gas distribution whereas studies of the gradient option made use of detailed HI and CO surveys. Recently, Melisse and Bloemen (1990) have therefore performed a correlation analysis of a 3-D empirical gas model of the Galaxy and the COS-B γ-ray survey, based on the same HI and CO surveys used in the *gradient model*, and made a quantitative comparison of the two CR distribution models.

Melisse and Bloemen (1990) investigated whether the COS-B intensity distribution of the Milky Way in the energy range 300 MeV – 5 GeV can be described by a relation of the form

$$I_\gamma = \epsilon_\gamma(\mathbf{r}_\odot) \int n_H(\mathbf{r}) \left(\frac{n_H(\mathbf{r})}{n_H(\mathbf{r}_\odot)}\right)^\alpha d\mathbf{r}, \quad \text{with} \quad \frac{n_{CR}(\mathbf{r})}{n_{CR}(\mathbf{r}_\odot)} \equiv \left(\frac{n_H(\mathbf{r})}{n_H(\mathbf{r}_\odot)}\right)^\alpha,$$

where $n_H(\mathbf{r})$ is the Galactic distribution of the gas density, $n_{CR}(\mathbf{r})$ that of the CR density, and $\epsilon_\gamma(\mathbf{r}_\odot)$ is the γ-ray emissivity (per nucleon) in the solar vicinity. The integration is along the line of sight. The HI and CO line velocities served as distance indicators, in combination with an assumed Galactic rotation curve; the distance-ambiguity problem was solved by taking into account average scale heights, which is sufficiently accurate for this work. The full velocity resolution of the surveys was used for the derivation of the gas densities, corresponding to length scales of typically 100 pc. If HI and H_2 are not coexisting (an assumption

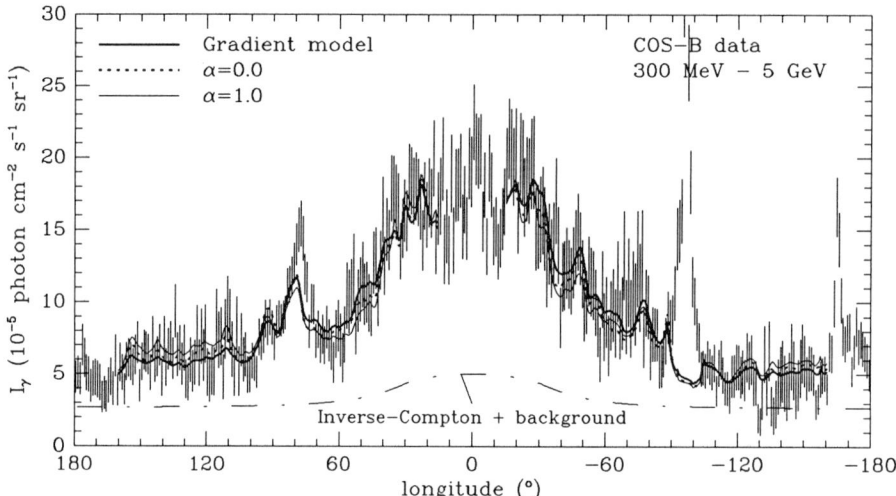

Fig. 4. Longitude profiles ($|b| \lesssim 5°$) of the observed γ-ray intensity distribution ($\pm 1\sigma$ error bars) and the predictions from the *gradient model* and the *coupling model* (model 2).

that mainly was made for technical reasons), then the above relation can be written as

$$I_\gamma = \epsilon_\gamma(\mathbf{r}_\odot) \int \left\{ n_{\text{HI}}(\mathbf{r}) \left(\frac{n_{\text{HI}}(\mathbf{r})}{n_{\text{HI}}(\mathbf{r}_\odot)} \right)^\alpha + 2XW_{\text{CO}}(\mathbf{r}) \left(\frac{2XW_{\text{CO}}(\mathbf{r})}{n_{\text{H}_2}(\mathbf{r}_\odot)} \right)^\alpha \right\} d\mathbf{r},$$

where W_{CO} is the CO brightness temperature integrated over a velocity channel ($T\Delta V$) and X is the average Galactic conversion factor between observed CO intensity and H_2 column density. The value of $n_{\text{HI}}(\mathbf{r}_\odot)$ was taken to be 0.5 H atom cm^{-3}. Estimates of $n_{\text{H}_2}(\mathbf{r}_\odot)$ depend on X; a relation of the form $n_{\text{H}_2}(\mathbf{r}_\odot) = 0.5(X/2.5 \times 10^{20})$ H atom cm^{-3} is consistent with various determinations. In order to enable a direct comparison with the *gradient model* of Strong et al. (1988), the same inverse-Compton predictions as used by these authors were added. Also, a (free) isotropic background level was included, so the actual model is of the form $I_\gamma + I_{\text{IC}} + I_b$, with I_γ as defined above (hereafter referred to as model 1). In addition, a model with CR-matter coupling for molecular clouds only was considered, i.e. $\alpha = 0$ in the HI term of the I_γ expression (hereafter model 2).

A maximum-likelihood technique was applied to estimate the values of the free parameters in the model: α, $\epsilon_\gamma(\mathbf{r}_\odot)$, X, and I_b. The region fitted covers the entire Galactic disk up to $|b| \approx 5°$; two $\sim 30°$ wide longitude intervals, centered on the Galactic center and anticenter directions, were excluded because of the poor kinematic distance information here, as well as regions around the well known strong γ-ray point sources. In order to compare the likelihood of the *coupling model* with that of the *gradient model*, the analysis of Strong et al. (1988) was repeated for the same area of the sky. Longitude profiles of the observed and modelled intensity distributions are shown in Fig. 4.

Fig. 5 shows the likelihood-ratio distributions of α. This figure indicates that a model with $\alpha \approx 0$ is to be preferred, by far. Also, it can be seen from

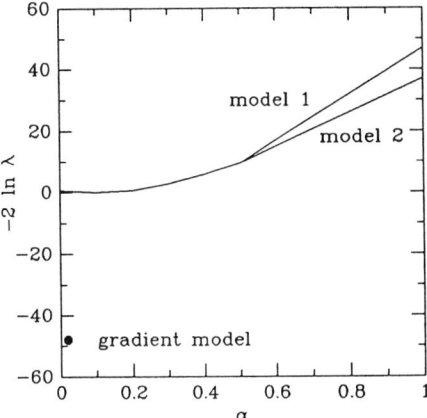

Fig. 5. Likelihood-ratio distribution of the α parameter in the *coupling model* (see Section 4). The preference for $\alpha \simeq 0$, and even more so for the *gradient model*, is evident.

this figure that the *gradient model* fits the data significantly better. Taking into account the difference in the number of degrees of freedom, this corresponds formally to a chance probability of $\lesssim 10^{-8}$. Fig. 4 indicates that the worse fit of the *coupling model* is due to large-scale deviations rather than small-scale aspects. The main reason turns out to be the small scale height of the γ-ray emitting disk in the *coupling model*, resulting from the concentration of cosmic rays in the Galactic disk. With increasing α this is to some extent compensated for in the fitting procedure by reducing the X value (because of the relatively small scale height of H_2), but this leads to a too small inner/outer Galaxy intensity contrast (as can be seen in Fig. 4 for model 2; the effect is even stronger for model 1), because H_2 is concentrated in the inner Galaxy. Invoking a radial CR gradient in the *coupling model* cannot compensate for this effect (i.e, α can still be expected to converge to 0 rather than 1). Hence, the main conclusion that can be drawn from these findings is that the scale height of the CR density distribution is significantly larger than that of the gas density distribution. This result is probably the most direct evidence from γ-ray astronomy for the existence of a thick-disk (or halo) distribution of 1 – 10 GeV CR protons (largely responsible for the high-energy γ-ray emission considered here) and confirms similar conclusions from radio synchrotron data for CR electrons (e.g., Baldwin 1976; Beuermann *et al.* 1985). If a CR halo is included in the modelling, weak CR-matter coupling in the Galactic disk ($\alpha \lesssim 0.5$) probably cannot be excluded.

4.2 Gamma Rays from violent Interstellar Events

Energetic events in the interstellar medium associated with regions of relatively high gas density, such as, supernova-cloud interactions and the collision of high-velocity clouds (HVC's) with the Galactic disk, may be visible as regions of enhanced γ-ray emission (see, e.g., the review by Morfill and Tenorio-Tagle 1983). The basic requirement is that the energy supplied is converted into the acceleration of particles with a reasonable efficiency; interstellar matter in the region of enhanced cosmic-ray (CR) density converts part of the CR energy into γ-rays. If spatially unresolved, such phenomena are potential γ-ray point sources, but particularly convincing would be the detection of extended γ-ray structures. Although the SAS-2 and COS-B γ-ray surveys indicate that a fairly homogeneous distribution of cosmic rays throughout the Galaxy can account in good detail for the observed γ-ray intensity distribution, several features are not accounted for.

The search for unexplained γ-ray excesses is closely related to the search for γ-ray point sources. In early analyses of the COS-B data, γ-ray point sources (apart from the Crab and Vela γ-ray pulsars) were simply defined as statistically significant peaks in the γ-ray intensity distribution with an angular shape consistent with the COS-B point-spread function, which lead to the so called 2CG catalogue (Swanenburg et al. 1981). With the availability of CO surveys (in addition to already existing HI 21 cm data) it became feasible to take the diffuse Galactic emission into account in the search for sources, which were then defined as point-like excesses superimposed on the model of the diffuse emission (Pollock et al. 1985ab; Simpson and Mayer-Hasselwander 1987, 1989). These analyses, however, were not particularly aimed at detecting extended γ-ray excesses. The following describes a simple alternative approach.

Fig. 6 presents a 'finding chart' of potential γ-ray sources and extended features for the 100 MeV – 5 GeV range (which is a good compromise between counting statistics and angular resolution). Basically, this map represents the γ-ray excesses that are left when the expected diffuse emission (from Strong et al. 1988) is subtracted from the observed γ-ray intensity distribution. It is not a pure subtraction. In order to correct for deviations between model and observations on scales of $10° - 20°$ in longitude direction, the modelled intensities were first re-adjusted to the observed intensities on this scale: for a given position, the predictions in a strip of 21 pixels in longitude and 1 pixel in latitude (pixel size is $0.5° \times 0.5°$), centered on this position, were scaled upwards or downwards such that for only 20% of the pixels the observed intensities are below the predicted intensities. This approach should reveal all indications for excesses, although the value of 20% is a rather arbitrary choice. Clearly, no formal statistical significance can be assigned to the excesses; 'finding chart' is probably the best denomination for the resultant map. To give the reader some indication of the significance, it may be useful to point out, on the basis of detailed studies of some selected regions, that even excesses with only two contours can be considered of real interest. Less attention should be paid to the precise shape of the (weak) excesses. Interestingly, however, some features are clearly extended, most pronounced near $\ell \approx 15° - 20°$ and $\ell \approx 330° - 340°$.

With a flux of $\sim 2 \times 10^{-6}$ photon cm^{-2} s^{-1} (100 MeV – 5 GeV), the excess near $\ell \approx 10°-15°$ would be among the most intense sources in the 2CG catalogue, if not extended. It coincides with two unique phenomena: 'Stockert's chimney', a (thermal) radio-continuum spur associated with the giant HII region S54 (Müller et al. 1987; Kundt and Müller 1987; see Fig. 7) and the crossing of the extension of the so called Complex C of HVC's with the Galactic disk (Fig. 8). Loop I (near $\ell \approx 30°$) seems not related. At a kinematic distance of ~ 3 kpc ($V_r \approx 30$ km s^{-1}), S54 is located at $z \approx 100$ pc and the radio spur extends up to $z \approx 400$ pc. Kundt and Müller (1987) argued that the spur is an outflow phenomenon driven by the HII region, consisting of a light relativistic pair plasma (with an admixture of

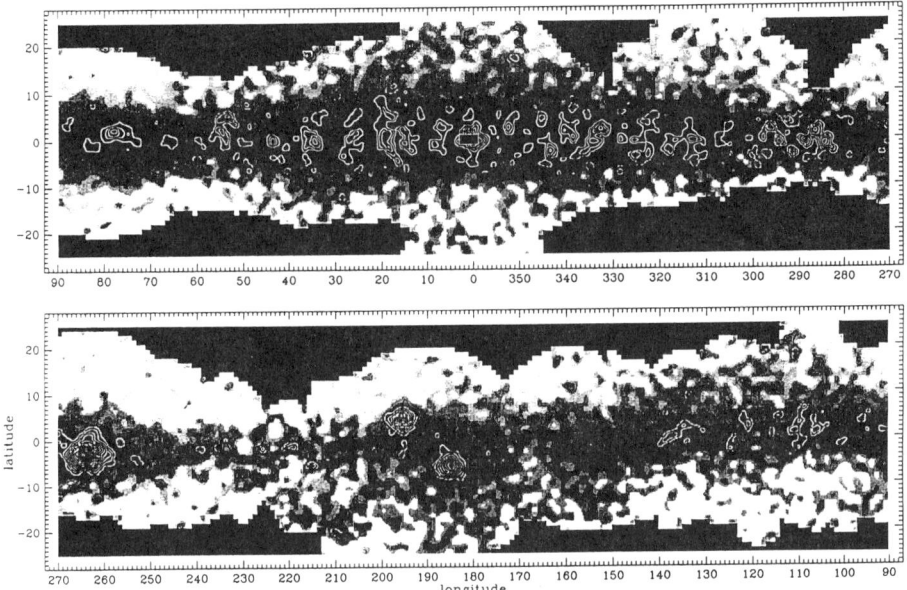

Fig. 6. 'Finding chart' of potential γ-ray point sources and extended excesses, derived from the COS-B γ-ray survey (100 MeV – 5 GeV). The underlying gray-scale map represents the observed γ-ray sky. The small box in the Galactic-center direction indicates a region that was excluded from the analysis; the feature surrounding it is probably an artifact. The strong excesses near $\ell \approx 185°$, $195°$, and $263°$ correspond to the Crab, Geminga, and Vela γ-ray sources. Contour values: 8, 13, 18, ...$\times 10^{-5}$ photon cm^{-2} s^{-1} sr^{-1}.

partially ionized hydrogen and cosmic rays), particularly for reasons of energy and absence of recoil and explosively receding fragments. Apart from the energy requirement ($E_k \approx 10^{54}$ erg), however, these arguments may not apply (i.e., the chimney may consist of 'ordinary' interstellar matter), if the HVC's and the intermediate-velocity gas seen in this region of the sky (see Fig. 9) are to be associated with S54. Shock acceleration of particles in the surroundings of S54 is then certainly plausible, whatever the precise energy source may be. A quite different interpretation of the findings may be the interaction of infalling HVC's with the Galactic disk. It has been suggested that shock ionization, formation of giant HII regions, and shock acceleration accompany such phenomena (e.g. Tenorio-Tagle 1981; Morfill and Tenorio-Tagle 1983). At a distance of 3 kpc, the CR enhancement (relative to the CR density of ~ 1 eV cm^{-3} in the solar vicinity) required to explain the excess γ-ray flux is about $10^7/M$, where M (in units of $M_\odot$) is the gas mass involved. If, for instance, the observed excess flux would originate largely from the HVC's with positive velocities surrounding S54 (Fig. 9c), with a mass of $\sim 5 \times 10^4 M_\odot$ at 3 kpc, then the average CR density would be enhanced by more than two orders of magnitude, and the total CR energy in this volume would exceed 10^{53} erg. A detailed analysis will be presented elsewhere.

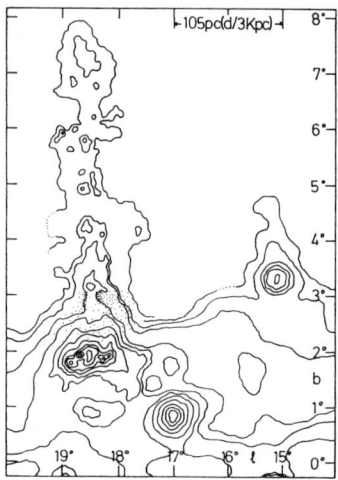

Fig. 7. Sketch of 'Stockert's chimney' at 2.7 GHz, with S54 located at $\ell \approx 18.5°$, $b \approx 2°$ (Müller et al 1987; Kundt and Müller 1987).

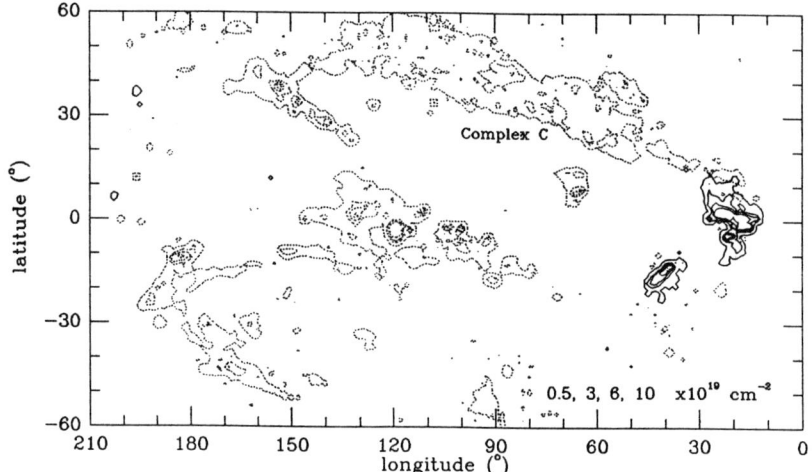

Fig. 8. HI column-density map of the HVC's in the catalogue of Hulsbosch and Wakker (1988), illustrating the possible connection between Complex C and the γ-ray excess near $\ell \approx 15°-20°$. Dotted: negative velocities. Full: positive velocities. The so-called Outer Arm complex has been excluded.

Although speculative sofar, the above clearly illustrates that the next generation of γ-ray telescopes may provide exciting new insight into energetic interstellar phenomena.

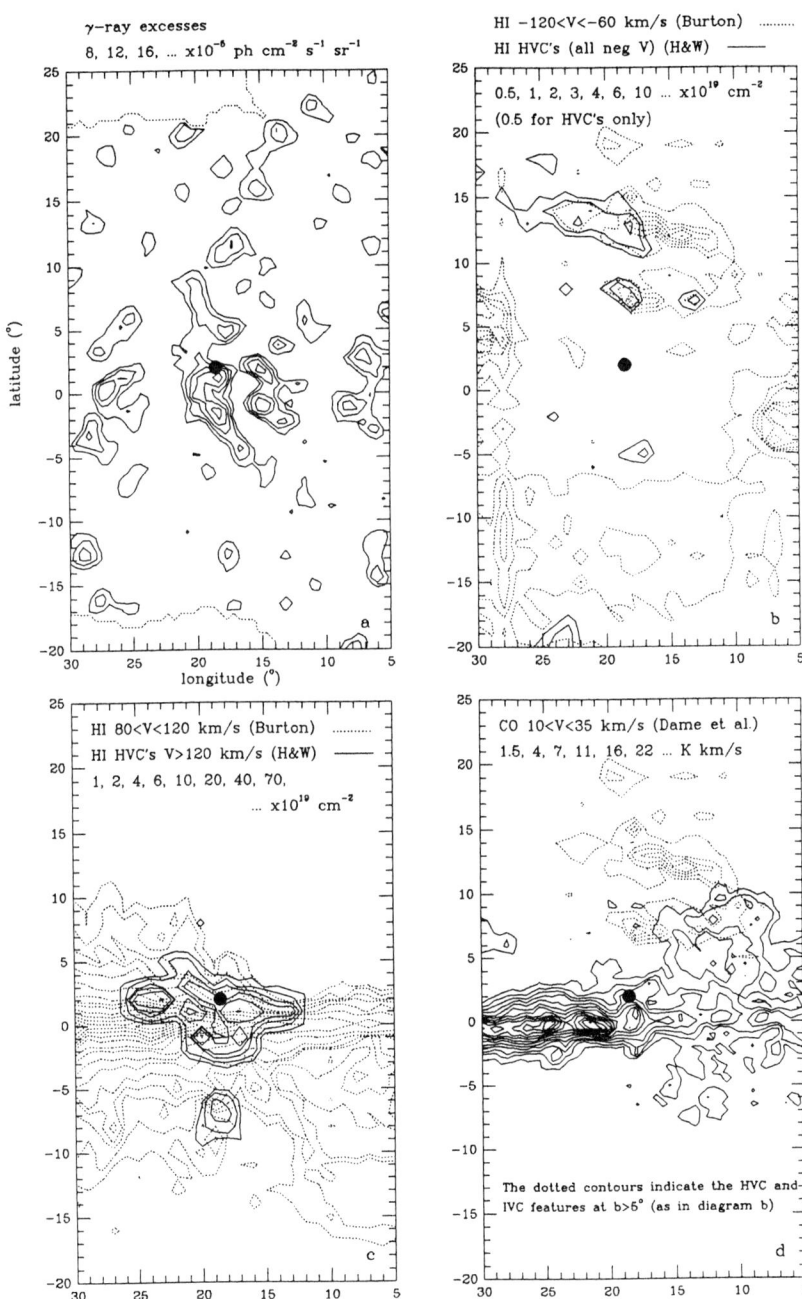

Fig. 9. A compilation of selected HI (b and c) and CO (d) observations around the extended γ-ray excess near $\ell \approx 15° - 20°$ (a). The black dot indicates the position of S54. In Fig. 9b, the extended low-intensity feature seen at $b < -5°$ is probably spurious (see Burton and te Lintel Hekkert 1985). Also, in Fig. 9c, the feature seen at $b < -10°$ (extending to $\ell \approx 350°$) needs confirmation.

5. PROSPECTS

As already mentioned in the Introduction, several new γ-ray satellites will be in orbit soon. The Soviet-French GAMMA-1 experiment and EGRET, next to COMPTEL, OSSE, and BATSE one of the four γ-ray telescopes aboard GRO, are operating in the high-energy part of the γ-ray spectrum and can be considered the direct follow-up's of SAS-2 and COS-B. GAMMA-1 will have the capability of generating images of small regions of the sky with an angular resolution of about 1°, but its sensitivity will be limited. EGRET will cover the energy range 20 MeV – 30 GeV, with over an order-of-magnitude increase in sensitivity compared to COS-B. It has an energy resolution of about 15% (FWHM) in the central part of the energy band covered, an angular resolution of typically 1° (FWHM), and an accuracy for point source location of about 10 arcmin. COMPTEL will explore the 1 to 30 MeV range with an energy resolution of 5 – 8%, an angular resolution of typically 2° – 3°, and an accuracy for point source location of typically 10 arcmin. Both EGRET and COMPTEL have a large field of view of approximately 1 steradian, which enables a full sky survey in the first year of the GRO mission, followed in later phases by indepth studies of selected regions. The OSSE experiment will provide observations in the low-energy γ-ray domain (0.1 – 10 MeV), with an energy resolution similar to that of COMPTEL, but limited imaging possibilities. BATSE will continuously monitor a large fraction of the sky in order to detect and study γ-ray bursts and other transient sources; the other GRO experiments wil provide additional burst capabilities. The French GRANAT/SIGMA telescope enables imaging at sub-MeV energies (0.1 – 2 MeV) with a location accuracy of about 20 arcmin, although with limited sensitivity and spectral resolution.

Particularly the EGRET and COMPTEL instruments should be able to advance drastically studies of the type described in this paper. COMPTEL will provide the important link to low energies, for which the information available is very limited at the moment, largely based on balloon-borne experiments (see, e.g., Sacher and Schönfelder 1984; Lavigne et al. 1986; Schönfelder et al. 1988). It will provide the first survey of the Milky Way at MeV energies, which is essential for an accurate separation of the different γ-ray components. In addition to the CR studies already discussed, spectral and spatial analyses of the COMPTEL data should provide information on the distribution and propagation of low-energy (< 100 MeV) electrons, which can only be studied in interstellar space by γ-ray observations at MeV energies.

Important progress is expected in the field of γ-ray line spectroscopy around 1 MeV. The ISM is responsible for two different components:
- Gamma-ray lines from the decay of radioisotopes produced by the interactions of low-energy CR nuclei with interstellar matter.

A variety of lines is expected (Ramaty et al. 1979), the most prominent ones being those of ^{12}C at 4.44 MeV and ^{16}O at 6.15 MeV. Measurements of these CR-induced decay lines enable studies of the distribution of low-energy (< 100 MeV/nucl) CR nuclei and of their role in the ionization of the interstellar medium. It is uncertain, however, whether planned missions will indeed be able to detect these lines (see Higdon 1987).
- Gamma-ray lines from the decay of (long-lived) radioisotopes produced during various nucleosynthesis processes in specific celestial objects (such as supernovae and novae) and distributed into interstellar space.

The 1.809 MeV line of the most prominent candidate, ^{26}Al, has already been observed in the general direction of the inner Galaxy (Mahoney et al. 1982, 1984; von Ballmoos et al. 1987). The question of the origin of the line can be expected to be solved by mapping the line emission along the Galactic disk. The OSSE and COMPTEL telescopes will provide important new insight into this problem. Another line produced in interstellar space is the 0.511 MeV line from the annihilation of positrons and electrons, the first cosmic γ-ray line ever detected. The production of this line is not related to a single chemical element, but to a variety of different processes which generate positrons. Because positrons survive typically about 10^5 years in interstellar space before they annihilate, a diffuse 0.511-MeV line component may even originate from positrons produced by relatively short-lived radioisotopes such as ^{56}Co, ^{22}Na, and ^{44}Ti. Cosmic-ray interactions with interstellar matter are a potential interstellar source of positrons. The line has been observed now at least a dozen times in the general direction of the Galactic center. Most of these measurements are consistent with the existence of a diffuse line component produced in interstellar space with an angular extent of the order of $\sim 60°$. In addition, there may be a 0.511 MeV point source at the position of the Galactic center which is variable in intensity (see Share et al., 1988). Again, like in the case of the 1.809 MeV ^{26}Al line, further measurements with high spectral and spatial resolution that generate maps in the light of the 0.511 MeV line are needed to clarify the origin of this line.

ACKNOWLEDGEMENTS

I acknowledge the receipt of a Fellowship from the Royal Netherlands Academy of Arts and Sciences (KNAW).

REFERENCES

Allen, R. J., Baldwin, J. E., Sancisi, R. 1978, *Astr. Ap.*, **62**, 397.
Arnaud, M., Cassé, M. 1985, *Astr. Ap.*, **144**, 64.
Axford, W. I. 1981, in *Origin of Cosmic Rays*, ed. G. Setti, G. Spada, A. W. Wolfendale, Dordrecht, Reidel, p. 339.
Axford, W. I. 1987, *Proc. Int. Cosmic Ray Conf., 20th, Moscow* **8**, 120.
Axford, W. I., Leer, E., Skandron, K. G. 1977, *Proc. Int. Cosmic Ray Conf., 15th, Plovdiv* **11**, 132.
Baldwin, J. E. 1976, in *The Structure and Content of the Galaxy and Galactic Gamma Rays*, ed. C. F. Fichtel, F. W. Stecker, Greenbelt, GSFC, p. 206.
von Ballmoos, P., Diehl, R., Schönfelder, V. 1987, *Ap. J.*, **318**, 654.
Beck, R., Biermann, P., Emerson, D. T., Wielebinski, R. 1979, *Astr. Ap.*, **77**, 25.
Bell, A. R. 1978, *M.N.R.A.S.*, **182**, 147.
Berezinskii, V.S., Bulanov, S.V., Ginzburg, V.L., Dogiel, V.A., Ptuskin, V.S. 1990, in *Cosmic Ray Astrophysics*, ed. V.L. Ginzburg, North Holland.
Beuermann, K., Kanbach, G., Berkhuijsen, E. M. 1985, *Astr. Ap.*, **153**, 17.
Bhat, C. L., Issa, M. R., Mayer, C. J., Wolfdendale, A. W. 1985, *Nature* **314**, 515.

Bignami, G. F., Fichtel, C. E. 1974, *Ap. J. (Letters)*, **189**, L65.
Blandford, R. D., Ostriker, J. P. 1978, *Ap. J. (Letters)*, **221**, L29.
Blandford, R. D., Ostriker, J. P. 1980, *Ap. J.*, **237**, 793.
Blandford, R. D., Cowie, L L. 1982, *Ap. J.*, **260**, 625.
Bloemen, J. B. G. M. 1985, *Astr. Ap.*, **145**, 391.
Bloemen, J. B. G. M. 1989, *Ann. Rev. Astr. Ap.*, **27**, 469.
Bloemen, J. B. G. M., Dogiel, V. A. 1990, Dorman, V. L., Ptuskin, V. S. *Astr. Ap.* , in prep.
Bloemen, J. B. G. M., Reich, P., Reich, W., Schlickeiser, R. 1988, *Astr. Ap.*, **204**, 88.
Brecher, K., Burbidge, G. R. 1972, *Ap. J.*, **174**, 253.
Brown, R. L., Marscher, A. P. 1977, *Ap. J.*, **212**, 659.
Brindle, C., French, D. K., Osborne, J. L. 1978, *M.N.R.A.S.*, **184**, 283.
Burbidge, G. R. 1974, *Phil Trans. Roy. Soc. Lond. A* **277**, 481.
Burbidge, G. R. 1983, in *Composition and Origin of Cosmic Rays* , ed. M.M. Shapiro, Dordrecht, Reidel, p. 245.
Burton, W. B., te Lintel Hekkert, P. 1985, *Astr. Ap. Suppl.*, **62**, 645.
Cassé, M., Goret, P. 1978 *Ap. J.*, **221**, 703.
Cesarsky, C. J. 1980, *Ann. Rev. Astr. Ap.*, **18**, 289.
Cesarsky, C. J. 1987, *Proc. Int. Cosmic Ray Conf., 20th, Moscow* **8**, 87.
Cesarsky, C. J., Völk, H. J. 1978, *Astr. Ap.*, **70**, 367.
Clifton, T. R., Frail, D. A., Kulkarni, S. R., Weisberg, J. M. 1988, *Ap. J.*, **333**, 332.
Cowsik, R. 1980, *Ap. J.*, **241**, 1195.
Cowsik, R. 1986, *Astr. Ap.*, **155**, 344.
Dogiel, V. A., Uryson, A. V. 1988, *Astr. Ap.*, **197**, 335.
Eichler, D. 1979, *Ap. J.*, **229**, 419.
Eichler, D. 1980, *Ap. J.*, **237**, 809.
Ekers, R. D., Sancisi, R. 1977, *Astr. Ap.*, **54**, L973.
Fichtel, C. E., Kniffen, D. A. 1984, *Astr. Ap.*, **134**, 13.
Ginzburg, V. L. 1953, *Usp. Fiz. Nauk.*, **51**, 343.
Ginzburg, V. L. 1987, *Proc. Int. Cosmic Ray Conf., 20th, Moscow* **7**, 7.
Ginzburg, V. L., Syrovatskii, S. I. 1964, *The Origin of Cosmic Rays* , Oxford, Pergamon Press.
Hayakawa, S. 1969, *Cosmic Ray Physics* , New York, Wiley-Interscience.
Heiles, C. 1987, in *Interstellar Processes* , ed. D. J. Hollenbach, H. A. Thronson, Dordrecht, Reidel, p. 171.
Higdon, J. C. 1987, *Proc. Int. Cosmic Ray. Conf., 20th, Moscow*, **1**, 60.
Hulsbosch, A. N. M., Wakker, B. P. 1988, *Astr. Ap. Suppl.*, **75**, 191.
Hummel, E., Sancisi, R., Ekers, R. D. 1984a. *Astr. Ap.*, **133**, 1.
Hummel, E., Smith, P., van der Hulst, J. M. 1984b. *Astr. Ap.*, **137**, 138.
Ipavich, F. 1975, *Ap. J.*, **196**, 107.
Jokipii, J. R. 1976, *Ap. J.*, **208**, 900.
Jones, F. C. 1979, *Ap. J.*, **229**, 747.
Klein, R., Urbanik, M., Beck, R., Wielebinski, R. 1983, *Astr. Ap.*, **127**, 177.
Klein, R., Wielebinski, R., Beck, R. 1984, *Astr. Ap.*, **133**, 19.
Kniffen, D. A., Fichtel, C. E. 1981, *Ap. J.*, **250**, 389.
Kóta, J., Owens, A. J. 1980, *Ap. J.*, **237**, 814.
Krymsky, GF. 1977 *Dok. Akad. Nauk. S.S.S.R* **234**, 1306.
Kundt, W., Müller, P. 1987, *Astr. Space Sci.*, **136**, 281.

Lavigne, J. M., Mandrou, P., Niel, M., Agrinier, B. et al. 1986, *Ap. J.*, **308**, 370.
Lawson, K. D., Mayer, C. J., Osborne, J. L., Parkinson, M. L. 1987, *M.N.R.A.S.*, **225**, 307.
Lerche, I., Schlickeiser, R. 1982a, *M.N.R.A.S.*. **201**, 1041.
Lerche, I., Schlickeiser, R. 1982b, *Astr. Ap.*, **107**, 148.
Mahoney, W. A., Ling, J. C., Jacobson, A. S., Lingenfelter, R. E. 1982, *Ap. J.*, **262**, 742.
Mahoney, W. A., Ling, J. C., Wheaton, W. A., Jacobson, A. S. 1984, *Ap. J.*, **286**, 578.
Marscher, A. P., Brown, R. L. 1978, *Ap. J.*, **221**, 583.
Mayer-Hasselwander, H. A. 1983, in *Kinematics, Dynamics, and Structure of the Milky Way* , ed. W. L. H. Shuter, Dordrecht, Reidel, p. 223.
Mayer-Hasselwander, H. A., Simpson, G. 1989, *Adv. in Space Research.* , in press.
Melisse, J., Bloemen, J. B. G. M. 1990, *Proc. Int. Cosmic Ray Conf., 21st, Adelaide* , OG 3.1-18.
Meyer, J.-P. 1985, *Ap. J. Suppl.*, **57**, 173.
Morfill, G. E. 1982a, *M.N.R.A.S.*, **198**, 583.
Morfill, G. E. 1982b, *Ap. J.*, **262**, 749.
Morfill, G. E., Tenorio-Tagle, G. 1983, *Space Sci. Rev.*, **36**, 93.
Müller, P., Reif, K., Reich, W. 1987, *Astr. Ap.*, **183**, 327.
Owens, A. J., Jokipii, J. R. 1977, *Ap. J.*, **215**, 677.
Phillipps, S., Kearsey, S., Osborne, J. L., Haslam, C. G. T., Stoffel, H. 1981, *Astr. Ap.*, **98**, 286.
Phillipps, S., Kearsey, S., Osborne, J. L., Haslam, C. G. T., Stoffel, H. 1981, *Astr. Ap.*, **103**, 405.
Piccinotti, G., Bignami, G. F. 1976, *Astr. Ap.*, **52**, 69.
Pikelner, S. B. 1953, *Doklady Acad. Sci. USSR* **88**, 229.
Pollock, A. M. T., Bennett, K., Bignami, G. F., Bloemen, J. B. G. M., Buccheri, R., et al. 1985, *Astr. Ap.*, **146**, 352.
Pollock, A. M. T., Bennett, K., Bignami, G. F., Bloemen, J. B. G. M., Buccheri, R., et al. 1985, *Proc. Int. Cosmic Ray Conf., 19th, La Jolla* **1**, 338.
Ptuskin, V. S. 1979, *Ap. Space Sci.*, **61**, 259.
Ramaty, R., Kozlovsky, B., Lingenfelter, R. E. 1979, *Ap. J. Suppl.*, **40**, 487.
Ramaty, R., Lingenfelter, R. E. 1979, *Nature,* **278**, 127.
Reich, P., Reich, W. 1988, *Astr. Ap.*, **196**, 211.
Reynolds, R.J. 1989, *Ap. J. Letters,* **339**, L29.
Rogers, M. J., Sadzinska, M., Szabelski, J., van der Walt, D. J., Wolfendale, A. W. 1988, *J. Phys. G,* **14**, 1147.
Sacher, W., Schönfelder, V. 1984, *Ap. J.*, **279**, 817.
Schönfelder, V., von Ballmoos, P., Diehl, R. 1988, *Ap. J.*, **335**, 748.
Share, G. H. et al. 1988, *Ap. J. Letters,* **326**, 717.
Shklovskii, I. S. 1952, *Astr. zh.* **29**, 418.
Shukla, P. G., Paul, J.A. 1976, *Ap. J.*, **208**, 893.
Simpson, G., Mayer-Hasselwander, H. A. 1987, *Proc. Int. Cosmic Ray Conf., 20th, Moscow* **1**, 89.
Simpson, J. A. 1983 *Ann. Rev. Nucl. Part. Sci.*, **33**, 323.
Sofue, Y., Fujimoto, M., Wielebinski, R. 1986, *Ann. Rev. Astr. Ap.*, **24**, 459.
Stecker, F. W. 1977, *Ap. J.*, **212**, 60.
Stecker, F. W., Jones, F. C. 1977, *Ap. J.*, **217**, 843.

Strong, A. W., Bloemen, J.B.G.M., Dame, T. M., Grenier, I., Hermsen, W., et al. 1988, *Astr. Ap.,* **207**, 1.
Swanenburg, B. N., Bennett, K., Bignami, G. F., Buccheri, R., Caraveo, P. A. et al. 1981, *Ap. J. (Letters),* **243**, L69.
Tenorio-Tagle, G. 1981, *Astr. Ap.,* **94**, 338.
Van der Walt, D. J., Wolfendale, A. W. 1988 *Space Sci. Rev.,* **47**, 1.
Völk, H. J. 1983, *Space Sci. Rev.,* **36**, 3.
Völk, H. J. 1984, in *High Energy Astrophysics* , ed. J. Audouze, J. Tran Thanh Van, Gif-sur-Yvette; Editions Frontières, p. 281.
Völk, H. J., Klein, U., Wielebinski, R. 1989, *Astr. Ap.* **213**, L12.
Wolfendale, A. W. 1988, in *Molecular Clouds in the Milky Way and External Galaxies* , ed. R. Dickman, R. Snell, J. Young, in press. Heidelberg, Springer-Verlag, p. 76.

THE ROLES OF COSMIC RAYS IN INTERSTELLAR DYNAMICS

T.W. HARTQUIST
Max Planck Institute for Physics and Astrophysics, Institute for
Extraterrestrial Physics, 8046 Garching, FRG.

ABSTRACT The energy density of cosmic rays is comparable to the thermal and magnetic energy densities in the interstellar medium. Hence, they often affect interstellar dynamics significantly. Work on cosmic ray moderation of steady shocks and of supernova remnants is reviewed. The role that cosmic rays play at present in supporting diffuse gas in the Galactic halo and their effects on accretion flows in evolving galaxies are described. Cosmic ray driven and modified instabilities are considered.

1. INTRODUCTION

The average thermal energy and average magnetic field energy densities in the interstellar medium are each roughly 1 eV cm^{-3}. Cosmic rays have a comparable average energy density, provided primarily by particles with energies of about 1 GeV. In this paper, I describe the often nonnegligible influence of cosmic rays on interstellar dynamics.

Sections 2 and 3 contain a short introduction to cosmic ray transport and acceleration theory. Second order Fermi or stochastic acceleration of high energy particles is most important in turbulent magnetic media which have only very small ordered large scale velocities. First order Fermi acceleration occurs in converging flows like shocks. A considerable literature treats the problem of first order Fermi acceleration in plane parallel shocks and the modification of the shock structure due to cosmic ray acceleration.

Supernova remnants drive the shocks of greatest importance for the global structure of the interstellar medium. Section 4 is a brief review of work done on the modification of supernova remnant structure by cosmic ray acceleration.

One can reasonably conjecture that because cosmic rays suffer no radiative losses that their pressure is the dominant one in diffuse material at large distances from energy sources such as stars or active galactic nuclei. Support for this conjecture comes from studies of the diffuse Galactic halo gas which probably is levitated by cosmic ray pressure as described in section 5.

In section 6, the influence of cosmic rays on the flow of cooling gas accreting onto young galaxies is considered. If the conjecture that cosmic rays dominate the pressure in diffuse material at large distances from energy sources is correct, they must have had a tremendous influence on the evolution of the Galaxy in its earliest phases.

Section 7 is a short exposition on the ways that cosmic rays drive or influence various instabilities in the interstellar medium. The Parker, thermal, and convective instabilities are all likely to be influenced by the presence of cosmic rays.

Section 8 is a brief summary.

2. THE COSMIC RAY TRANSPORT EQUATION

For one of the clearest presentations of a derivation of the cosmic ray transport equation, the reader should refer to the review article by Blandford and Eichler(1987). The papers by Skilling(1975) and by Völk(1975) and some of the references in them treat parts of the derivation more elegantly or more completely. However, to my knowledge no "textbook quality" discussion of the derivation of the transport equation and the coefficients in it exists.

Here, I simply give the form of the transport equation for the idealized case that the waves are propagating in one direction only along the largescale magnetic field which is assumed to be in the $\hat{z}$ direction and to have a strength B_o. Further, the waves are assumed to be small amplitude Alfvén waves with a power spectrum $P(k_z)$ where k_z is the wavenumber. The transport equation is

$$\frac{\partial f}{\partial t} + \nabla_x \cdot ((\mathbf{u} + v_A \hat{z})f) - \frac{\partial}{\partial z}(K_{zz}\frac{\partial f}{\partial z}) = \frac{1}{3p^2}(\nabla_x \cdot (\mathbf{u} + v_A \hat{z}))\frac{\partial}{\partial p}(p^3 f) \quad (1)$$

where $f(x,p,t)$ is the isotropic part of the distribution function, $\mathbf{u}$ is the velocity of the medium, and v_A is the Alfvén speed. The diffusion coefficient is given roughly by

$$K_{zz} \sim \frac{1}{3}(\frac{B_o}{\delta B})^2 c r_g \quad (2)$$

where c and r_g are the speed of light and the gyroradius, and $(\delta B)^2/8\pi \sim k_z P(k_z)$ The term on the right hand side leads to acceleration or deceleration of the cosmic rays depending on whether the relative motion of the waves is converging or diverging; this term is responsible for first order Fermi acceleration.

If waves moving both parallel and antiparallel to the large scale field are considered, an additional acceleration term appears. It accounts for a process which gives rise to acceleration even in the absence of convergence. The process is called stochastic or second order Fermi acceleration and arises because small particles gain energy from scattering repeatedly on massive bodies (i.e. waves) having a nonsingular velocity distribution. Consider, for instance, one set of very heavy particles with a thermal distribution and one set of initially cold, light particles; the evolution to equipartition occurs as the light particles are accelerated to speeds which are large compared to those of the very heavy particles.

3. COSMIC RAY ACCELERATION IN PLANE PARALLEL SHOCKS

A number of authors(Axford, Leer, and Skadron 1977; Krymsky 1977; Bell 1978a,b; Blandford and Ostriker 1978) independantly and nearly simultaneously started to address explicitly the question of how cosmic rays are accelerated by the first order Fermi process in steady plane-parallel shocks propagating parallel to the large scale magnetic field.

Consider initially the case in which the reaction of the cosmic rays back onto the flow of the thermal fluid is ignored. Also take $u \gg v_A$ and assume that

$$\mathbf{u} = u_s \hat{z} \qquad z < 0 \qquad (3a)$$

$$\mathbf{u} = u_+ \hat{z} \qquad z > 0 \qquad (3b)$$

The time independant cosmic ray transport equation for $u \gg v_A$ is

$$\frac{\partial}{\partial z}(uf - K_{zz}\frac{\partial f}{\partial z}) = \frac{\partial u}{\partial z}\frac{1}{3p^2}\frac{\partial}{\partial p}(p^3 f) \qquad (4)$$

which can be rewritten as

$$\frac{\partial}{\partial z}(-K_{zz}\frac{\partial f}{\partial z} - u\frac{p}{3}\frac{\partial f}{\partial p}) = -u\frac{\partial}{\partial z}\frac{\partial}{\partial p^3}(p^3 f) \qquad (5)$$

Hence, one must solve equation (4) for $z < 0$ and for $z > 0$; clearly, the right hand side of the equation is zero for both regions. The boundary conditions at $z=0$ require that f is continuous which together with equation (5) implies that $K_{zz}\frac{\partial f}{\partial z} + u\frac{p}{3}\frac{\partial f}{\partial p}$ is also continuous. One finds that

$$f(z,p) = f_-(p) + [f_+(p) - f_-(p)]\exp\{\int_0^z \frac{u_s}{K_{zz}}dz\} \qquad z < 0 \qquad (6a)$$

$$f(z,p) = f_+(p) \qquad z > 0 \qquad (6b)$$

$$\frac{\partial f_+}{\partial \ln p} = \frac{3u_s}{u_s - u_+}(f_+ - f_-) \qquad (6c)$$

Consequently,

$$f_+(p) = qp^{-q} \int_0^p f_-(p')p'^{q-1}dp' \qquad (7a)$$

with

$$q = \frac{3u_s}{u_s - u_+} \qquad (7b)$$

Thus, the passage of plasma through a steady plane-parallel shock propagating parallel to the large scale magnetic field leads to a power law distribution in momentum of the cosmic rays, if the lengthscale over which the flow velocity in the thermal fluid changes is very short compared to the cosmic ray mean free path, which is of the order of $r_g(\frac{B_0}{\delta B})^2$

In fact, because the cosmic rays have an energy density that can be a substantial fraction of the shock ram pressure, their interaction with the waves can affect the flow in the thermal fluid on lengthscales comparable to or larger than their mean free path. The analysis given in the above paragraph must be extended so that a self-consistent calculation of the flow dynamics and the cosmic ray dynamics can be performed.

One approach which has been adopted is that of generating, by taking momentum moments of the cosmic ray transport equation, fluid equations that describe the cosmic ray behavior (McKenzie and Völk 1982; Völk, Drury, and McKenzie 1984).

Before considering the set of fluid equations with which plane parallel steady shocks moving parallel to the large scale field can be studied, the reader should bear in mind that just as energy can be transferred from waves to cosmic rays, the streaming of cosmic rays can transfer energy to waves. The energy transfer rate, Q_W, per unit volume to waves is just the integral over momentum of the term on the right hand side of equation (5) times $-4\pi p^4/\gamma m$. If the cosmic rays are very relativistic, then

$$Q_W = -(u+w)\frac{dP_c}{dz} \qquad (8)$$

where the cosmic ray pressure, P_c, for very relativistic particles is

$$P_c \equiv \frac{4\pi}{3}\int_0^\infty \frac{p^4}{\gamma m}fdp \qquad (9)$$

and w is v_A or $-v_A$ depending on the direction of wave propagation. Even though its validity has not always been rigorously established, the assumption that the wave energy is dissipated locally as heat, rather than convected away, generally is made. That is

$$Q_W = L_W \qquad (10)$$

where L_W is the rate per unit volumn at which heating occurs, and L_W can depend on the wave pressure,

$$P_W \equiv \frac{1}{8\pi} \int P(k_z) dk_z \tag{11}$$

The fluid equations which describe the flow of the thermal gas are

$$\rho u = a_1 \tag{12a}$$

$$\rho u^2 + P_T + P_c + P_W = a_2 \tag{12b}$$

$$\rho u \left(\frac{u^2}{2} + \frac{\gamma}{\gamma-1} \frac{P_T}{\rho} + F_c + F_W \right) = a_3 \tag{12c}$$

where P_T is the thermal pressure and a_1, a_2, and a_3 are constants, and radiative losses have been neglected. $\gamma = 5/3$ for ionized gas. F_c and F_W are the cosmic ray and wave energy fluxes which are treated below. From (1), one can find that in a steady flow,

$$\frac{dF_c}{dz} = (u+w)\frac{dP_c}{dz} \tag{13a}$$

with

$$F_c = \frac{\gamma_c}{\gamma_c - 1}(u+w)P_c - \frac{\bar{K}}{\gamma_c - 1}\frac{dP_c}{dz} \tag{13b}$$

where $\gamma_c = 4/3$ for very relativistic cosmic rays, and

$$\bar{K} \equiv \frac{\int_0^\infty dp \frac{p^4}{\gamma m} \bar{K}_{zz} \frac{\partial f}{\partial z}}{\int_0^\infty dp \frac{p^4}{\gamma m} \frac{\partial f}{\partial z}} \tag{13c}$$

Finally,

$$F_W = 2P_W \left(\frac{3}{2}u + w\right) \tag{14}$$

Equations(8)-(14) constitute a set with which a two fluid model of cosmic ray moderated, plane-parallel, steady flows moving parallel to the largescale magnetic field can be constructed. By adopting a fluid description of the cosmic rays, one loses information about the cosmic ray spectrum in order to deal with coupled ordinary differential equations only. In some applications (see e.g. Hartquist and Morfill 1986), L_W is a specified function and P_W is calculated, and $\bar{K}$ is expressed as a function of P_W. In other applications, P_W is assumed to be dynamically negligible, and $\bar{K}$ is a specified constant. In adiabatic (but not in radiative) steady flows, the lengthscale over which variations in the fluid parameters occur depends on the constant value specified for $\bar{K}$, but the final downstream conditions corresponding to a given set of upstream conditions do not. (Note, for instance, that equation (7) is independant of K_{zz}.) In general, L_W and $\bar{K}$ are unknown.

The actual cosmic ray pressure at the shock front is a further unknown. One might assume that only those cosmic rays which exist in the distant upstream gas are accelerated. However, some particles from the high energy tail of the thermal distribution of particles are probably also accelerated in the vicinity of the viscous subshock in the thermal plasma. The means by which particles are injected as cosmic rays is not understood, even though the existence of the cosmic rays themselves imply that nearly thermal particles must be turned into cosmic rays in some environment. Hence, while the constant value assumed for $\bar{K}$ does not affect steady shock cosmic ray acceleration calculations significantly, the injection efficiency does.

Völk, Drury, and McKenzie(1984) have used the two fluid equations to construct steady shock models. $\bar{K}$ was taken to be constant and no injection of cosmic rays was assumed. Results were obtained for a number of values of the parameters $\beta \equiv 8\pi P_{T_1}/B_o^2$, $N \equiv \frac{P_{c_1}}{P_{c_1}+P_{T_1}}$, and Mach number where the subscript 1 indicates preshock values. The fraction of the dissipated kinetic energy which is converted into cosmic ray energy generally exceeds about one half when the Mach number is greater than about 5 or 6, $N > 1/2$, and $\beta \geq 1$.

A number of other approachs have been adopted to study the cosmic ray moderation of shocks.

Eichler(1979,1984) made use of the fact that if K_{zz} is a strongly varying function of p then the independant variables on which f depends can be taken to be p and u rather than p and z, and $f(p,u)$ can be approximated by

$$f(p,u) \approx f_+(p)\theta[u_c(p) - u] \tag{15}$$

where θ is the step function and u_c is a function of p only. He then proceeded to develop an elegant analytic treatment of the structure of steady cosmic ray moderated shocks.

Ellison and Eichler(1984) have constructed models of cosmic ray moderated shocks by using Monte Carlo techniques to study the position dependance of $f(p,z)$. They solved the cosmic ray transport equation

$$\mu u \frac{\partial f}{\partial z} = \frac{\partial f}{\partial t}|_c \tag{16}$$

where μ is the cosine of the pitch angle (the angle between the momentum vector and the large scale magnetic field) and $\frac{\partial f}{\partial t}|_c$ is a collision term. The collision term was determined by positing a distribution of scattering centers assumed to be present for each momentum. The trajectories of roughly 10^5 particles of different momenta were followed. At each point in the shock the cosmic ray pressure was evaluated from the integral over the spectrum obtained from the Monte Carlo calculation. The cosmic ray pressure was then included in the hydrodynamic equations which govern the flow of the thermal fluid, and the system was solved self-consistently. Some attempt was made to address the problem of cosmic ray injection, as scattering centers for particles at thermal energies were also included.

Recently, Bell(1987) and Falle and Giddings(1987) have used numerical partial differential equation solvers to investigate the structures of model plane parallel shocks propagating parallel to the large scale magnetic field and governed by the cosmic ray transport equation and the time dependant fluid equations for the thermal gas. Sufficiently long integrations in time yield the structures of steady shocks. Bell permitted cosmic rays which obtain $p \geq 100 m_p c$ (where m_p is the proton mass) to escape the system. (Otherwise, in steady shocks, all of the energy would go into a small number of very high energy cosmic rays.) He assumed that $K_{zz} \propto p^{1/2}$, the shock speed to be $0.01c$, and the Mach number to be 100. He calculated the ratio of cosmic ray pressure to total pressure in the distant postshock gas as a function of the ratio of cosmic ray pressure to total pressure in the distant upstream gas. He found that the former ratio increases rapidly to values above 0.5 when the latter ratio attains values around 3×10^{-4} For unmoderated shocks, (7b) gives $q = 4$. Bell found that at high energies q remains 4 in the moderated shocks but that the spectrum is softer (i.e. $q > 4$) at some lower energies. The softening at lower energies is greatest for the largest preshock ratios of cosmic ray pressure to total pressure. Note that all of the calculations described in this section have applied to adiabatic shock models.

As mentioned above, several important physical quantities are unknown for shocks. One of the most important unknown quantities is the value of $\bar{K}$. Clearly, semiempirical studies in which spectral diagnostics of real interstellar shocks are used to infer the values of important quantities such as $\bar{K}$ are required. Boulares and Cox(1988) have performed such an investigation. They considered spectral data for the Cygnus Loop supernova remnant and used models like those of Völk (1984) to interpret it. Specifically, they addressed the problem of how many types of shocks are present in the Cygnus Loop. Regions of strong optical line emission arise as shocks propagate through dense clumps in the Loop, whereas soft X-ray emission originates in more diffuse interclump regions which have been shocked. Balmer line emission is also observed from the Loop and on the basis of ordinary unmoderated shock models, such Balmer line emission would not be expected to be associated with either of these two types of shocked regions. Boulares and Cox argued that the observed Balmer line photons are emitted in cosmic ray heated precursors of the shocks which induce the X-ray emission if $\bar{K} \leq 2 \times 10^{25} \mathrm{cm}^2 \mathrm{s}^{-1}$ (c.f. $\bar{K}$ is generally thought to be roughly $10^{28} \mathrm{cm}^2 \mathrm{s}^{-1}$ in the Solar neighborhood.) The observed SII photons, whose origins have eluded explanation probably are emitted in the same precursors. Boulares and Cox also argued that cosmic ray pressure may be responsible for the measured discrepancy between ram pressure and distant postshock thermal pressure in the gas emitting strongly in the optical (Raymond et al. 1988), but magnetic pressure may also play a role. An additional point made by Boulares and Cox is that when cosmic ray moderation is included, models of the observed shocks do not require, as unmoderated shock models do, that the preshock number densities for some cases lie in the range of 0.15 cm^{-3} to 15 cm^{-3}. This range is thought to be excluded by simple thermal instability arguments.

4. SPHERICALLY SYMMETRIC SUPERNOVA REMNANTS

As mentioned in the previous section, models of evolving plane parallel cosmic ray moderated shocks exist(Bell 1987; Falle and Giddings 1987). One aim of the scientists developing those models was the investigation of the evolution of the cosmic ray spectrum with time. The only studies of cosmic ray moderated spherical supernova remnants were based on fluid descriptions of the cosmic ray dynamics, and conclusions concerning the spectral evolution were not possible.

Chevalier(1983) has found similarity solutions for the evolution of adiabatic remnants in which the cosmic ray diffusion coefficient is so small that diffusion terms are ignored. The equations governing the remnant evolution were taken to be

$$\frac{\partial \rho}{\partial t} + u\frac{\partial \rho}{\partial r} + \rho\frac{\partial u}{\partial r} + \frac{2\rho u}{r} = 0 \tag{17a}$$

$$\frac{\partial u}{\partial t} + u\frac{\partial u}{\partial r} + \frac{1}{\rho}\frac{\partial P_T}{\partial r} + \frac{1}{\rho}\frac{\partial P_c}{\partial r} = 0 \tag{17b}$$

$$\frac{\partial P_T}{\partial t} + u\frac{\partial P_T}{\partial r} - \frac{5P_T}{3\rho}\left(\frac{\partial \rho}{\partial t} + u\frac{\partial \rho}{\partial r}\right) = 0 \tag{17c}$$

$$\frac{\partial P_c}{\partial t} + u\frac{\partial P_c}{\partial r} - \frac{4P_c}{3\rho}\left(\frac{\partial \rho}{\partial t} + u\frac{\partial \rho}{\partial r}\right) = 0 \tag{17d}$$

For similarity solutions $u_s \propto t^{-3/5}$, where u_s is the speed of the leading shock, when the ambient medium is uniform. The results depend on ϕ, the ratio of the cosmic ray pressure to the total pressure at the shock front. Table I contains results from Chevalier's work, r_s is the radius of the remnant. Note that even for a small value of ϕ, the thermal pressure approachs zero at the origin.

TABLE I Results for the Similarity Solutions Found by Chevalier

$\phi =$	0.00	0.01	0.50	0.99
$r = 0$				
$P_c/\rho_1 u_s^2$	0.00	0.23	0.30	0.33
$P_T/\rho_1 u_s^2$	0.23	0.00	0.00	0.00
$r = 0.5 r_s$				
$P_c/\rho_1 u_s^2$	0.00	0.0093	0.27	0.33
$P_T/\rho_1 u_s^2$	0.23	0.23	0.031	0.0003
$r = 0.9 r_s$				
$P_c/\rho_1 u_s^2$	0.00	0.0057	0.31	0.42
$P_T/\rho_1 u_s^2$	0.41	0.40	0.24	0.0023

Dorfi(1989) has used the time dependant version of the two fluid equations given by McKenzie and Völk(1982) to study the evolution of adiabatic spherical remnants expanding into uniform surrounding media following the uniform deposition of 10^{51} ergs of thermal energy within a radius of 10^{12} cm. The entire high density atmosphere was assumed to extend to 10^{14} cm. In Dorfi's models, $T_1 = 8000$K, $\rho_1 = 5 \times 10^{-25}$g cm^{-3} and $P_{c_1} = P_{T_1}$.

He assumed that cosmic rays are also injected at the shock front. The injection efficiency, η_{inj}, is defined by

$$P_c^{inj} \equiv \frac{1}{\gamma_c - 1} \frac{\eta_{inj}}{2} \rho_1 u_1 \triangle u^2 \qquad (18)$$

where P_c^{inj} and $\triangle u^2$ are the pressure of the injected cosmic rays and $\triangle u^2$ is the jump in the value of u^2 at the shock front. Dorfi followed remnant evolution for differing values of $\bar{K}$, η_{inj} and γ_c. The model remnants each had a free expansion phase during which a reverse shock existed. Following the termination of that phase, a remnant entered the normal "adiabatic" expansion phase. For $\bar{K} = 10^{27}$ cm^{2}s^{-1}, the cosmic ray pressure was relatively constant throughout the remnant and roughly equal to the preshock pressure during the adiabatic phase. In contrast, for $\bar{K} = 10^{25}$ cm^{2}s^{-1}, the cosmic ray pressure was peaked around $r = r_s$ at several to about ten times the value in the ambient medium and decreased inwardly to several times less than the preshock value at $r = 0$ The cosmic ray pressure for a model with $\bar{K} = 10^{23}$ cm^{2}s^{-1}, also peaked at several to about ten times the value in the ambient medium but dropped by many orders of magnitude towards the remnant center. Results were somewhat dependant on γ_c and η_{inj}, but the sensitivity to $\bar{K}$ was greater; perhaps, the values, 0. and 10^{-3}, adopted for η_{inj} for the remnants with smaller diffusion coefficients were too small (c.f. the discussion of the paper by Bell(1987) given in the previous section) compared to the range of values in which the postshock cosmic ray pressure depends sensitively on η_{inj}. The ratio of cosmic ray pressure to total pressure grew with time. The growth of the ratio with time was in part due to the dropping of the total pressure in the remnant, with time, relative to the ambient cosmic ray pressure. From (6a), it follows that the cosmic ray acceleration time is given roughly by $\bar{K}/u^2$ and for the smaller values of $\bar{K}$ used by Dorfi, the acceleration time is shorter than the remnant age. Yet throughout much of the remnants' lives, the computed postshock cosmic ray pressures are only small fractions of the thermal pressures; this would seem to contrast with the result found by Völk (1984) that cosmic ray pressures are large behind shocks with Mach numbers a bit above 5 or 6. However, as pointed out by Drury, Markiewicz, and Völk (1989) very high Mach numbers (>>10 say) and no injection would be expected to give rise to low ratios of postshock cosmic ray pressure to total pressure in any treatment including one like that of Völk et al. (1984). It would seem that Dorfi's results do not show high ratios of cosmic ray pressure to thermal pressure in the remnants because he chose small η_{inj}s for his small $\bar{K}$ models and did not follow the remnants' histories to low enough Mach number. It would be of interest to compare a small $\bar{K}$ model (of the type constructed by Dorfi) with a larger η_{inj} and the corresponding similarity solution found by Chevalier.

Drury, Markiewicz, and Völk(1989) have studied cosmic ray acceleration in supernova remnants with a model based on the assumptions that the postshock region at the front of the remnant is thin, that the interior of the remnant has a high pressure but is massless, and that the expansion rate of the outer boundary of the cosmic ray precursor is proportional to the mean cosmic ray diffusion coefficient and inversely proportional to the thickness of the precursor. These assumptions permit one to write ordinary differential equations, with time as the independant variable, governing the mass in the precursor, the mass in the postshock shell, the energy in the precursor, the energy in the remnant interior, and the motions of the interfaces (e.g. the shock front) between the different regions. Only moments of the cosmic ray distribution function were calculated, and no spectral information was obtained. Cosmic ray diffusion terms were included to describe transport at the interfaces between the different regions. Various physical arguments were presented to justify the assumed behaviors of parameters including the cosmic ray injection efficiency at the shock front, the diffusion coefficient, and the heating rate of the plasma by the dissipation of Alfvén waves. The solution of the ordinary differential equations led the authors to conclude that both the thermal pressure and the cosmic ray pressure are significant in the remnant interior.

5. COSMIC RAY SUPPORT OF THE HALO GAS

Ipavich(1975) and Jokipii(1976) investigated models of a cosmic ray driven Galactic wind. Attempts to understand the ionization structure of gas in the Galactic halo have stimulated more recent interest in the role that cosmic rays play in supporting halo gas. Specifically, Chevalier and Frannson(1984) and Hartquist, Pettini, and Tallant(1984) concluded that extragalactic photons can induce sufficient ionization to produce the C^{3+} and Si^{3+} detected in absorption against halo stars and extragalactic objects (e.g. Savage's review in this volume) and located at several kiloparsecs above the disk only if the gas in which those species exist is of low density (10^{-3} cm^{-3}), is relatively uniformly distributed, and has a large filling factor. Since it was thought that the gas is not hot ($\leq 10^5$K) and not moving very rapidly (Pettini and West 1982), cosmic ray pressure was proposed to be responsible for the support of this supposedly pervasive plasma (Chevalier and Frannson 1984; Hartquist 1984). More recently acquired data (Savage's review) suggest that the Southern hemisphere gas is fairly dynamically quiescient, but the Northern hemisphere gas is more active. The Southern hemisphere data continue to provide evidence for the importance of cosmic rays in the Galactic halo.

Since cosmic ray support of the C^{3+} and Si^{3+} bearing plasma was mentioned in the context of a specific model of the ionization structure, there has perhaps been some confusion that the idea that cosmic rays are of prime dynamical importance in the diffuse halo gas is somehow tied to photoionization models of the ionization structure. It is not and is, in fact, most strongly suggested, for reasons discussed more fully in the introduction of a paper by Hartquist and Morfill(1986), by the point concerning the low temperatures and low velocities mentioned above. Further, the conjecture that cosmic rays provide the dominant pressure in plasmas far from energy sources, such as stars, is attractive because cosmic rays experience no radiative losses.

Hartquist and Morfill(1986) studied two fluid models of static cosmic ray supported gas in the Galactic halo. The equations used were similar to (8) through (14) with $u = 0$ and $w = v_A$. The large scale magnetic field was assumed to be constant and perpendicular to the disk. Equation (12b) was replaced by

$$\frac{dP_c}{dz} = -\rho g \qquad (19)$$

where g is the gravitational acceleration. The thermal balance of the gas was not calculated, and, consequently, the equation corresponding to (12c) was not included. L_W was taken to be proportional to P_W^2/P_B (where P_B is the magnetic pressure) as one would expect for a second order wave decay mechanism (such as that due to the three wave interaction if the amplitudes of the Alfvén waves propagating towards the disk are proportional to the amplitudes of Alfvén waves propagating away from the disk). $\bar{K}$ was taken to be proportional to P_w^{-1}. The choice of the constants of proportionality in the expressions for L_w and $\bar{K}$ are discussed by Hartquist and Morfill. Two of the most interesting features of the models are: 1) Convective stability as determined by the criterion that $-(3P_c/4)dP_c/dz < -(1/\rho)d\rho/dz$ obtains only in those models in which the cosmic ray energy production rate, S, per unit area in the disk obeys the inequality, $S/8v_{A_o}P_{co} < 2.0$, where the subscript o indicates that the quantity is evaluated at the base of the halo. 2) The heating rate per unit volume due to the dissipation of waves generated by cosmic ray streaming in the halo is given by $g\rho v_A$.

Breitschwerdt, McKenzie, and Völk(1987) have constructed two fluid models of a Galactic wind driven by cosmic rays. The finding that the outflow speed can be low at heights less than 10 kpc, say, yet pass through a sonic point at many tens of kiloparsecs is of great interest. Even if data were to show conclusively that the outflow speed is low near the disk, they would not constrain the speed of the wind at the terminal shock which constitutes the interface between the interstellar medium and the intergalactic medium. Because the cosmic ray acceleration time to very high energies is long, the most energetic cosmic rays must be accelerated in a long lived shock like the terminal shock of the Galactic wind (Jokipii and Morfill 1985).

6. ACCRETION FLOWS ONTO GALAXIES

Böhringer and Morfill(1988) have constructed two fluid models of cooling flows in the vicinity of large galaxies. Such an accretion flow probably existed around the Galaxy in its earliest phases of evolution. Specific models were developed for the Perseus cluster and for M87. Spherical symmetry was assumed.

For the M87 model, the integration was started at an inner boundary of $r = 5$ kpc. The adopted gravitational potential was given by a King function for $M = 7.7 \times 10^{12} M_\odot$ and $a = 20$ kpc. The accretion rate was set equal to 10 $M_\odot$ yr^{-1}, and the Alfvén speed, $\bar{K}$, and hydrogen nucleus number density at 5 kpc were taken to be 3 km s^{-1}, 10^{28} cm^{2}s^{-1}, and 3×10^{-2} cm^{-3}. The total cosmic ray energy production rate was taken to be 4.5×10^{41} erg s^{-1}. For several kiloparsecs above $r = 5$ kpc, the density dropped with height.

Böhringer and Morfill claimed that the inversion layer is Rayleigh-Taylor unstable. Their arguments are based on the application of the standard one fluid analysis of the instability, an application which they stated is appropriate in the limit that the diffusion coefficient is small. Perhaps, such instability triggers cloud formation.

In addition, the authors identified, for two values of the Alfvén speed, regions in cosmic ray production rate - diffusion coefficient parameter space for which the cosmic ray pressure would affect substantially the structure of the M87 inflow near the inner boundary.

For the Perseus cluster model, a cosmic ray production rate and an infall rate which were about 30 times larger than those used in the M87 model were assumed. Distributed, as well as central, cosmic ray energy production was included. An inversion layer was also found to exist for that model.

7. COSMIC RAYS AND INSTABILITIES

In the previous two sections, convective and Rayleigh-Taylor instabilities were mentioned. Thermal instability is probably important in the Galactic halo gas and in the cooling flows in M87 and the Perseus cluster; Galactic high latitude, high velocity clouds and warm(10^4K) optical line emitting regions in the cooling flows are generally thought to form due to this instability. Those three instabilities need to be investigated in the context of two fluid models. For instance, a two fluid study of the interplay of the thermal and convective instabilities analogous to that performed by Balbus(1988) for a one fluid system should be executed.

Drury and Falle(1986) have analyzed a cosmic ray induced instability which was first discovered in numerical calculations of the time dependant two fluid shock models (Dorfi and Drury 1985). Sound waves in a plasma containing cosmic rays will grow if the scale height of the cosmic rays is less than about $|P_c(1 + \phi)|/a\rho$ where a is the sound speed and $\phi \equiv \partial ln\bar{K}/\partial ln\rho$.

The growth rate of the Parker(1966) instability may be affected if the cosmic ray sources are included. The initial stationary state is generally assumed to consist of plane parallel stratified plasma supported against a constant gravitational field by magnetic pressure and cosmic ray pressure, though no sources or sinks of cosmic rays are assumed to exist. The initial magnetic field is taken to be perpendicular to the gravitational field. One could imagine such an initial configuration with embedded cosmic ray sources; since it is an equilibrium configuration, the sources must be balanced by losses, which may be due to the generation of waves which then dissipate and heat the thermal gas which in turn cools by radiative processes. When perturbations are introduced on the initial state, cosmic ray losses may no longer balance sources; for instance, the path length along a field line between two sources increases leading to a reduction of the cosmic ray pressure gradient component parallel to the field and, hence, of the wave generation rate. An imbalance may result in greater cosmic ray streaming along the field lines which bow out of the plane in places. Streaming into the regions of bowed field lines could suppress the Parker instability, the growth of which normally occurs because gas can slip towards the plane along the field lines where they bow, by levitating the gas.

The presence of cosmic rays probably greatly affects the evolution of fragments formed in old supernova remnants. Hartquist and Morfill(1983) have argued that some chemical abundances in the neutral material in the Per OB2 association provide evidence of the presence of a sufficient number density of 2 MeV cosmic rays to produce a dynamically significant pressure. Such low energy cosmic rays do not enter a neutral cloud freely if it has a large enough column density; ionization losses in the cloud produce a cosmic ray anisotropy on either side that drives the production of waves which hinder the cosmic ray streaming (Skilling and Strong 1976; Cesarsky and Völk 1978). Hartquist, Oppenheimer, and Elmegreen(1979) have argued that the pressure differences due to the exclusion of the cosmic rays can lead to the thermal instability of clouds with certain column densities.

8. SUMMARY

The pressure due to cosmic rays in the interstellar medium is significant, and at large distances from energy sources it may dominate because they do not experience rapid radiative losses. Cosmic rays are accelerated in shocks, and the shocks may be modified greatly by the presence of cosmic rays. Cosmic rays probably support gas at great distances from galaxies, but a thorough analysis of the stability of cosmic ray supported configurations is required.

ACKNOWLEDGEMENT

Dr. S. B. Charnley's instruction on and help with the use of T_EX are greatly appreciated.

REFERENCES

Axford, W. I., Leer, E., and Skadron, G. 1977, In *Proc. 15th Intern. Cosmic Ray Conf.(Plovdiv)*, **11**,132.
Balbus, S. A. 1988, *Ap. J.*, **328**, 395.
Bell, A. R. 1978a, *M. N. R. A. S.*, **182**, 147.
Bell, A. R. 1978b, *M. N. R. A. S.*, **182**, 443.
Bell, A. R. 1987, *M. N. R. A. S.*, **225**, 615.
Blandford,R. and Eichler, D. 1987, *Phys. Reports*, **154**, 1.
Blandford, R. D. and Ostriker, J. P. 1978, *Ap. J.*, **221**, L29.
Böhringer, H. and Morfill, G. E. 1988, *Ap. J.*, **330**, 609.
Boulares, A. and Cox, D. P. 1988, *Ap. J.*, **333**, 198.
Breidtschwerdt, D., McKenzie, J. F., and Völk, H. J. 1987, *In: Interstellar Magnetic Fields: Observations and Theory*, eds. R. Beck and R. Grave, (Springer Verlag: Heidelberg) p131.
Cesarsky, C. J. and Völk, H. J., 1978, *Astron. Astrophys.*, **70**, 367.
Chevalier, R. A. 1983, *Ap. J.*, **272**, 765.
Chevalier, R. A. and Fransson, C. 1984, *Ap. J.*, **279**, L43.
Dorfi, E. 1989, preprint.
Dorfi, E. and Drury, L. O'C. 1985, In *Proc 19th Intern. Cosmic Ray Conf.(La Jolla)*, **3**, 121.
Drury, L. O'C., Markiewicz, W. J., and Völk, H. J. 1989, *Astron. Astrophys.*, submitted.
Drury, L. O'C. and Falle, S. A. E. G. 1986, *M. N. R. A. S.*, **223**, 353.
Eichler, D. 1979, *Ap. J.*, **229**, 419.
Eichler, D. 1984, *Ap. J.*, **277**, 429.
Ellison, D. C. and Eichler, D. 1984, *Ap. J.*, **286**, 691.
Falle, S. A. E. G. and Giddings, J. 1987, *M. N. R. A. S.*, **225**, 399.
Hartquist, T. W. and Morfill, G. E. 1983, *Ap. J.*, **266**, 271.
Hartquist, T. W. and Morfill, G. E. 1986, *Ap. J.*, **311**, 518.
Hartquist, T. W., Oppenheimer, M., and Elmegreen, B. 1979, *Astron. Astrophys.*, **75**, 137.
Hartquist, T. W., Pettini, M., and Tallant, A. 1984, *Ap. J.*, **276**, 519.
Ipavich, F. M. 1975, *Ap. J.*, **196**, 107.
Jokipii, J. R. 1976, *Ap. J.*, **208**, 900.
Jokipii, J. R. and Morfill, G. E. 1985, *Ap. J.*, **290**, L1.
Krymsky, G.F. 1977, *Sov. Phys. Dol.*, **23**, 327.
McKenzie, J. F. and Völk, H. J. 1982, *Astron. Astrophys.*, **116**, 191.
Parker, E. N. 1966, *Ap. J.*, **145**, 811.
Pettini, M. and West, K. A. 1982, *Ap. J.*, **260**, 561.
Raymond, J. C., Hester, J. J., Cox, D., Blair, W. P., Fesen, R. A., and Gull, T. R. 1988, *Ap. J.*, **324**, 869.
Savage, B. D. 1989, this volume.
Skilling, J. 1975, *M. N. R. A. S.*, **172**, 557.
Skilling, J. and Strong, A. W. 1976, *Astron. Astrophys.*, **53**, 253.
Völk, H. J. 1975, *Rev. Geophys. Space Phys.*, **13**, 547.
Völk, H. J. Drury, L. O'C., and McKenzie, J. F. 1984, *Astron. Astrophys.*, **130**, 19.

Section III

Evolutionary Processes in the Interstellar Medium

TURBULENT STRIPPING OF INTERSTELLAR CLOUDS BY INTERACTION WITH SUPERNOVA REMNANTS

Richard I. Klein
University of California, Lawrence Livermore National Laboratory and
Department of Astronomy, University of California at Berkeley

Christopher F. McKee
Departments of Physics and Astronomy, University of California at Berkeley

Philip Colella
University of California, Lawrence Livermore National Laboratory and
Department of Mechanical Engineering, University of California at Berkeley

ABSTRACT

We present results for the interaction of a supernova remnant blast wave with a small interstellar cloud, using for the first time high resolution local adaptive mesh refinement techniques with an underlying second order accurate 2-D Godunov hydrodynamic scheme. The cloud shock is assumed to be strong enough that it is non-radiative. We follow the morphological evolution of the cloud in great detail as it undergoes a series of complex shock-shock interactions and Kelvin-Helmholtz and Rayleigh-Taylor instabilities. We demonstrate, over a large range of cloud densities and shock strengths, that clouds are efficiently destroyed in a few cloud crushing times (essentially, the Rayleigh Taylor time) by a combination of instabilities and large scale shear flow. We investigate the scaling properties of clouds of different densities and shocks of different strengths. Our results have uncovered the development of copious supersonic vortex rings produced in the shear flow layer of the interaction. These vortex rings may wrap up ambient magnetic fields, enhancing the synchrotron emission and possibly explaining the compact radio hot spots seen in Cas A. Our calculations have, for the first time, carried the cloud-shock interaction well into the fragmentation regime. We demonstrate that similar calculations with standard fixed grid hydrodynamic schemes would require one to two orders of magnitude more computational time than is needed with our adaptive mesh approach to achieve comparable results.

INTRODUCTION

The interaction between supernova remnants (SNRs) and interstellar clouds in the galaxy is known to play a major role in determining the structure of the interstellar medium (ISM). We know that the ISM is highly inhomogeneous, consisting of both diffuse atomic clouds (T~100K) and dense molecular clouds (T~10K) surrounded by a low density warm ionized gas(T~10^4K) and by a very hot coronal gas (T~10^6K). Next to radiation directly from stars, supernova explosions represent the most important form of energy injection into the ISM; they determine the velocity of interstellar clouds, accelerate cosmic rays, and can compress clouds to gravitational instability, possibly spawning a new generation of star formation. The shock waves from supernova remnants can compress, accelerate, disrupt and render hydrodynamically unstable interstellar clouds, thereby ejecting mass back into the intercloud medium. Thus, while the interaction of the SNR blast wave with cloud inhomogeneities can clearly alter the appearance of the ISM, the cloud inhomogeneities can similarly have a profound effect on the structure of the SNR.

Recent observations of SNR of enhanced emission in the Balmer line filaments show evidence of cloud-shock interactions for Tycho (Braun, 1988). Velusamy (1987) finds evidence of the remnant cloud interaction in his radio observations of W28 and W44 taken at 327 MHz. These observations clearly show the distortion of the radio shell as the remnant begins to wrap around a dense cloud. The observations of the SNR IC443 by Braun and Strom (1986) show the later evolution of the cloud shock interacting with the outer layers of the cloud stripped off at high velocity.

Given the importance of the interaction of the supernova shocks with clouds for understanding the structure and the dynamics of the ISM as well as the potential importance of the interaction as a means of triggering new star formation, the problem has been studied both analytically and numerically over the past decade. Even when idealized as the interaction of a strong shock with a spherically symmetric cloud embedded in a less dense intercloud medium, the problem represents an extremely complex non-linear hydrodynamic flow encompassing a rich family of shock-shock interaction phenomena. The multi-dimensional nature of the evolution of the disrupted cloud is such as to make a detailed analytic calculation intractable. The first serious attempt to follow the interaction numerically was made by Woodward (1976), who used a combined Eulerian-Lagrangian approach to follow the interaction of the shock from a spiral density wave with a galactic cloud. These results showed the start of both Rayleigh-Taylor and Kelvin Helmholtz type instabilities; however, the calculation was not carried out far enough in time and lacked the spatial resolution to ascertain the final fate of the cloud. A subsequent attempt to investigate this problem by Nittman et al. (1982) used a flux-corrected transport approach and was very unresolved. Recently, Tenorio-Tagle and Rozyczka (1986) attempted to follow the evolution with a second order accurate hydrodynamic scheme, but again the calculation was under-resolved and clearly showed the effects of strong numerical diffusion at the interface of the cloud boundary and the intercloud medium. This made it impossible to disentangle the mixing of cloud-intercloud matter due to physical instabilities from mixing due to numerical effects. All of the previous work on this important problem leave unanswered several questions of key importance: What is the ultimate fate of clouds that have been impacted by SNR shocks? What is the total momentum delivered to the cloud? How much mass is lost from the cloud? What are the mechanisms by which clouds are disrupted and to what extend does disruption take place? How does cloud morphology scale with cloud density, shock Mach number and cloud size? Is the cloud driven to gravitational instability or is the cloud destroyed? What is the effect of the interstellar magnetic field on the evolution? What are the observable consequences of the interaction?

We have recently found (Klein, Colella and McKee, 1989a) that highly complex shock-shock interactions play a major role in determining the morphology of the cloud. Instabilities and shear flow motions are crucial to track accurately. Small scale structure in the flow may contribute significant mass loss back to the ISM and must be well resolved. These physical phenomena place an enormous constraint on the capabilities of most conventional numerical methods for solution of the 2-D equations of hydrodynamics. Even high order accurate approaches such as PPM with fixed Eulerian grids would require at least 10^6 grid points to follow the evolution accurately enough to answer the questions posed above. Clearly, one has a great need to evolve 2-D hydrodynamics with a great enough accuracy to deal with physical constraints and at the same time do so economically in both storage and time.

METHODOLOGY

To address these difficulties, we have used the local adaptive mesh refinement techniques with second order Godonov methods developed by Berger and Colella 1989 (cf. Klein, Colella, and McKee, 1989a). This first important problem will be the forerunner of a broad-based program we are developing to use adaptive mesh refinement to study astrophysical gas dynamics. We employ a second order finite difference solution of the 2-D Euler equations on a square grid in a cylindrically symmetric geometry. The numerical integration of the Euler equations is accomplished using an operator split version of a second order Godunov method (Colella and Woodward, 1984). The Godunov method conserves mass, momentum and total energy. We assume that the cloud and intercloud gas are both adiabatic, although we allow the cloud and intercloud medium to have different values of the adiabatic index γ. The resulting method is second order accurate in space and time, and captures shocks and other discontinuities with minimal numerical overshoot and dissipation.

From the point of view of being able to resolve detailed complex physical structures with reasonable amounts of supercomputer time and memory, the most important feature of our code is that it employs a dynamic regridding strategy known as local Adaptive Mesh Refinement (AMR) to dynamically refine the solution in regions of interest or excessive error. This is effected by placing a finer grid over the region in question with the grid spacing reduced by some even factor (typically) in each spatial dimension. The boundary of the refined grid is always chosen to coincide with cell edges of the coarser grid. Multiple levels of grid refinement are possible with the maximum number of nested grids supplied as a parameter in the calculation. Typically our calculations employ two nested grids over the initial coarse grid. A level 3 grid has 256 cells for each cell in a level 1 grid. In our present work, we determine those regions which require refinement by estimating the local truncation error in the density and refining those regions where the error is greater than some initially specified amount. In addition, we require the maximum level of refinement in the neighborhood of all cells containing cloud material. Special care is taken to ensure the correct fluxes across boundaries between and fine grids. This dynamic adaptive gridding approach is a crucial factor in our ability to economically resolve important features in the cloud shock interaction.

CLOUD SIZE SCALES

As the SNR expands through the ISM, it drives a shock into any cloud it encounters. Assuming that these are strong shocks, the pressure behind the blast wave and the pressure behind the transmitted cloud shock are comparable, and one finds that (McKee and Cowie, 1975)

$$v_s \approx (\rho_i/\rho_c)^{1/2} v_b , \qquad (1)$$

where v_s and v_b are the cloud shock and blast wave velocities and ρ_c and ρ_i the initial cloud and intercloud densities, respectively. Following McKee (1988), we define characteristic timescales for the cloud-shock interaction. Let $\chi \equiv \rho_c/\rho_i$ be the density contrast and assume that $\chi \gg 1$. Assume that the cloud is a sphere with radius a at a distance R_b from the supernova explosion. The blast wave in the Sedov-Taylor phase will expand as $R_b \propto t^{2/5}$, so the age of the SNR is,

$$t \equiv \frac{dR_b}{dt} = \frac{2}{5} \frac{R_b}{v_b} . \qquad (2)$$

The blast wave in the intercloud medium crosses the cloud in a time

$$t_{ic} \equiv \frac{2a}{v_b}, \qquad (3)$$

whereas the cloud shock crushes the cloud in a time

$$t_{cc} \equiv \frac{a}{v_s} = \frac{\chi^{1/2} a}{v_b} . \qquad (4)$$

The cloud crushing time t_{cc} is of the order of the sound crossing time in the crushed cloud; it is also about the timescale for the growth of large scale Rayleigh-Taylor instabilities. Finally, the cloud accelerates up to the velocity of the intercloud gas in a characteristic drag time t_d defined by $\rho_i v_b t_d = \rho_c a$,

or

$$t_d = \frac{\chi a}{v_b} = \chi^{1/2} t_{cc}. \tag{5}$$

In this paper, we will consider only clouds that can be characterized as "small", so that the SNR does not evolve significantly during the time for the cloud to be crushed:

$$t > t_{cc} \Rightarrow a < \frac{0.4R}{\chi^{1/2}}. \tag{6}$$

Indeed, we shall focus on the case in which the cloud is "very small", so that $t \gg t_d$, and $a \ll 0.4R/\chi$. In either case, we have $a \ll R$ so that the blast wave may be treated as a planar shock. In the opposite limit of a shock interaction with a large cloud, the SNR blast wave will undergo substantial weakening over the time it takes to cross the cloud. We expect substantial disruption for the small clouds, but only impulsive effects for large clouds.

CLOUD EVOLUTION

a. Cloud Crushing

Since there are no intrinsic scales in the problem, it is parameterized by the Mach number of the SNR blast wave M and the density ratio χ. Our calculations assumed 2-D axisymmetry for an inviscid fluid with no magnetic field. Two cases were considered for the cloud: $\gamma = 1.1$ and $\gamma = 5/3$. The intercloud gas was assumed to have $\gamma = 5/3$. Several calculations have been made for Mach numbers in the range 10-1000 and density ratios 10-400.

It is useful to follow the morphological evolution of the cloud through several cloud crushing times to obtain a sense of the different stages of development. We present the time-development of the isodensity contours of the cloud for the case γ (cloud) = γ (intercloud) = 5/3, $\chi=10$, $M=10$. At $t=0.84\ t_{cc}$ (Fig. 1), the transmitted shock is compressing the cloud from the front, secondary shocks have enveloped the sides of the cloud as the blast wave passes over the cloud, and a reflected bow shock moves upstream into the intercloud medium. The reflected shock becomes a standing bow shock and eventually a weak acoustic wave carrying away a small amount of energy from the supernova shock (Spitzer, 1982). At $t=1.05 t_{cc}$ (Fig. 2) the blast wave behind the cloud reflects off the axis giving rise to a Mach reflected shock back into the cloud. After $t=1.26 t_{cc}$ (Fig. 3), behind the cloud, a

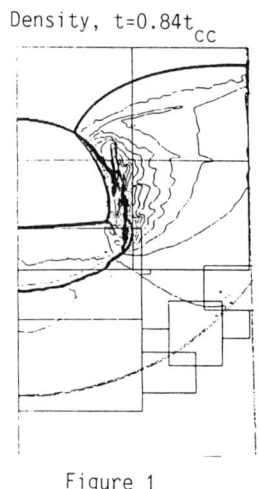

Figure 1

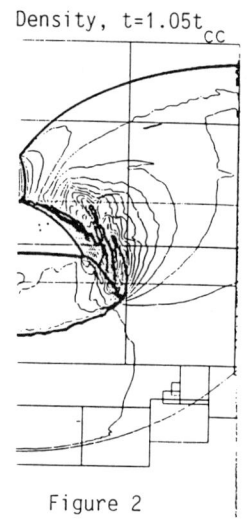

Figure 2

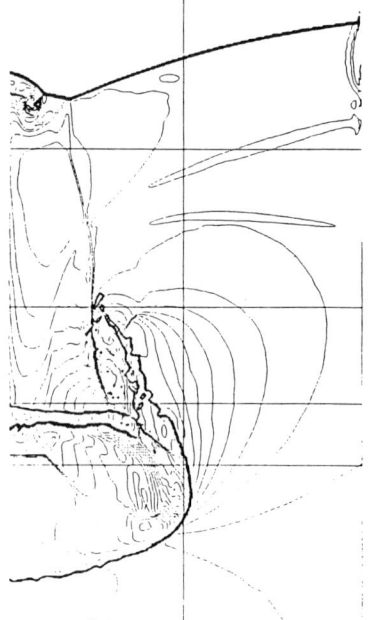

Figure 3

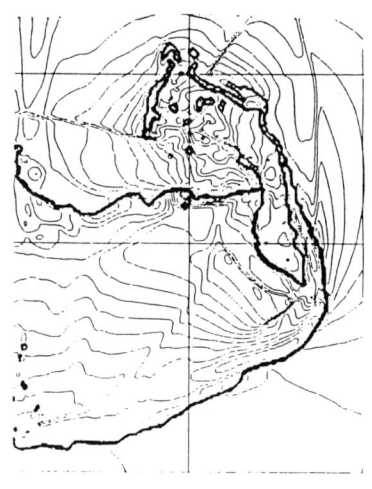

Figure 4

Isodensity contours of cloud and intercloud matter at different times.

double Mach reflection with the appearance of two triple point interactions occurs. This classic oblique shock interaction (Hornung, 1986; Glaz et.al. 1985) shows evidence of a strong supersonic vortex ring far behind the cloud. The vortex ring may have interesting observational consequences for SNRs (see below), but plays no role in the continued dynamical evolution of the cloud. The reflected shock and the transmitted shock undergo a strong interaction at $t=1.68 t_{cc}$ (Fig. 4) resulting in a initial flattening of the cloud. We also note the beginning of a strong shear flow. Substantial flattening of the cloud is observed at $t=2.1 t_{cc}$ from the strong shocks which have squeezed it like a vise. The pressure maximum on the nose of the cloud exceeds the pressure minimum on the sides and the cloud begins to expand laterally (Fig. 5). We note the growth of Richtmyer-Meshkov instabilities (Richtmyer, 1960) on the cloud nose which grow more slowly than the classic Rayleigh Taylor modes. At $t=2.5 t_{cc}$, we see evidence of Kelvin Helmholtz instabilities, on the sides of the cloud; weak shocks still residing in the cloud interior dissipate their energy (Fig. 6).

b. **Shear Flow and Vortex Production**

At $3.78 t_{cc}$ a prominent shear layer exists due to the motion of the cloud through the ICM. The shear produces copious vortex rings along the shear flow layer and leads to substantial Kelvin-Helmholtz instabilities which break up the arms (Fig. 7). The cloud consists of a distorted unstable axially flattened core component and a severely disrupted halo of cloud material. Over 70% of the original cloud mass is in small fragments which, in the absence of cooling, should merge with the intercloud medium. The unstable break up is dominated by large scale differential shear. At $t=9.7\ t_{cc}$, the cloud is completed destroyed (Fig. 8) and consists of several thousand fragments. At $4.2\ t_{cc}$ the strong supersonic vortex rings align along the shear flow layer produced in the dominant arm of cloud material that has been pulled from the main core of the cloud as well as along a second substantially fractured mass of cloud that has been fragmented from the arm. In Fig. 9 we show the associated flow field alongside of isodensity contours of the cloud and intercloud gas at $t=4.2\ t_{cc}$. It is clear that regions of strong circulation (high vorticity, numbered 1-5) are associated with positions along the shear flow layer where the cloud has undergone severe fragmentation. As vortex rings are formed in the shear layer and move away from the initial cloud are, the vortex rings are broken off. The process is called vortex shedding. It is suggestive of the possibility that the vorticity in the intercloud matter is acting to enhance the cloud break-up along the differential shear layer, thus acting as a mix-master aiding the development of the Kelvin-Helmholtz instabilities. This interesting possibility is worth further study.

To gain more insight into the development of the vortex rings we develop an equation for the time dependent change of the vorticity $\omega \equiv \nabla \times u$ where u is the fluid velocity. The equation of motion with the inclusion of body forces F is

$$\frac{\partial u}{\partial t} + (u \cdot \nabla) u = \frac{-\nabla P}{\rho} + \frac{F}{\rho}. \tag{7}$$

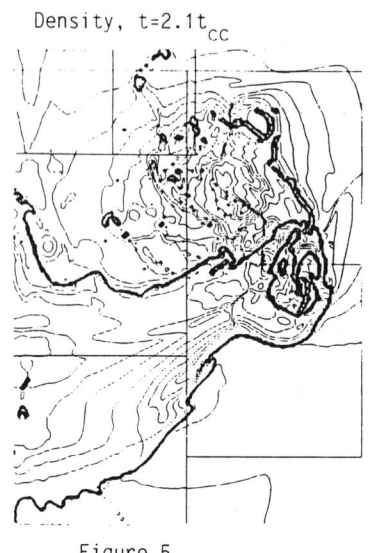

Figure 5

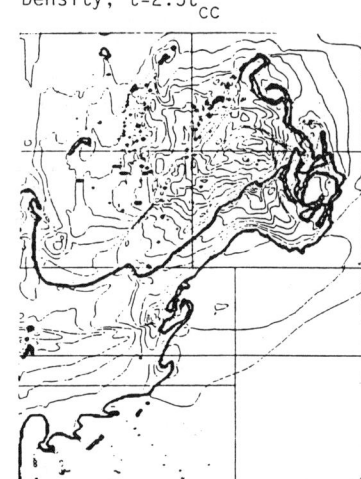

Figure 6

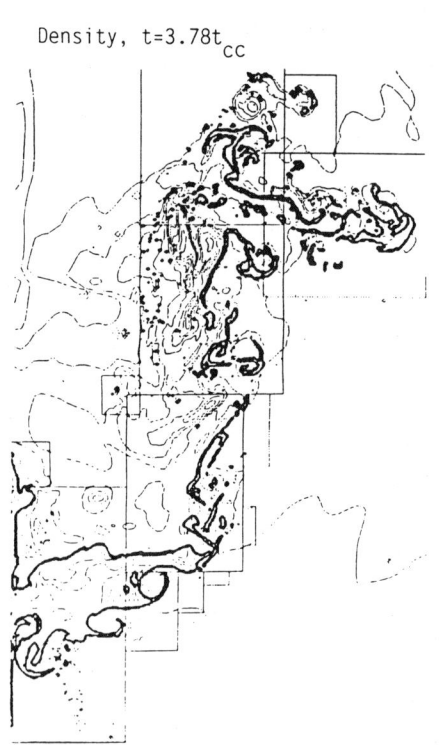

Figure 7

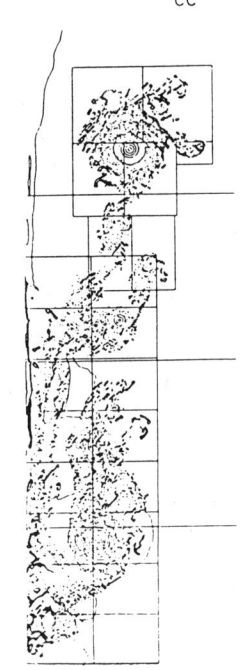

Figure 8

Supernova Stripping of Clouds 125

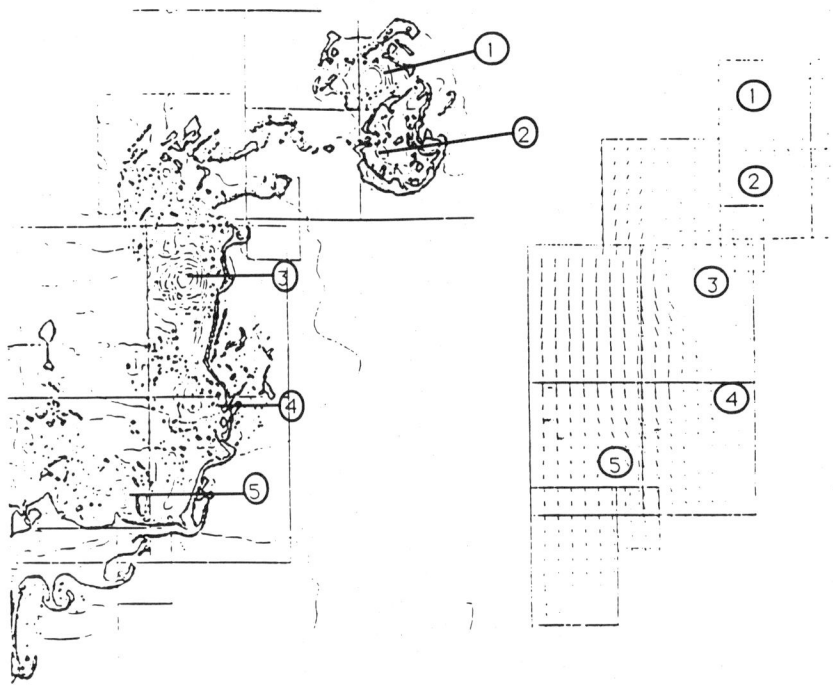

Figure 9 Isodensity contours (on left) at t=4.2 t_{cc}, flow field (on right). Numbers are sites of vorticity maximums.

Using the vector identity

$$(\nabla \times u) \times u = (u \cdot \nabla)u - \frac{1}{2}\nabla(u^2) \tag{8}$$

we find

$$\frac{\partial u}{\partial t} + \omega \times u = \frac{-\nabla P}{\rho} + \frac{F}{\rho} - \frac{1}{2}\nabla u^2 \tag{9}$$

Now taking the curl of each side and using the identity

$$\nabla \times (\omega \times u) = \omega \nabla \cdot u + (u \cdot \nabla)\omega - \omega \cdot \nabla u, \tag{10}$$

we derive the material derivative of the time rate of change of the vorticity

$$\frac{d\omega}{dt} = \omega \cdot \nabla u - \omega \nabla \cdot u - \frac{\nabla P \times \nabla \rho}{\rho^2} + \nabla \times (F/\rho). \tag{11}$$

We see that the vorticity depends upon four processes. The first term is a shear term which represents vortex-tube stretching. The second term represents the effects of compression: If the vorticity filaments bunch together the magnitude of the vorticity component increases in the direction of the vortex filament. If these were the only two processes, then the circulation $\int \omega \cdot dA = \oint v \cdot d\ell$ would be conserved. The next term is a baroclinic term which is the major source of vorticity in the cloud-shock interaction. The shock impinges on the cloud obliquely and produces surfaces of constant pressure that are not coincident with surfaces of constant density at the interface of the cloud and intercloud matter. This gives a non-zero cross product of gradients. The vorticity in the ICM is greater than that in the cloud because of the higher velocities in the lower density material. Our calculations show that most of the vorticity remains concentrated near the cloud boundary, where it originated. The fourth term is important if the fluid is viscous. If the force F/ρ is frictional, it can be represented as $F/\rho = \nu\nabla^2 u$ where ν is the viscosity; then $\nabla \times (F/\rho) \sim \nu\nabla^2\omega$. This represents the diffusion of vorticity from regions of high to low concentration. It is proportional to the amount of numerical viscosity in the finite difference approximations. Given the importance of vorticity as a possible observational diagnostic of the remnant cloud interaction as well as its possible role in the cloud fragmentation, it is of great importance to demonstrate that numerical viscosity does not play a role in determining the amount of vorticity production. We have computed the time evolution of the cloud for four increasingly resolved initial grids, doubling the number of cells in both Δr and Δz with each increase in resolution. We have shown that the time evolution of the vorticity for even the coarsest mesh ($\Delta x=.08$ in code units, corresponding to about half of the initial cloud radius) tracks to a remarkable degree of accuracy the vorticity of the finest resolution, which is equivalent to a 7×10^6 zone calculation for a fixed grid method. This clearly establishes that numerical viscosity, which is proportional to grid resolution, does not affect the production of vorticity for the adaptive grid techniques we are using. This type of calculation is a powerful check on the conservation of vorticity.

Let us consider the characterization of the evolution of the interstellar cloud in more detail. In Table 1, we display the results of adiabatic calculations for three models in which $\gamma = 5/3$ in both the cloud and ICM. The calculations are done for two models ($M=10$ and 100) for density contrast $\chi =10$ and one model ($M=100$) for density contrast 100. The first entry in the table is the time normalized to the intercloud crossing time. The second entry gives the time normalized to the cloud crushing time and the drag time, $t_d = \chi^{1/2} t_{cc}$. The next column is the sound speed behind the cloud shock normalized to the blast wave velocity. The shocked intercloud gas moves at a velocity $(3/4)\, v_b$ relative to the cloud for $\gamma = 5/3$, so the next entry measures the ratio of the current cloud/intercloud relative velocity Δv to its initial value; in the frame of the shocked intercloud gas, this is a measure of cloud deceleration. The next column is a characterization of the cloud's aspect ratio in the radial and axial direction weighted by its half mass distribution. Here $r_{1/2}$ is the radial half-mass distance and $Z_{1/2}$ is the axial half-mass distance. The last column gives the radial $\dot{r}_{1/2}$ and axial $\dot{Z}_{1/2}$ expansion velocities of the cloud. These velocities are computed by using the half mass distance distributions at the two final times in the calculation.

Several conclusions can be drawn from these results. Comparing the results at the same normalized "final" time $t = 4.2 t_{cc}$ for clouds of the same density $\chi = 10$, but subjected to blast waves of different Mach number, 10 and 100, we note that both clouds have decelerated to about 0.15 of their initial velocities. Thus, these clouds have almost stopped, leading to a small pressure differential between the front of the cloud surface and the sides so that there is little force driving further radial expansion; hence the clouds have a radial expansion velocity $\dot{r}_{1/2} \approx 0$. The strong shear flow in the cloud is still dominant, however, and both clouds are supersonically shearing apart at about the same axial expansion velocity $\dot{Z}_{1/2}$ of 3 times the cloud velocity. The physical extent of the stretching in both the radial and axial direction

$$\frac{r_{1/2}(t)}{r_{1/2}(0)}, \frac{Z_{1/2}(t)}{Z_{1/2}(0)}$$

is essentially the same for the two cases. The remarkable agreement of these features of the clouds and their similar morphological structure leads one to suspect that the cloud evolution may scale similarly with the Mach number of the SNR shock. This Mach scaling can be clearly seen if we scale the time, velocity and pressure as $t' = t/M$, $v' = vM$ and $P' = PM$. Substituting these scaled quantities into the Euler equations, we find that Euler equations are invariant under this transformation. Thus, we find that for fixed γ and density contrast χ, the morphological evolution is a function of t/t_{cc} only, in the limit of large M.

Clouds with greater density contrasts χ show greater expansion in both the radial and axial directions, as shown both by the results in Table 1 and by Figures 10 and 11, which portray the state of a shocked cloud with $\chi = 100$ at 4 t_{cc} and $\chi = 400$ at 2 t_{cc}. This follows from the fact that the characteristic expansion time for the cloud is the sound crossing time (which, as remarked above, is about t_{cc}), whereas the time for the cloud to decelerate is the drag time $t_d = \chi^{1/2} t_{cc}$. The lateral expansion of the cloud is due to the lower pressure on the sides of the cloud caused by the Venturi effect (Nittman et al. 1982). This pressure difference decays on the drag time; by the time shown in Figure 11, this expansion has stopped. At $t = 4\ t_{cc}$, the axial expansion velocity is a substantial fraction of v_b for both $\chi = 10$ and $\chi = 100$; since t_{cc} is larger for $\chi = 100$, the length of the cloud is greater in this case. We expect the axial expansion of the cloud to stop within a few drag times. This has been verified for the $\chi = 10$ case, but not the $\chi = 100$ case.

c. <u>Cloud Fragmentation</u>

At late times (several t_{cc}) the cloud is rendered to a turbulent flow with several fragments reduced to a foam on the scale of grid resolution. It is of great interest to follow the mass loss of the cloud as it fragments, and to understand how the fragmentation scales with varying cloud density. In Fig. 12, we show the mass of the cloud core as a function of time for clouds with density contrasts $\chi=10,100,400$. The cloud core is defined to be the most massive cloud fragment. The mass loss vs time has been fitted with a exponential to determine the fragmentation time t_f, defined as the time for each cloud to be left with 1/e of its original mass. We find for $\chi=10$ that the cloud fragments initially into two roughly equal mass fragments. The mass fragments then begin a series of further fragmentation stages into smaller pieces due to combined Rayleigh Taylor and Kelvin Helmholtz instabilities. We have developed an algorithm that performs pattern recognition of fragments on our hierarchical grid structure and sorts their masses to produce a mass spectrum for fragmentation. Fig. 13 shows the fragment mass plotted against the mass in each bin for the $\chi=10$, $M=10$ case at $t=3.78\ t_{cc}$. We note that the cloud has broken into two fragments of essentially the same mass comprising about 78% of the mass of initial cloud, two additional fragments comprising ~9% of the initial mass and ~580 fragments accounting for the remaining 12% of the initial mass. In Fig. 14 we blow-up the low mass end of the spectrum (12% of the initial mass) where we see the distribution of the more massive of these low mass fragments. Mass fragmentation spectra such as these can provide a distinct "fingerprint" of the way in which a cloud undergoes fragmentation and are a powerful tool to establish scaling properties for different clouds with varying conditions. We are in the process of determining which aspects of this fragmentation are physical and which are affected by numerical resolution. In Fig. 8 we show isodensity contours of the cloud at $t=9.67\ t_{cc}$ where the cloud is completely destroyed. The final fate

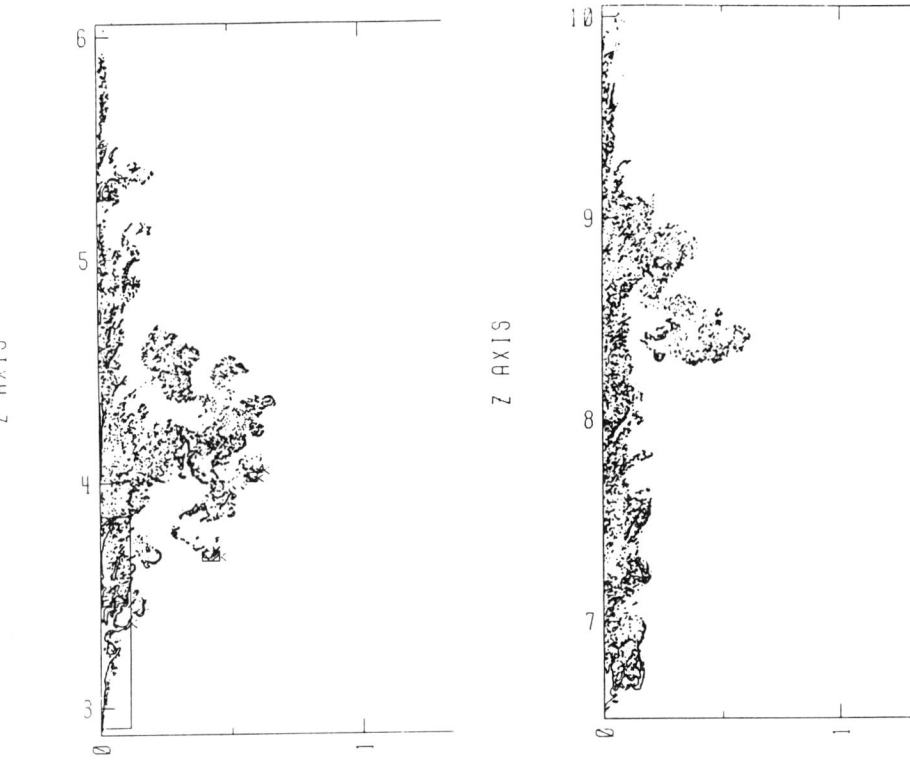

Figure 10 Isodensity contours for $\chi=100$, $M=100$ at $t=4.0\ t_{cc}$

Figure 11 Isodensity contours for $\chi=400$, $M=100$ at $t=2.0\ t_{cc}$. Note morphology of cloud consisting of a dense "head" followed by a trail of several thousand fragments with an aspect ratio of 20 to 1.

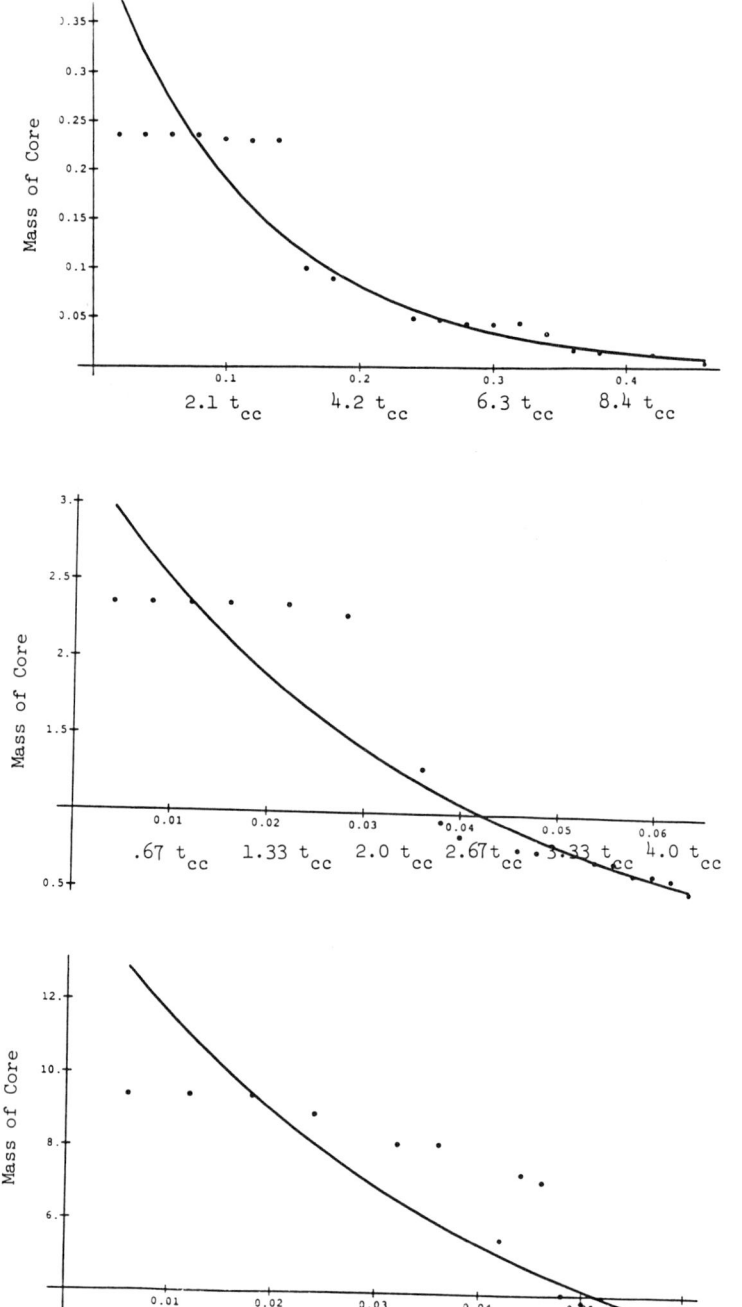

Figure 12 Core mass vs time for χ=10, 100, 400

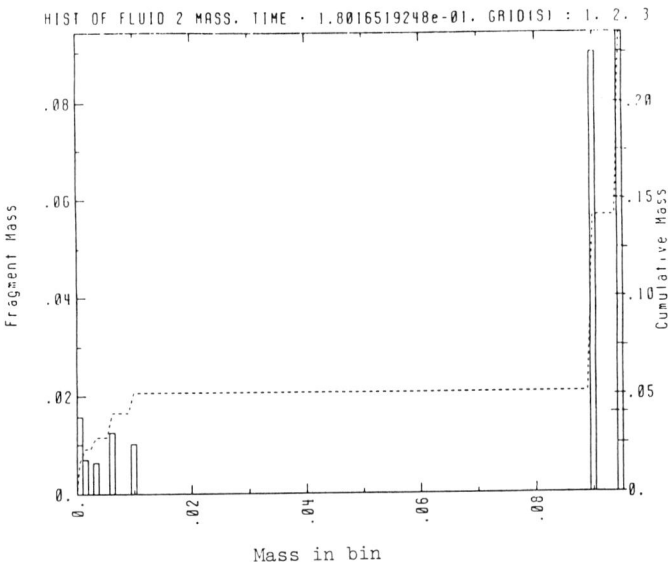

Figure 13 Fragment mass vs mass in bin for $\chi=10$, M=10 at t=3.78 t_{cc}

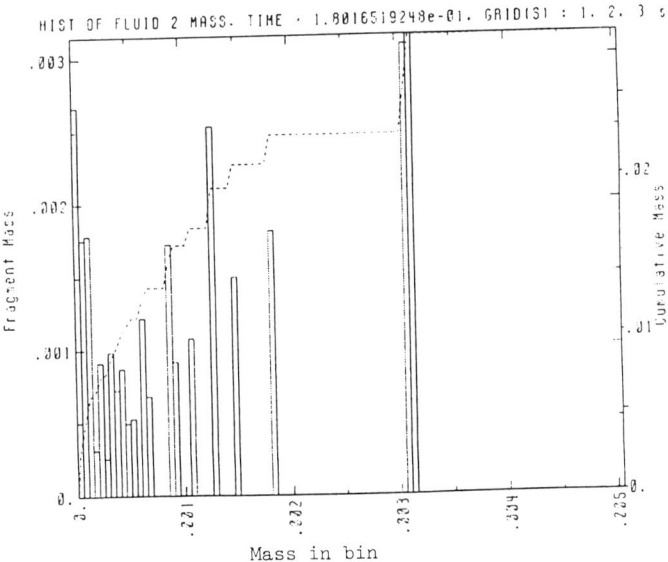

Figure 14 Fragment mass vs mass in bin for $\chi=10$, M=10 at t=3.78 t_{cc} at low mass end of the fragmentation spectrum.

of this cloud consists of a quasi-static halo of fragments of which 50% of the mass resides in an axially elongated distribution stretched out 5-6 times its initial shape, and the rest of the mass resides in a multitude of fragments much less dispersed.

For clouds with $\chi \gg 10$, the stripping process proceeds differently. For $\chi=400$, the cloud fragments gradually, with a continuous erosion by loss of small fragments (cf. Fig. 11 and Fig. 12). Since small fragments rapidly become comoving with the intercloud medium whereas the cloud core decelerates gradually, mall fragments trail far behind the massive cloud core until the core itself is destroyed by Kelvin-Helmholtz instabilities as it drags through the intercloud medium. In Fig. 11 the cloud core mass at $t = 2t_{cc}$ is 26% of the original cloud and we see that the cloud has the distinct morphology of a dense cloud core trailed by a multitude of fragments in a narrow tail.

Our results show that clouds are fragmented in a time $t_f \sim (1.5 - 4)\, t_{cc}$ as χ ranges from 400 to 10; recall that t_{cc} is of order the Rayleigh Taylor timescale. The numerical coefficient is smaller for the higher density contrasts, presumably because the relative velocity of the cloud remains greater.

This conclusion is consistent with that of Nittman et al. (1982), who concluded that the cloud would be destroyed in a time $\sim 3\, t_{cc}$ due to the combined effect of lateral expansion and strong fluid rotation behind the cloud. Because of the increase in the cross section of the cloud due to the lateral expansion, the time for the fragmented cloud to accelerate up to a velocity comparable to that of the shocked ICM is several times less than the initial drag time $t_D = \chi a/v_b$. Thus for $\chi \lesssim 10^2$, the fragmentation time and the acceleration time are comparable; on this point, our conclusion differs from that of Nittman et al. In an analytic study of the related problem of the stripping of gas from a galaxy moving through an intracluster medium, Nulsen (1982) concluded that the stripping time is of the order of the drag time t_d; in the absence of gravitational effects (which he found to be generally small), our results indicate that stripping will occur substantially more rapidly.

We have performed calculations for several similar models for $\gamma=1.1$ in the cloud. This softer equation-of-state is more representative of clouds that are radiative, although it should be pointed out that truly radiative clouds can get rid of their stored energy efficiently, and we would expect substantially more shock compression than the models considered here. We note that these "radiative" clouds move substantially more rapidly than their $\gamma=1.67$ counterparts. These clouds are significantly more radially compressed, and thus experience far less drag than the $\gamma=1.67$ clouds. This can again be understood by consideration of the sound speed in these clouds. We find that the scaling of sound speed c_c with γ is such that $c_c(\gamma=1.1) \ll c_c(\gamma=5/3)$, so that these "radiative" clouds expand laterally more slowly. We note that the high density "radiative" cloud is still experiencing large supersonic axial shearing. As with the previous $\gamma=5/3$ models Mach scaling appears to be established.

OBSERVATIONAL CONSEQUENCES

An outcome of these calculations that may be potentially very important for observations of SNR is the discovery of the copious production of vortex rings distributed along the strong shear flow layer (Fig. 9). Approximating the rotation of these vortices by rigid body rotation, we can relate the vorticity ω in an individual vortex ring to the pressure differential across the vortex ΔP, and we find that $\omega = (8\Delta P/\rho)^{1/2}/r$. This appears to be an excellent approximation when compared to our detailed calculations. Those rings with large aspect ratio may be subject to non-axisymmetric instabilities and break up into yet smaller vortex structures (Saffman and Baker, 1979). "Fat" rings, with small aspect ratio, are likely to remain intact. Recent high resolution radio observations of the Cas A SNR (Tuffs, 1986) have revealed several hundred intense compact radio emission peaks distributed throughout the remnant. We have demonstrated that strong shear flows associated with shock-cloud interactions result in the production of many supersonic vortex rings. These vortex rings can be expected to wind up ambient magnetic fields present in the interstellar clouds until equipartition between the energy in the field and the vortex is achieved. It is quite possible that the resulting intense wound up magnetic field and the associated betatron acceleration could account for the synchrotron emission of electrons, thus explaining the observations in Cas A. Equipartition magnetic fields are often invoked in astrophysics to explain non-thermal emission; our results suggest that the fields may indeed reach equipartition, but only in a small fraction of the volume. Chevalier (1976) postulated the presence of turbulent vortices, acting as magnetic scattering centers in SNRs to explain particle acceleration by a second-order Fermi mechanism. We conjecture that the radio hot spots may indeed be indirect observational evidence of the presence of vortex rings produced behind the shocked clouds. The vortex rings would have low density and pressure at the center, thus appearing weak in optical, UV and x-ray emission.

Finally, the cloud morphology itself is a important signature. The clouds can be expected to be elongated structures with aspect ratios ~5-6, with multitudes of fragments trailing behind the cloud core. A possible example of this has been observed by Braun and Strom (1986) in IC443.

CONCLUSIONS

We have performed second order accurate, high resolution local adaptive mesh refinement calculations of the interaction of a supernova shock with interstellar clouds. These extremely powerful hydrodynamic techniques have enabled us to calculate exceedingly complex flows much more rapidly, much more accurately, and much further in time than previous work with standard fixed grid hydrodynamics. We have followed the evolution of interstellar clouds well into the regime of fragmentation. Our calculations have demonstrated high accuracy with 80,000 grid cells that would only be achievable with fixed grid high order accurate hydrodynamic schemes with >1,000,000 grid cells. We find:

1) Small non-radiative interstellar clouds are efficiently destroyed in a few cloud crushing times times by combined Rayleigh-Taylor and Kelvin Helmholtz instabilities dominated by large scale shear flow. Clouds that have the same density but are enveloped by strong shocks of differing Mach number exhibit scaling behavior in their morphological evolution.

2) Small clouds are highly fragmented by non-radiative shocks. Cloud fragments will most likely eventually feed their mass back into the ISM by thermal evaporation.

3) Small adiabatic clouds fragment to such an extent that it is unlikely that fragments large enough to become gravitationally unstable and form stars will survive. The cloud destruction proceeds more rapidly than the free fall time.

4) Clouds evolve toward a elongated structures with aspect ratios of five to six, consisting of multitudes of fragments.

5) Our calculations indicate the copious production of supersonic vortex rings. These vortex rings may be effective in winding up the ambient magnetic field in clouds, increasing the magnetic field strength and enhancing the synchrotron emission of cosmic ray electrons. This could explain the recent observations of numerous compact radio hot spots in Cas A.

In the future, we shall use adaptive mesh refinement hydrodynamic techniques to investigate a broad range of astrophysical gas dynamical phenomena.

ACKNOWLEDGEMENTS:

The calculations presented in this paper were performed on the XMP416 and YMP832 at the Lawrence Livermore National Laboratory. This work was performed under the Auspices of the U.S. DOE by LLNL under Contract W-7405-Eng-48, supported in part by the Applied Mathematical Sciences Program of the office of energy research, and was also performed in part under the auspices of a special NASA astrophysics theory program which supports a joint Center for Star Formation Studies at NASA Ames Research Center, University of California, Berkeley, and University of California, Santa Cruz. The work of CFM is supported by NSF grant AST 86-15177.

Table 1

	t/t_{ic}	t/t_{cc} t/t_{drag}	c_c/v_b	$\frac{4}{3}(\Delta v / v_b)$	$r_{1/2}(t)/r_{1/2}(0)$ $Z_{1/2}(t)/Z_{1/2}(0)$	$\dot{r}_{1/2}/v_b$ $\dot{Z}_{1/2}/v_b$
$\frac{\chi=10}{M=10}$	6.7	4.2 1.3	0.18	0.16	1.8 3.2	~0.0 0.35
	15.3	9.66 3.0		0.074	2.38 5.69	~0.0 ≤0.045
$M=100$	6.7	4.2 1.3	0.18	0.14	2.0 2.6	~0.0 0.32
$\frac{\chi=100}{M=100}$	21.3	4.3 0.43	.056	0.25	3.7 8.4	~0.0 0.42

BIBLIOGRAPHY

Berger, M. J., and Colella, P., 1989, to appear in J. Comp. Phys.

Braun, R., 1988, IAU Coll. 101, Supernova Remnants and the Interstellar Medium, Ed. R. S. Roger and I. L. Landecker, Cambridge Univ. Press, 227.

Braun, R. and Strom, R.G., 1986, Astronomy and Astrophysics, 164, 193.

Chevalier, R. A., 1976, Ap.J., 207, 450.

Colella, P. and Woodward, P., 1984, J. Comp. Phys. 54, 174.

Glaz, H. M., Colella, P., Glass, I.I., Deschambault, R. L., 1985, Proc. Roy. Soc. Lond. A 398, 117.

Hornung, H., 1986, Ann. Rev. Fluid Mech., 18, 33.

Klein, R. I., Colella, P., and McKee, C. F., 1989a, "The Physics of Compressible Turbulent Mixing International Workshop", Princeton University, Ed. W. Dannevik, 1989, Springer Verlag, New York Inc., Lecture Notes Series.

Klein, R. I., Colella, P. and McKee, C. F., 1990, Ap. J. in preparation.

McKee, C. F., 1988, IAU Coll.101, Supernova Remnants and the Interstellar Medium, Ed. R. S. Roger and I. L. Landecker, Cambridge Univ. Press, 205.

McKee, C. F. and Cowie, L. L., 1975, Ap.J., 195, 715.

Nittman, J., Falle, S., and Gaskell, P., 1982, M.N.R.A.S., 201, 833.

Nulsen, P.E.J. 1982, M.N.R.A.S., 198, 1007.

Richtmyer, R.D., 1960, Comm. Pur. Appl. Math., 13, 297.

Saffman P. G., Baker, G. R., 1979, Ann. Rev. of Fluid Mech. 11, 95.

Spitzer, L., 1982, Ap.J., 262, 315.

Tenorio-Tagle, G., and Rozyczka, M., 1986, Astron. Astrophys., 155, 120.

Tuffs, R. J., 1986, M.N.R.A.S., 219, 13.

Velusamy, T., 1988, IAU Coll 101, Supernova Remnants and the Interstellar Medium, Ed. R. S. Roger and I. L. Landecker, Cambridge Univ. Press, 265.

Woodward, P., 1976, Ap.J., 207, 464.

WINDS FROM HOT STARS

JOHN H. BIEGING
Radio Astronomy Laboratory, 601 Campbell Hall, University of California,
Berkeley, CA 94720

ABSTRACT The winds of hot stars are significant sources of energy and momentum, as well as of nuclear-processed matter, for the interstellar medium. This paper reviews the relevant properties of hot stars and their winds and discusses our current understanding of both the large-scale effects of these winds on the galaxy, and the local effects on the evolution of H II regions and supernova remnants.

INTRODUCTION

In this paper I review the role of stellar winds in the deposition of energy and matter in the interstellar medium. I will consider only the winds of the hottest stars (spectral types O and B, and the Wolf-Rayet stars). Winds from pre-main sequence stars and from cool evolved stars are covered in reviews by Rodriguez and Knapp elsewhere in this volume.

The importance of stellar winds from the hottest stars has been fully appreciated only in the last decade or so. Two instruments have been essential to the development of our knowledge of these winds: the International Ultraviolet Explorer (IUE) and the Very Large Array (VLA). The IUE has made possible systematic studies of the spectra of a large number of OB and Wolf-Rayet (WR) stars. The UV spectra of these stars typically show P Cygni line profiles indicative of mass loss. Model analyses of the spectra have been used to determine mass loss rates and wind velocities. The VLA has provided complementary data on mass loss, especially for the most luminous hot stars, by detecting the (generally very weak) free-free emission from the ionized stellar wind. The VLA has sufficient sensitivity and angular resolution to measure the flux density of stellar wind thermal emission unambiguously (i.e., free of source confusion problems which limit even the largest single-dish radio telescopes), for a relatively large sample of stars.

Hot-star winds are astrophysically important for at least two reasons. First, they are a significant source of energy and of nuclear-processed matter to the interstellar medium. This aspect of these winds is the subject of this review. Second, the prodigious mass loss rates developed in these winds may have a profound effect on the evolution of the star. Evolutionary effects will not be discussed here; the reader is referred to the review by Chiosi and Maeder (1986) for an overview of the effects of mass loss on stellar evolution of OB and WR stars.

This review is divided into three parts: (1) the parameters of stellar winds for hot stars; (2) the calculation of global input rates of matter and energy into the ISM; and (3) local effects of stellar winds on both photoionized nebulae (H II regions) and on the development of supernova remnants.

STELLAR WIND PARAMETERS OF HOT STARS

In this section I review the relevant properties of mass-losing hot stars, including the types of stars which lose mass, mass loss rates, wind velocities, and correlations with other stellar properties.

Stars Which Lose Mass

The domain of mass loss in the HR diagram has been discussed by Abbott (1982a), who demonstrates that all stars with initial masses greater than 15 $M_\odot$ show evidence for mass loss in their visible and UV spectra. This criterion implies that all O-stars and all B0 dwarfs are expected to have winds throughout their lifetimes. (In this paper the terms "mass loss" and "stellar wind" are used more or less interchangeably, since the mass loss process occurs through a wind driven by the star. Because the wind velocity is greater than the stellar escape velocity–see below–the wind material is indeed lost from the stellar system.) Spectroscopic evidence also implies that virtually all B and A supergiants are also losing mass through winds. Finally, all Wolf-Rayet stars appear to have strong winds. In fact, the existence of a dense, massive wind may be a necessary condition for producing the characteristic emission line spectra which are a defining property of the WR stellar category.

Mass Loss Rates

Two approaches have been used extensively for determining mass loss rates from hot stars. The first method uses the continuum intensity of the free-free emission from the wind, together with the distance and wind velocity, to derive the mass loss rate, $\dot{M}$ (cf. Wright and Barlow 1975; Panagia and Felli 1975). The continuum intensity is best measured at microwave frequencies with the VLA, although in principle the IR excess (over the stellar photosphere) due to the wind can also be used. In this case, since the IR free-free emission is produced mainly from a region close to the star, the wind velocity as a function of distance from the star must be estimated or assumed to derive $\dot{M}$ (cf. Castor and Simon 1983). In contrast, at microwave frequencies, the thermal wind emission arises several hundred stellar radii from the star, where the wind can be assumed to have reached its terminal velocity. Moreover, the photospheric contribution at microwave frequencies is completely negligible, in contrast to the IR emission, where a photospheric contribution must be estimated and subtracted from the observed emission to obtain the wind contribution. For these reasons the mass loss rates derived from microwave flux density measurements are less model-dependent and therefore more reliable, than those derived from IR observations.

Mass loss rates derived from radio observations of a complete sample of OB stars have been published by Bieging, Abbott, and Churchwell (1989–see also references therein). Individual objects of particular interest have been studied by White and Becker (1982, 1983) and Becker and White (1985) and by Persi et al. (1985), who derive mass loss rates and other wind properties (e.g., wind temperature) from radio observations. Radio-derived mass loss rates of WR stars have been obtained by a number of workers (Seaquist 1976; Florkowski and Gottesman 1977; Dickel, Habing, and Isaacman 1980; Hogg 1982, 1985; Felli and Panagia 1982; and Bieging, Abbott, and Churchwell 1982). A complete sample of WR stars within 3 kpc of the sun and north of declination -47° was surveyed with the VLA by Abbott et al. (1986).

A complicating factor in the use of radio flux densities to determine mass loss rates of hot stars has become apparent over the past several years. Roughly one-quarter of all OB and WR stars observed show evidence–either in the radio spectral index, angular size, or time variability–for nonthermal radio emission, possibly from relativistic electrons accelerated by shock processes in the stellar wind. The distinction between thermal and nonthermal emission is essential to derive accurate mass loss rates and requires more than a single observation at a single frequency. For a more complete discussion, see Abbott et al. (1986) and Bieging, Abbott, and Churchwell (1989).

A second method which has been extensively employed to derive mass loss rates for OB stars is the analysis of UV spectra of C IV and N V, principally from IUE observations (e.g., Garmany et al. 1981; Garmany and Conti 1984). This method is most reliable for stars with low mass loss rates, since for high rates the UV lines become saturated. In this sense the radio continuum and UV spectral analyses are complementary. Abbott (1985) has shown that mass loss rates derived by these methods form a continuous sequence when plotted against bolometric luminosity, for values of L from 10^5 $L_\odot$ to 3×10^6 $L_\odot$. Garmany and Conti (1984) found, from a combination of UV and radio-derived values, that mass loss rates scale as a power of the stellar luminosity, with a best-fit relation

$$\dot{M} = 1.35 \times 10^{-7} (L/10^5\ L_\odot)^{1.62}\ M_\odot\ \mathrm{yr}^{-1}$$

The scatter of the data around this fit is about a factor of 3 either way in $\dot{M}$.

This dependence of $\dot{M}$ on L is consistent with the predictions of radiation-driven wind theory, as formulated by Castor, Abbott, and Klein (1975) and subsequently refined (Abbott 1982b; Friend and Abbott 1986; Pauldrach, Puls, and Kudritsky 1986; Puls 1987). Attempts to parametrize the mass loss rate in terms of other stellar parameters–i.e., a "Reimers"- type law involving mass, radius, or effective temperature–give no obviously better fit. Van Buren (1985) finds no correlation between $\dot{M}$ and T_{eff}. The power-law correlation between $\dot{M}$ and L is generally regarded as convincing evidence that the winds of OB stars are driven by radiation pressure acting on absorption lines of the ions in the stellar atmosphere. However, the factor of 3 scatter in mass loss rates about the best-fit power law is not understood, but is probably larger than expected solely from observational uncertainties in the radio flux density, distance, and wind velocity.

Cassinelli and van der Hucht (1986) have reexamined the case for radiation-driven winds from WR stars. Based on revised surface abundances (but meager statistics) they derive mass loss rates which scale as $L^{2.0}$, or slightly steeper than for OB stars. There is however a serious "momentum problem" for the WR winds. The ratio of wind momentum, $\dot{M}V_\infty$, to radiative momentum, L/c, is much greater than unity for all WR stars with measured $\dot{M}$. This momentum excess is too large to be explained by the radiation-driven wind model, even with multiple scattering of photons. An (unknown) alternative mechanism may be the driving source of the WR winds.

Distances for OB stars are most reliably determined through membership in associations. For stars not in associations, the distance uncertainty, which enters as the 1.5 power in the calculation of $\dot{M}$, is typically worse than for cluster members. The WR stars in general have rather poorly known distances, which are generally the largest source of uncertainty in deriving $\dot{M}$.

Stellar Wind Velocities

The maximum velocity, V_∞, attained by the stellar wind must be known or estimated to determine both the mass loss rate, as discussed above, and to calculate the kinetic energy and momentum flux carried by the wind. V_∞ is generally determined in practice in one of two ways. For OB stars, the blue-shifted absorption edge of UV P Cygni lines which are optically thick throughout the zone of acceleration, gives V_∞ directly with typical uncertainties of about 10%. For WR stars, the reddening is often too large to permit good UV spectra. In such cases the width of optically thick emission lines can be used to obtain V_∞, with uncertainties typically 10 - 20%, depending on the star (see Abbott et al. 1986 for a more complete discussion).

For OB stars, there is a rather well-defined relationship between V_∞ and V_{esc}, the escape velocity from the stellar surface. Abbott (1982) found that

$$V_\infty = aV_{esc}$$

where

$$V_{esc} = (2GM(1-\Gamma)/R_*)^{1/2}$$

is the escape velocity and $\Gamma = \sigma_e L_*/4\pi GMc$ corrects the surface gravity for the effect of radiation pressure. The factor a depends on the stellar effective temperature, and varies between 1 for cooler stars (e.g., B0) and 3.5 for the hottest stars. This relation yields V_∞ to an accuracy of about 20% for OB stars. Van Buren (1985) found an alternative relation between V_∞ and T_{eff} only, with a rather larger dispersion of about 50% in V_∞.

For WR stars, there is considerable variation in values of V_∞ for stars of the same spectral type, as noted by Abbott et al. (1986). Thus estimates of V_∞ for individual WR stars may have errors of order 25% if based solely on average values by spectral type.

Wind Energies of OB and WR Stars

For OB stars, measured mass loss rates range up to $\sim 1 \times 10^{-5}$ $M_\odot$ yr^{-1}. These rates are expected to occur over the main sequence lifetime of a few million years, so the effect on stellar evolution can be profound (cf. Chiosi and Maeder 1986) for stars of spectral type O3 to O6. Since the mass loss rate scales approximately as $L^{1.6}$, the evolutionary effect is less severe for late O and B stars.

Measured wind velocities, V_∞, for O-stars lie generally in the range 1500 - 3000 km s^{-1}. The total kinetic energy carried by the wind can be estimated as the wind power, $(\dot{M}V_\infty^2)/2$, times the main sequence lifetime. For the most massive OB stars, the total kinetic energy carried by the wind is $\leq 10^{51}$ erg. Thus an early-O star may generate as much kinetic energy in its wind as that in a supernova remnant.

The wind velocities of WR stars are measured to be typically in the range 1500 - 4000 km s^{-1}. Abbott et al. (1986) found that mass loss rates of their sample of 40 WR stars within 3 kpc of the sun all lay within the relatively narrow range of 0.8 - 8.0 × 10^{-5} M$_\odot$ yr^{-1}, with an average value of 2 × 10^{-5} M$_\odot$ yr^{-1}. Van der Hucht, Cassinelli, and Williams (1986), using revised values of T_{eff} and atmospheric abundances, found mass loss rates which are systematically a factor of 2 to 3 higher than those of Abbott et al. (1986), i.e., $\dot{M} \sim$ 3 to 10 × 10^{-5} M$_\odot$ yr^{-1}. Using these higher rates, the wind power of a typical WR star is $\sim 10^{38}$ erg s^{-1}. If the WR phase lasts for 4 × 10^5 years, as evolutionary calculations indicate, then the total kinetic energy of a WR wind is $\sim 10^{51}$ ergs, also comparable to the kinetic energy of a supernova remnant.

GLOBAL INPUT OF MATTER AND ENERGY BY STELLAR WINDS

Input Data

Estimates of the total rate of kinetic energy and matter injected, per unit surface area of the galactic disk for the solar vicinity, have been made by Abbott (1982a), Abbott et al. (1986), Van Buren (1985), and van der Hucht, Williams, and Thé (1987). In order to determine such global averages, the following information is required:

(1) Catalogs of hot stars. For OB stars, the catalog of Garmany, Conti, and Chiosi (1982) is ~90% complete to a distance of 3 kpc. For the WR stars, the work of van der Hucht et al. (1981, 1988) is also essentially complete to about 3 kpc. These catalogs can be used to derive the surface density of stars in each spectral type and luminosity class.

(2) Evolutionary models. Models are generally used to infer the initial masses, main sequence lifetimes, luminosities, effective temperatures, and surface abundances for the individual stars from the catalogs cited above. There has been much progress recently in evolutionary calculations for massive stars. It has been realized that mass loss itself can significantly alter the evolution of a massive star (cf. Chiosi and Maeder 1986). Together with improvements in treatment of physical processes like convection (e.g., "overshooting"), an optimist can believe that stellar parameters can be reliably inferred from observational data coupled with evolutionary models.

(3) A calibration of $\dot{M}$ against other stellar parameters. It was noted above that the best correlation seems to be between $\dot{M}$ and L, and that other parameterizations, in terms of M_*, R_*, etc., give no better fit to the observations.

(4) A calibration of V_∞ against other stellar parameters, e.g., V_{esc} and T_{eff}, as discussed above.

Calculations of Input Rates

Abbott (1982a) and Abbott et al. (1986) have used these four categories of information to compute the mass, kinetic energy, and momentum fluxes from winds for the catalogued OB and WR stars within 3 kpc of the sun, using the data of Garmany, Conti, and Chiosi (1982) and van der Hucht et al. (1981), and

have derived the average rates per kpc² for these quantities. A comparison of energy, momentum, and mass inputs by stellar category showed that:
 1. The O-stars dominate both the radiative luminosity and the ionizing flux–not a surprising result.
 2. The WR star winds contribute about half of the mass, momentum, and kinetic energy deposited in the ISM by stellar winds, a conclusion which was not previously appreciated.
 3. The stellar wind kinetic energy return is dominated by the most luminous stars (early-O and WR). This result is mainly a consequence of the rather steep luminosity dependence of $\dot{M}$, though there is a corollary tendency for the most luminous stars to have the highest wind velocities. (There are notable exceptions, however, such as the very luminous B1 star P Cyg, which has a very low wind velocity of 220 km s^{-1}.)

Van Buren (1985) has also determined mass and energy return rates for winds of OB stars. He redetermined the luminosity function and derived an IMF, including a correction for extinction, which was significantly flatter than that of Garmany, Conti, and Chiosi (1981). Applying this IMF and a somewhat different calibration of $\dot{M}$ and V_∞ in terms of stellar parameters, Van Buren (1985) obtained mass and kinetic energy return rates for OB stars which were significantly higher than Abbott's (1982a) values, as summarized in Table 1.

Table 1: Comparison of Mass and Kinetic Energy Return Rates Calculated for OB Stars

Quantity	Van Buren (1985)	Abbott (1982)	
Mass	2.6×10^{-4}	8.6×10^{-5}	$M_\odot$ yr^{-1} kpc^{-2}
Kinetic Energy	5.4×10^{38}	1.0×10^{38}	erg s^{-1} kpc^{-2}

The discrepancy is mainly a consequence of Van Buren's relatively flat IMF, which implies a higher density of the most massive, and therefore most luminous, stars, which dominate the mass and energy return. There is at present no obvious resolution of this discrepancy, though the OB star catalog of Garmany, Conti, and Chiosi (1981) is probably missing no more than 10% of all O-stars within 3 kpc of the sun. It seems possible, therefore, that Van Buren's extinction correction may have somewhat overestimated the density of the most luminous stars.

The survey of Abbott et al. (1986) confirmed the importance of WR winds as sources of enriched matter and of kinetic energy to the ISM. However, van der Hucht, Cassinelli, and Williams (1986) have argued that revised surface abundances of WR stars (especially the WC spectral type) predicted by Prantzos et al. (1986) should be used to compute WR mass loss rates and wind composition. These revised abundances, with substantial increases in CNO abundances over those assumed by Abbott et al. (1986), yield significantly higher mass loss rates, by factors of ~2 for the WN type and ~3 for the WC type. As a consequence, van der Hucht, Cassinelli, and Williams (1986) find that the WR stars are even more important as sources of matter and energy for the ISM than was previously believed. Table 2 summarizes the estimated return rates by stellar category: O-stars, WR stars, and B and A supergiants. The rates for O-stars and BA supergiants are from Abbott (1982) while the WR star rates are the revised values of van der Hucht, Cassinelli, and Williams (1986).

Table 2: Comparison Of Stellar Inputs By Category (averages for D < 3 kpc)

Quantity	%-contribution by			Total Rate (kpc^{-2})
	WR	O	BA	
Mass	77%	19%	4%	1.6×10^{-4} M$_\odot$ yr^{-1}
K.E.	70	28	2	3.2×10^{38} erg s^{-1}
Momentum	74	23	3	2.3×10^{30} g cm s^{-1}
Rad. Lumin.	8	65	27	3.5×10^{40} erg s^{-1}
Ion. Photons	5	93	2	3.0×10^{50} s^{-1}

Sources:
(1) O stars, B and A supergiants: Abbott 1982
(2) WR stars: van der Hucht, Cassinelli, & Williams 1986

Element Enrichment
As indicated in Table 2, the derived rate of mass returned by hot star winds to the ISM is $\sim 1.6 \times 10^{-4}$ M$_\odot$ yr^{-1} kpc^{-2}, with an uncertainty of at least a factor of two. Some part of this material is enriched in nucleosynthesis products. Because of their highly evolved state, the WR stars dominate the chemical enrichment of the ISM by the hot star winds. Following Abbott (1982) but using the revised abundances of van der Hucht, Cassinelli, and Williams (1986), we can estimate the yield of He and the CNO elements by the WR star winds. The "yield", Y_M, is defined here as the mass of element M by which the ISM is enriched per unit mass of the ISM which is formed into stars. If X_M is the relative mass fraction of element M, then we calculate

$$Y_M = [(\Delta X_M(\text{WN}) \times \dot{M}_{\text{WN}}) + (\Delta X_M(\text{WC}) \times \dot{M}_{\text{WC}})]/SFR$$

with separate terms for the WN and WC spectral types, and where $\Delta X_M = X_M(\text{WR}) - X_M(\text{ISM})$. With the abundances and mass loss rates of van der Hucht, Cassinelli, and Williams (1986) and a star formation rate, SFR, of 4×10^{-3} M$_\odot$ yr^{-1} kpc^{-2} (probably close to a lower limit–see Jura 1987), the yields of He, C, N, and O for the WN and WC spectral types are summarized in Table 3. For the assumed values, the table indicates that almost 1 $M_\odot$ of CNO elements is returned to the ISM for every 100 $M_\odot$ converted to stars. Compared with the enrichment yields expected for supernovae (cf. Chiosi and Maeder 1986), the WR star winds are estimated to produce $\geq 60\%$ of the total yield of He and $\geq 15\%$ of the yield of the CNO elements. Thus, the WR star winds are the dominant source of helium and a non-negligible source of C, N, and O for the enrichment of the interstellar medium.

Table 3:   Element Enrichment by Wolf-Rayet Star Winds

Element	Relative Mass Fractions			Yield
	X(WN)	X(WC)	X(ISM)	
He	0.95	0.32	0.22	1.4×10^{-2}
C	.00037	.39	.0034	5.2×10^{-3}
N	.021	.00	.0012	3.3×10^{-4}
O	.0011	.25	.0082	3.1×10^{-3}
CNO	.022	.64	.013	8.6×10^{-3}

Note: mass loss rates and mass fractions for the WN and WC spectral categories are from van der Hucht, Cassinelli, and Williams (1986)

"LOCAL" EFFECTS OF STELLAR WINDS ON THE ISM

Standard Model of Wind-Blown Bubbles

The standard picture of the interaction of a high-velocity wind with the ambient medium has been developed by a number of workers. Avedisova (1972) proposed a model, based on the Sedov solution for supernova shocks, to explain certain ring nebulae associated with WR stars (e.g., NGC 6888). Related work was published by Steigman, Strittmatter, and Williams (1975) and by Castor, McCray, and Weaver (1975), who considered an adiabatic similarity solution for wind-blown shells expanding into a uniform medium. These models predict the radius and velocity of the shell as a function of time. An improved analytical model was developed by Weaver et al. (1977) to include the effects of radiative losses at later times. The basic prediction of the model is that the high-velocity stellar wind will drive a shock wave into the ambient gas to produce three concentric zones: (1) the stellar wind itself, with a velocity of ~ 2000 km s^{-1}, density falling as $1/r^2$, and a temperature of $\sim 10^4$ K; (2) outside this region, a shocked stellar wind with nearly constant density and a decelerating velocity field, and with a temperature of 10^6 - 10^7 K; and (3) an outer shell of high-density shocked interstellar gas, with $T \approx 10^4$ K.

These analytical results are in good agreement with recent numerical hydrodynamic calculations by Tenorio-Tagle *et al.* (1989) for a "standard model" wind-blown shell, which develops precisely the structure predicted by the analytic model of Weaver *et al.* (1977).

A number of wind-blown bubble candidates have been identified. The X-ray "supershell" in Cygnus (Cash *et al.* 1980) has been suggested as a bubble blown by the collective effect of winds from members of the Cyg OB2 association (Abbott, Bieging, and Churchwell 1981). The total thermal energy content of the bubble is $\sim 6 \times 10^{51}$ ergs, while the wind luminosity of stars in Cyg OB2 (on which the bubble is centered) is $> 2 \times 10^{52}$ erg Myr^{-1}. The energetics and estimated age of the bubble are consistent with a wind-blown origin.

Some of the giant H I shells catalogued by Heiles (1979) are probably produced by stellar winds. The size and energy requirements of such shells can be explained by one or a few early-O star winds. Heiles has noted a possible association of the H I shells with young clusters.

Finally, a number of ring-shaped nebulae may be formed, or strongly influenced by, stellar winds. Obvious examples include the Rosette and Bubble (NGC 7635) Nebulae, and ring nebulae associated with Wolf-Rayet stars (Avedisova 1972; Chu 1981).

Effects of a Cloudy Medium

The standard model for wind-blown bubbles is unrealistic, since it assumes that the wind expands into a homogeneous medium. In fact, the ISM is full of clouds and density variations. The effects of such structure on wind-blown bubbles can be substantial. McKee, Van Buren, and Lazareff (1984) considered a cloudy medium around an O-star, with a cloud mass spectrum $N(M) \sim M^{-2}$. They argue for the importance of two physical effects in the development of a bubble: (1) "photoevaporation", by which photoionization removes material from neutral clouds, thereby increasing the density and mass of the H II region (cf. Elmegreen 1976, who modelled an ensemble of uniform clouds); and (2) the "rocket effect", which displaces the clouds away from the ionizing star. These two effects result in a "homogenization" of the cloudy medium out to a radius R_h which depends on the initial mean density of the ambient gas, n_m, as $R_h = 56 n_m^{-0.3}$ pc. Remarkably, R_h is virtually independent of the stellar lifetime. The mass displaced (i.e., the mass of the swept-up shell) is predicted to be $M_h = 2.4 \times 10^4 n_m^{0.1}$ M$_\odot$, nearly independent of the initial mean density. The effect of the cloudy medium on the bubble depends on the mechanical luminosity, L_W ($\equiv \dot{M} V_\infty^2 / 2$) of the wind. If

$$L_W < L_{crit} = 1.3 \times 10^{36} (S_{49}/n_m)^{1/3} \text{ erg s}^{-1}$$

(where S_{49} is the stellar ionizing flux in 10^{49} photons s^{-1}, and n_m is in cm^{-3}), then the bubble expands into the homogenized medium around the star as described by the standard model discussed above. If $L_W \geq L_{crit}$, the bubble expands until it reaches the homogenization radius, R_h. At this point cloud evaporation "poisons" the bubble: rapid radiative cooling reduces the bubble to a momentum conserving expansion and the bubble stalls. Thereafter, the expansion of the bubble is controlled by cloud evaporation, so that $R_{bubble} \approx R_h$.

McKee, Van Buren, and Lazareff (1984) argue that a classification scheme suggested by Chu (1981) to categorize WR nebulae, and used also by Lozinskaya

(1982) for O-star nebulae, can be explained as an evolutionary sequence of wind-blown bubbles in cloudy media. The stages of development are:

(1) Wind blown shell. This is the earliest stage, with the shell size comparable to that of the H II region.

(2) Amorphous H II region. A thick shell is dominated by the photoionization of clouds. The effects of this process on the clouds have recently been calculated by Bertoldi (1989) and Bertoldi and McKee (1990), who conclude that small clouds are entirely evaporated, while large clouds are first imploded, then accelerated away from the star by the rocket effect. Observational examples of such imploded clouds may be the globules seen in many optical H II regions (e.g., Rosette, Gum).

(3) Ring-like H II regions. When the homogenization radius exceeds the Strömgren radius, the nebula is a thin shell of H II filled by the (low-density) wind bubble.

(4) Invisible. For late-type O-stars near the end of the main sequence lifetime, the emission measure of the bubble is too low to be readily detectable.

(5) Stellar ejecta. After the star leaves the main sequence, the nebular gas near the star is expected to be entirely wind material. Examples include the ring nebulae associated with WR stars studied by Chu, Treffers, and Kwitter (1983 and references therein).

Supernova Remnants in Wind-Blown Bubbles

All early-type stars capable of producing stellar wind-blown bubbles should end their evolution as type II supernovae. What is the effect of the pre-existing bubble on the evolution of the supernova remnant (SNR)? This is an important point, because the initial ambient conditions of the remnant expansion are very different for a wind-blown bubble ($n \sim 0.01$ cm^{-3}, $T \sim 10^4 - 10^6$ K) as compared with the usually assumed values (e.g., $n = 1$ cm^{-3}, $T = 100$ K). McKee, Van Buren, and Lazareff (1984) point out that in such a case, the SNR will interact only with the stellar wind ejecta out to $R \approx 20$ pc. Since the mass of the SNR should be comparable to the mass of the stellar ejecta within this radius, the SNR remains essentially in free expansion out to $R \approx 20$ pc, and does not go into the Sedov-Taylor phase.

Recent one-dimensional numerical hydrodynamic calculations by Tenorio-Tagle *et al.* (1989) support this picture. These authors have modelled a supernova event of kinetic energy 10^{51} ergs occurring within a wind-blown shell, for a wind with $V_\infty = 1000$ km s^{-1}, $\dot{M} = 3 \times 10^{-5}$ M$_\odot$ yr^{-1}, and a bubble age of 2×10^5 yr. These wind parameters are reasonable for a WR wind (though with a V_∞ somewhat lower than is typical). They describe two model cases:

(1) Mass in the wind-blown shell is comparable to that of the SN ejecta. In this case, the transmitted shock outruns the wind-blown shell and proceeds into the undisturbed gas, forming a second outer shell. The predicted X-ray size of the SNR should then exceed the Hα size (from the shocked shell). The SNR expands freely at first, to reach the Sedov phase in only ~20% of the time required for the standard (non-wind) case.

(2) Mass in the wind-blown shell is much larger than the mass of the SNR. In this case the SN shock compresses and accelerates the wind shell, but radiative losses prevent the remnant from entering the Sedov phase at all. Instead, the shell goes directly into the momentum-conserving phase. The SN

shock is trapped in the massive outer wind shell and cannot move ahead into the undisturbed gas.

The results of Tenorio-Tagle et al. (1989) imply these consequences for SNR evolution in a wind-blown bubble:

—The remnant develops multiple-shell structure, due to the presence of reverse shocks.

—The SN shock quickly expands to reach the wind shell.

—As a result, the ages of SNR's derived from the assumption of a long Sedov phase may be overestimated by a factor of several.

—Wind-blown bubbles may account for the early optical emission from SNR's.

SUMMARY

This paper has reviewed the properties of winds from hot (OB and WR) stars and examined the effects of these winds both on the global input of energy and matter into the ISM, and on the local effects of wind-blown bubbles on photoionized nebulae and on supernova remnant evolution. Significant conclusions are:

(1) The total kinetic energy injected into the ISM by the wind of a luminous OB or WR star is comparable to that from a Type II supernova.

(2) WR star winds dominate both the element enrichment and the kinetic energy returned by all hot star winds, in a global average.

(3) The standard model of stellar wind-blown bubbles (e.g., Weaver et al. 1977) evidently requires substantial modification in a cloudy ISM.

(4) Wind-blown bubbles will significantly affect the evolution of a subsequent supernova remnant, in particular by greatly shortening (or eliminating) the Sedov phase.

ACKNOWLEDGEMENTS

Research at the U.C. Berkeley Radio Astronomy Laboratory is supported by NSF grant AST87-14721.

REFERENCES

Abbott, D.C. 1982a, *Ap. J.*, **263**, 723.
——— 1982b, *Ap. J.*, **259**, 282.
——— 1985, in *Radio Stars*, ed. R.M. Hjellming and D.M. Gibson (Dordrecht: Reidel), p. 61.
Abbott, D.C., Bieging, J.H., and Churchwell, E.B. 1981, *Ap. J.*, **250**, 645.
Abbott, D.C., Bieging, J.H., Churchwell, E.B., and Torres, A.V. 1986, *Ap. J.*, **303**, 239.
Avedisova, V. 1972, *Sov. Astr. A.J.*, **15**, 708.
Becker, R.H., and White, R.L. 1985, in *Radio Stars*, ed. R.M. Hjellming and D.M. Gibson (Dordrecht: Reidel), p. 139.
Bertoldi, F. 1989, *Ap. J.*, **346**, in press.

Bertoldi, F., and McKee, C.F. 1990, *Ap. J.*, , in press.
Bieging, J.H., Abbott, D.C., and Churchwell, E.B. 1982, *Ap. J.*, **263**, 207.
─────── 1989, *Ap. J.*, **340**, 518.
Cash, W., Charles, P., Bowyer, S., Walter, F., Garmire, G., and Riegler, G. 1980, *Ap. J. (Letters)*, **238**, L71.
Cassinelli, J.P., and van der Hucht, K.A. 1986, in *Instabilities in Luminous Early Type Stars: Proc. of a Workshop in Honour of C. de Jager*, ed. H. Lamers and C. de Loore (Dordrecht: Reidel).
Castor, J.I., Abbott, D.C., and Klein, R.I. 1975, *Ap. J.*, **195**, 157.
Castor, J.I., McCray, R., and Weaver, R. *Ap. J. (Letters)*, **200**, L107.
Castor, J.I., and Simon, T. 1983, *Ap. J.*, **265**, 304.
Chiosi, C., and Maeder, A. 1986, *Ann. Rev. Astr. Ap.*, **24**, 329.
Chu, Y-H. 1981, *Ap. J.*, **249**, 195.
Chu, Y-H., Treffers, R., and Kwitter, K. 1983, *Ap. J. Suppl.*, **53**, 937.
Dickel, H.R., Habing, H.J., and Isaacman, R. 1980, *Ap. J. (Letters)*, **238**, L39.
Elmegreen, B.G. 1976, *Ap. J.*, **205**, 405.
Felli, M., and Panagia, N. 1981, *Astr. Ap*, **102**, 424.
Florkowski, D.R., and Gottesman, S.T. 1977, *M.N.R.A.S.*, **179**, 105.
Friend, D.B., and Abbott, D.C. 1986, *Ap. J.*, **311**, 701.
Garmany, C.D., Conti, P.S., and Chiosi, C. 1982, *Ap. J.*, **263**, 777.
Garmany, C.D., and Conti, P.S. 1984, *Ap. J.*, **284**, 705.
Garmany, C.D., Olson, G.L., Conti, P.S., and Van Steenberg, M.E. 1981, *Ap. J.*, **250**, 660.
Heiles, C. 1979, *Ap. J.*, **229**, 533.
Hogg, D.E. 1982, in *IAU Symposium 99, Wolf-Rayet Stars: Observations, Physics, and Evolution*, ed. C.W.H. de Loore and A.J. Willis (Dordrecht: Reidel), p. 221.
─────── 1985, in *Radio Stars*, ed. R. Hjellming and D. Gibson (Dordrecht: Reidel), p. 117.
Jura, M. 1987, in *Interstellar Processes*, ed. D.J. Hollenbach and H.A. Thronson Jr. (Dordrecht: Reidel), p. 3.
Lozinskaya, T.A. 1982, *Astrophys. Space Sci.*, **87**, 313.
McKee, C.F., Van Buren, D., and Lazareff, B. 1984, *Ap. J. (Letters)*, **278**, L115.
Panagia, N., and Felli, M. 1975, *Astr. Ap*, **39**, 1.
Pauldrach, A., Puls, J., and Kudritsky, R.P. 1986, *Astr. Ap*, **164**, 86.
Persi, P., Ferrari-Toniolo, M., Tapia, M., Roth, M., and Rodruiguez, L.F. 1985, *Astr. Ap*, **142**, 263.
Prantzos, N., Doom, C., Arnould, M., and de Loore, C. 1986 *Ap. J.*, **304**, 695.
Puls, J. 1987, *Astr. Ap*, **184**, 227.
Seaquist, E.R. 1976, *Ap. J. (Letters)*, **203**, L35.
Steigman, G., Strittmatter, P., and Williams, 1975, *Ap. J.*, **198**, 575.
Tenorio-Tagle, G., Bodenheimer, P., Franco, J, and Różyczka, M. 1989, *M.N.R.A.S.*, in press.
Van Buren, D. 1985, *Ap. J.*, **294**, 567.
van der Hucht, K.A., Conti, P.S., Lundström, I., and Stenholm, B. 1981, *Space Sci. Rev.*, **28**, 227.
van der Hucht, K.A., Hidayat, B., Admiranto, A.G., Supelli, K.R., and Doom, C. 1988, *Astr. Ap*, **199**, 217.
van der Hucht, K.A., Cassinelli, J.P., and Williams, P.M. 1986, *Astr. Ap*, **168**, 111.

van der Hucht, K.A., Williams, P.M., and Thé, P.S. 1987, *Quart. Jour. R.A.S.*, **28**, 254.
Weaver, R., McCray, R., Castor, J, Shapiro, P, and Moore, R. 1977, *Ap. J.*, **218**, 377.
White, R.L., and Becker, R.H. 1982, *Ap. J.*, **262**, 657.
——— 1983, *Ap. J. (Letters)*, **272**, L19.
Wright, A.E., and Barlow, M.J. 1975, *M.N.R.A.S.*, **170**, 41.

THE TOTAL RATE OF MASS RETURN TO THE INTERSTELLAR MEDIUM FROM RED GIANTS AND PLANETARY NEBULAE

G. R. KNAPP and K. P. RAUCH
Princeton University Observatory, Princeton, NJ08544

E. M. WILCOTS
Astronomy Department, University of Washington, Seattle, WA 98195

ABSTRACT High luminosity post main sequence stars are observed to be losing mass in large amounts into the interstellar medium. The various methods used to estimate individual and total mass loss rates are summarized. Current estimates give $M_T \sim 0.3 - 0.6\ M_\odot\ yr^{-1}$ for the whole Galaxy.

INTRODUCTION

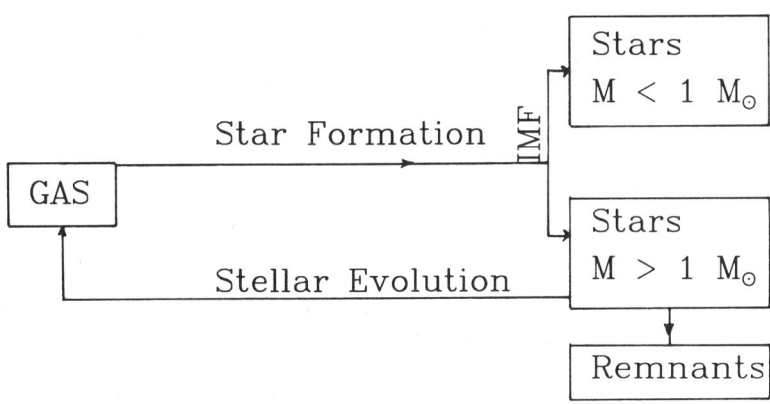

Fig. 1. Cycling of material between the interstellar gas and stars.

Stars lose mass throughout their lifetimes: during the star formation process as the star settles to the main sequence, on the main sequence, during the ascent of the red giant and asymptotic giant branches, and even during the post-AGB phase, as the star evolves towards a compact white dwarf. There are copious observational data on almost all of these phases, but as yet no complete theory of what causes the mass loss in the first place. This review concentrates on one aspect of the complex star- interstellar medium interaction, the return of mass to the interstellar medium (ISM) by the processes of stellar evolution.

The flow of material between the stars and the interstellar gas is sketched in Figure 1. The mass return to the ISM can be broken down into two main areas. First, there is the question of the efficiency of star formation, i.e. what fraction of the gaseous mass in star forming regions actually ends up in main sequence stars. Mass is observed to be driven out of star forming regions throughout the star formation process, and large amounts are ejected during the pre-main sequence phase, as shown by the presence of bipolar outflows from young stars (Rodriguez 1989). Second, once the stars have settled to the main sequence and emerged from their formative molecular clouds, mass is lost, in greater or lesser amounts, throughout the remainder of their evolution.

The fraction of the mass in a coeval group of main sequence stars which will be returned to the ISM has been estimated by Tinsley (1980). The relevant timescale is the main sequence lifetime of a 1 $M_\odot$ star, which is about the Hubble time. The fraction of mass which forms stars of $\leq$ 1 $M_\odot$ can be calculated for a given initial mass function (IMF). Next, for the stars evolving on this timescale, the fraction of their mass which ends up as inert remnants (white dwarfs, black holes, or neutron stars) is estimated. The leftover mass is the fraction returned to the ISM, R. The Miller-Scalo (1979) IMF gives R = 0.17. Because of the steepness of the IMF, well over half of this mass is ejected by stars at the low mass end of the evolving stars, 1 to 8(?) $M_\odot$, i.e. the stars which end their lives as white dwarfs (Tinsley 1975).

A summary of mass loss at all points on the Herzsprung-Russell diagram is given by de Jager *et al* (1988). Aside from the pre-main-sequence processes, the sources of mass return which are significant on a galactic scale appear to be (1) winds from massive main sequence and post-main-sequence (OB and Wolf-Rayet) stars (Bieging 1989) (2) winds from evolved stars on the red giant and asymptotic giant branch stars, discussed here and (3) supernova ejecta (Chevalier 1989). Winds from low mass stars on the main sequence are probably negligible (the solar mass loss rate is only 10^{-13} $M_\odot$ yr^{-1}).

Since the bulk of the mass lost from red giant and AGB stars comes from stars in the 1 to 8 $M_\odot$ range, these stars are an important source of the CNO elements in the Galaxy. The abundances of the CNO elements are altered by processing on the main sequence and by mixing of triple-α products to the surface of some stars on the AGB (Iben 1987). And since these stars are cool ($T_{eff} \leq 3000$ K) the material is shed in molecular, dusty form. The formation of silicate dust is observed in stars with n(O) > n(C) ("oxygen" stars) and of carbonaceous dust in stars with n(C) > n(O) (carbon stars). Mass loss on the AGB is thus a source, and perhaps the source, of interstellar dust.

The low effective temperatures of red giant stars means that their atmospheres are molecular (Tsuji 1976; Johnson *et al.* 1975) and hence their winds are primarily in molecular form. They can thus be studied by observations of radio- and millimeter-wavelength molecular lines. The detection of extended circumstellar molecular envelopes around these stars in molecular line emission allows the measurement of the mass loss rates from the central objects. The molecular emission lines have characteristic parabolic shapes; that the circumstellar envelopes are destined to become planetary nebulae upon further evolution of the central star is demonstrated by the detection of circumstellar molecular emission associated with young planetary nebulae

(Mufson, Lyon and Marionni 1975; Huggins and Healy 1989). These observations show that both oxygen and carbon AGB stars become planetary nebulae, with both oxygen-rich (e.g. Vy2-2 and M1-92, Davis et al. 1979) and carbon-rich (e.g. NGC7027, NGC2346) envelopes being seen (cf. Zuckerman and Aller 1986). These observations show that carbon and oxygen stars are not different stages of AGB evolution, but that there is a bifurcation of chemical properties during AGB evolution.

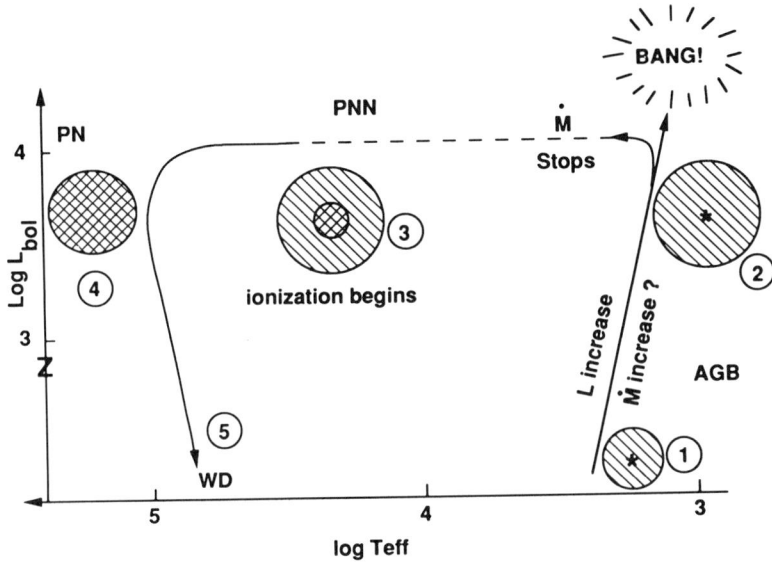

Fig. 2. Post main sequence evolution of a star of mass 1 to 8 $M_\odot$. (1) and (2) As the stellar luminosity increases on the AGB, an extended cool molecular circumstellar envelope is formed by mass loss. (3) After mass loss stops, the temperature of the central star increases, causing the ionization of the circumstellar envelope. (4) the envelope is completely ionized and is now a planetary nebula. (5) The planetary nebula expands into the ISM as the central star evolves to the white dwarf stage.

Figure 2 shows a sketch of the upper (high-luminosity) part of the HR diagram, accompanied by cartoons of the appearance of the circumstellar material at each stage. As the luminosity of the central star rises as the star ascends the AGB, mass loss takes place (stage 1). It is generally thought that the mass loss rate steadily increases with luminosity (cf. Kwok 1989; van der Veen et al. 1987); while this makes physical sense, there is little observational evidence to support this idea, and, as mentioned above, no theoretical framework. Indeed, observed mass loss rates at this stage cover more than four orders of magnitude. In any case, the star develops an extensive cold molecular circumstellar envelope due to mass loss. The effective radius of the envelope is limited by photodestruction of the molecules as the mass flows into the ISM (cf. Mamon, Glassgold and Huggins 1988), and the crossing time for these envelopes (about 10^4 years) is shorter than the total mass loss lifetime, so that observation of a sample of these envelopes gives a reasonable snapshot of the current mass loss rate. It is difficult to tell if a given star is on the red giant or asymptotic branches, especially for stars which are heavily dust enshrouded. However, since we wish to address the total rate of mass return from red giant stars, a census of all mass losing stars will give the correct answer whether there is one mass losing episode or two.

If the total mass lost is sufficient to remove the envelope before core instability and explosion occurs, the hydrogen-shell burning is disrupted and mass loss stops (Schönberner 1983). The central star then begins to contract and heat up, and the UV flux begins to ionize the circumstellar material to produce a planetary nebula (stage 3, Figure 2). After about 10^4 years all the material is ionized (though cf. Spergel et al. 1983) and the circumstellar shell is observed as a planetary nebula (stage 4, Figure 2). The planetary nebula then expands and diffuses into the ISM as the central star evolves to the white dwarf stage (stage 5, Figure 2).

Figure 2 then shows that we have four different views of the same phenomenon, i.e. four different opportunities to measure the mass lost on the red giant branches. (a) First, we can measure the total number (or luminosity) of the red giant branch stars in some region. This gives an indication of the current number of mass losing stars; if we can assume a mean mass loss rate per giant, the total mass loss rate M_T can be estimated. (b) The mass loss rates themselves can be measured for a volume limited sample of evolved giants and summed. (c) The total mass of a volume limited sample of planetary nebulae can be measured and summed; together with an estimate of the lifetime of the planetary nebula phase this gives M_T. (d) The total mass rate can be estimated from the white dwarf formation rate and measurements of the mass lost prior to the white dwarf stage. It is likely that (a) and (b) will provide a somewhat more complete census of the total mass loss. If mass is lost on the red giant as well as asymptotic branches, the mass lost during the red giant phase may have drifted too far from the star by the time it becomes a planetary nebula to be readily observable (though cf. Chu 1989). Also, some stars may copiously lose mass on the red giant and asymptotic branches before exploding as supernovae (indeed the evolution of the radio light curves for SN suggests that this is often the case, cf. Chevalier 1989) and will obviously not be found in planetary nebula/white dwarf counts.

This review will attempt to evaluate the results for each of these evolutionary stages. The subject is unfortunately not in very good shape, for it is one of those in which any work, to be useful, must get an answer within a factor two or three of the correct one, and it is not yet at that stage. Both red giants and planetary nebulae are high-luminosity objects and so are easily seen to large enough distances to make the compilation of a good sample of these rare objects possible; however, for neither type of object can the individual distances be measured with any reliability. This review will summarize our knowledge of post main sequence mass loss by each of the four methods outlined above; no attempt will be made to tackle the interesting question of chemical evolution. It will also concentrate on information about the solar neighborhood and its extrapolation to the rest of the galactic disk; the rôle of mass losing stars in the galactic center region is not discussed.

LUMINOSITY AND MASS LOSS RATE FOR AGB STARS

Red giant stars in the Galaxy are observed to be losing a large amount of mass, and mass loss rates in the range several times 10^{-8} M$_\odot$ yr^{-1} to several times 10^{-4} M$_\odot$ yr^{-1} are observed. The assumption of a mean mass loss rate per luminous giant star is therefore extremely uncertain. Part of the reason for this wide range in mass loss rates is likely to be the range in the luminosities and progenitor masses of stars which are currently losing mass in the Galaxy.

A somewhat better estimate of the total mass loss rate can be obtained from the luminosities if it is assumed that the mass loss rate is proportional to the luminosity. The observational evidence is fairly strong that, whatever causes the mass loss in the first place, the winds are at least driven to their terminal velocities by radiation pressure on dust grains (e.g. Jura 1988). The mass loss rate can then be estimated by conservation of momentum:

$$\dot{M} V_o \simeq \tau L_*/c \qquad (1)$$

where V_o, the wind terminal velocity, is usually within a factor of two of 15 km s^{-1}, the typical escape velocity from a red giant star, and τ, the effective optical depth of the envelope, is about 1. However, application of this equation to solar neighborhood red giants would greatly overestimate the total mass loss rate, since observed mass loss rates are much less than the value given by (1) for the great majority of evolved stars (Knapp 1986), i.e. $\tau \ll 1$.

Inference of the total mass loss rate from the stellar luminosity can probably be more reliably made for elliptical galaxies. In these systems, there is generally little star formation and most of the visible and IR radiation is emitted by evolving stars in an old, single- age population (Tinsley 1980). Estimates of the mean mass loss rate per giant star of mass 1 M$_\odot$ can be used to calculate the rate of injection of mass into the ISM of these galaxies. Faber and Gallagher (1976) find:

$$\dot{M} = 1.5 \times 10^{-11} \, L_B/L_\odot \; M_\odot \, yr^{-1} \qquad (2)$$

using the emission at visual wavelengths. More recent work has concentrated on the mid-IR emission because of the availability of measurements of a very large number of ellipticals by the *IRAS* satellite (Neugebauer et al. 1984). The shortest *IRAS* wavelength is 12 μ; it is likely that this is mainly radiated by the photospheres of red giants and by dust in the mass-loss circumstellar envelopes around them (Soifer et al. 1986; Impey et al. 1986; Jura et al. 1987). Jura et al. (1987) estimate that

$$\dot{M} = 4.3 \times 10^{-30} \, L_{12} \; M_\odot \, yr^{-1} \qquad (3)$$

where L_{12} is the 12μ luminosity in erg s^{-1} Hz^{-1}. The derivation of equation (3) assumes that about half of the 12μ luminosity comes from circumstellar material and that the typical outflow speed is 15 km s^{-1}. The resulting mass loss values agree well with those estimated using equation (2). Further observations of the mass loss rates from low-mass (1 $M_\odot$) stars in the solar neighborhood may well lead to the refinement of these estimates.

DIRECT MEASUREMENT OF MASS LOSS RATES

Since mass loss on the AGB causes the formation of an extended, dusty circumstellar envelope the mass loss rate can be estimated from observations of emission from molecular lines or from warm circumstellar dust. All of these estimates are subject to uncertainties in the abundance of the observed species relative to that of hydrogen. HI observations have succeeded in detecting a small number of circumstellar envelopes (e.g. Bowers and Knapp 1988) but show that in most envelopes the hydrogen is primarily in molecular form. Even were it not, observations of the HI 21 cm line would be of limited use for this problem because of the ubiquity of strong emission from the abundant interstellar HI (Lockman 1989).

Mass loss rates are primarily estimated from three types of observation: (1) thermal emission in the rotational lines of the abundant CO molecule; (2) the linear sizes of the emitting regions in OH/IR stars; and (3) the circumstellar infrared excess. Each of these methods will be briefly reviewed.

CO Observations

CO emission is readily detectable from circumstellar envelopes. The line is usually fairly optically thick, and if the outflow velocity is constant and the envelope much smaller than the beam used to observe it, the line shape is parabolic (Morris 1975). As an example, the CO(2-1) emission from the bright carbon star CIT6 is shown in Figure 3. The full width of the line at zero power is twice the terminal outflow velocity; the mean envelope velocity gives the stellar systemic radial velocity and the line intensity is roughly proportional to the loss rate of CO molecules. The mass loss rate can thus be evaluated using models of the CO line formation (Morris 1980). In these models, the outflow is assumed to be spherically symmetric with constant outflow velocity and constant mass loss rate. The envelope is assumed to be truncated at an effective radius r_e given by the photodissociation model of Mamon, Glassgold and Huggins (1988):

$$r_e \sim 6.3 \times 10^{16} \left(\frac{\dot{M}}{10^{-6}}\right)^{0.6} \left(\frac{10}{V_o}\right)^{0.4} \left(\frac{f_{CO}}{3 \times 10^{-4}}\right)^{0.5} \text{ cm} \quad (4)$$

where $f = n(CO)/n(H_2)$. The CO rotational levels are excited by microwave background photons, by collisions with H_2 and He and by infrared emission from

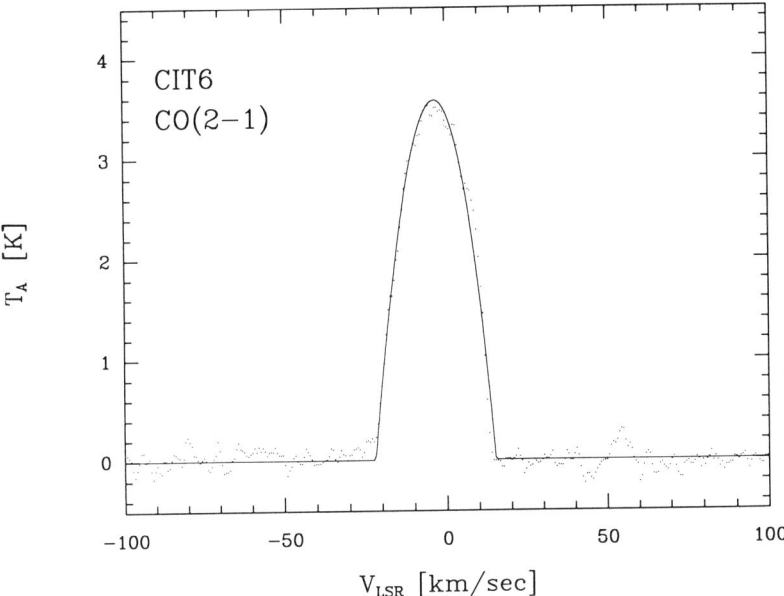

Fig 3. CO(2-1) line emission from the carbon star CIT6, observed with the 10.4 meter telescope of the Caltech Submillimeter Observatory (Phillips et al. 1989). The least-squares parabolic profile fit is shown.

the inner regions of the envelope at 4.6 μ, the wavelength of the v = 1 - 0 transition of CO.

These models give:

$$\dot{M} \sim \frac{T_A^* V_o^2 D^2}{Af} \; M_\odot \; yr^{-1} \qquad (5)$$

(Knapp and Morris 1985) where A is a numerical constant which depends on the transition observed and the telescope beamwidth, T_A^* is the observed main beam brightness temperature and D is the distance. Abundances for carbon stars of f = $7-10 \times 10^{-4}$ and for oxygen stars of f = $2-4 \times 10^{-4}$ are assumed; these are roughly the values found under the assumptions that, respectively, all of the oxygen and all of the carbon are in CO. Aside from distance uncertainties, these models probably give values of $\dot{M}$ good to about a factor of three in most cases (though cf. Sahai 1987). Both carbon and oxygen stars are observed to be sources of CO line emission. The mass loss rates found this way from CO observations range from values like 5×10^{-8} $M_\odot$ yr^{-1} for T Cas to 1 to 3 $\times 10^{-4}$ $M_\odot$ yr^{-1} for supergiant stars like CRL2688, IRC+10420 and NML Cyg. A note of caution in applying these models is suggested by recent, more sensitive observations of circumstellar envelopes; some evolved stars have been observed to have fast, non-spherically symmetric outflows in addition to the slower flows (Kawabe et al. 1987; Morris et al. 1987; Tsuji et al. 1988; Gammie et al. 1989; Phillips et al. 1989).

OH Observations

Several hundred luminous evolved stars are observed to be powerful sources of maser emission in lines of OH, H_2O and SiO (Engels 1979;.te Lintel Hekkert and Habing 1989). The characteristic maser emission seen in the ground state OH line at 1612 MHz has a double-horned shape produced by emission from a shell at a fairly large distance (typically 10^{16} cm) from the star. The OH in this shell is produced by photodissociation of the saturated species H_2O as the gas flows into the interstellar medium. The shell radius thus depends on the strength of the interstellar UV radiation field, on the gas- to-dust ratio in the envelope, on the relative abundance of H_2O, on the outflow velocity of the wind and on the mass loss rate. The outflow velocity can be measured from the observed line profile; if the first three quantities can be assumed to be constant from star to star, measurement of the OH shell radius gives the mass loss rate. This method of measuring $\dot{M}$ of course works only for oxygen (OH/IR) stars; in carbon stars the oxygen is bound in CO. By all indications, though, about 90% of all stars on the AGB are oxygen stars (e.g. Thronson et al. 1987; Hacking et al. 1985).

Measurements of shell angular radii in the 1612 MHz OH line can be made with high spatial resolution interferometry, using either the VLA (Bowers, Johnston and Spencer 1983; Herman 1983) or the MERLIN array (Diamond et al. 1985). These observations show that the large majority of OH/IR stars have circumstellar envelopes which are roughly spherically symmetric. Together with a distance to the star, these observations give the linear radius of the envelope. The radius can also be directly measured by the phase lag between the blue (near side) and red (far side) maser emission, since the maser is pumped by IR emission from the variable central star (Jewell et al. 1981; Herman 1983).

Stellar mass loss rates can then be estimated from the observed radii by calibrating the radius-mass relationship for stars with known mass loss rates (Bowers, Johnston and Spencer 1983) or by using models of the H_2O photodissociation (Huggins and Glassgold 1982; Netzer and Knapp 1987). For mass loss rates higher than about 10^{-7} $M_\odot$ yr^{-1}, Netzer and Knapp (1987) find

$$r_{OH} = A \times 10^{16} \left(\frac{\dot{M}}{10^{-5}}\right)^{0.7} V_o^{-0.4} \text{ cm} \qquad (6)$$

where A, which has a value of about 5, depends on the strength of the radiation field, and $n(H_2O)/n(H_2) = 3 \times 10^{-4}$. Both the CO and OH detections are heavily biased towards stars with high mass loss rates.

Circumstellar Dust

Luminous evolved stars have infrared excesses which are readily attributable to the presence of circumstellar dust, condensed in the outflowing wind (Gehrz and Woolf 1971). The characteristic emission spectra of this dust show that its likely composition is silicates in oxygen stars and graphite / amorphous carbon in carbon stars (e.g. Hacking et al. 1985; IRAS Science Team 1986). Indeed, comparison of the SiO abundances in the inner and outer regions of the envelopes around oxygen stars suggests that 90 - 99 % of the silicon in these envelopes is condensed into grains (Morris et al. 1979).

The spectral effect of the dust is to absorb the starlight at optical and near IR wavelengths and to reradiate this energy at mid and long IR wavelengths. The long wavelength color excess is then a direct measure of the amount of circumstellar dust and hence of the dust loss rate. If the gas to dust ratio is constant from star to star (of a given chemical type) the color excess gives the mass loss rate (Gehrz and Woolf 1971).

That the gas to dust ratio is fairly constant can be shown by comparison of the dust and gas loss rates. For a sample of about seventy carbon and oxygen stars whose dust loss rates are modeled by Rowan-Robinson and Harris (1983a,b), Knapp (1985) showed that the gas to dust ratio is roughly constant and further has approximately the interstellar value. This is not unexpected; the interstellar dust contains almost all of the heavy elements and the atmospheric abundances in AGB stars have approximately the normal galactic values.

Figure 4 shows the energy distribution (F_λ versus λ) for two carbon stars with very different mass loss rates. The data are taken from the compilation by Gezari, Schmitz and Mead (1987). RY Dra has a modest mass loss rate as measured by CO observations, about 3×10^{-7} $M_\odot$ yr^{-1}, and most of the flux is radiated at the characteristic wavelength of the cool photosphere, 1μ (T $\sim$ 2900 K). There is also a long wavelength excess over that expected for a black body. IRC+10216, on the other hand, has a high mass loss rate, about 6×10^{-5} $M_\odot$ yr^{-1} and, as Figure 5 shows, the photospheric radiation is almost completely absorbed, while the radiation from the source peaks at 4 to 5μ (T $\sim$ 750 K). The ratio of the flux densities at 5μ and 1μ is thus likely to give a fairly good estimate of the mass loss rate.

This approach has been taken by several authors, using data from the large scale sky surveys (at 2.2μ, Neugebauer and Leighton 1969) and at the IRAS wavelengths of 12μ, 25μ and 60μ. For example, Jura (1987) finds:

$$\dot{M} = 1.7 \times 10^{-7} \left(\frac{V_o}{15}\right) D^2 \left(\frac{L}{10^4}\right)^{-0.5} S_{60} \left(\frac{\lambda}{10}\right) \quad M_\odot \text{ yr}^{-1} \qquad (7)$$

where L is the bolometric luminosity of the source, λ its mean wavelength and S_{60} the flux density at 60μ. Similar analyses have been made by Skinner and Whitmore (1988a,b), van der Veen et al. (1987) and Thronson et al. (1987) using the IRAS flux density ratios such as S_{12}/S_{25}. A slightly different though related approach is to calibrate the IR flux density ratios using observations of stars with known mass loss rates (van der Veen et al. 1987; Thronson et al. 1987; Knapp et al. 1989). Figure 5 shows the color $S_{60}/S_{2.2}$ versus $\dot{M}$ for a sample of oxygen stars whose mass loss rates are measured by CO or OH observations, as outlined above - this plot can be used to transform the IR colors to mass loss rates.

A source of uncertainty in these calculations is the value of the outflow speed (cf. Equation (7)). However, while this is observed to vary from 3 to 50 km s^{-1}, the great majority of the observed stars have V_o within a factor two of 15 km s^{-1}.

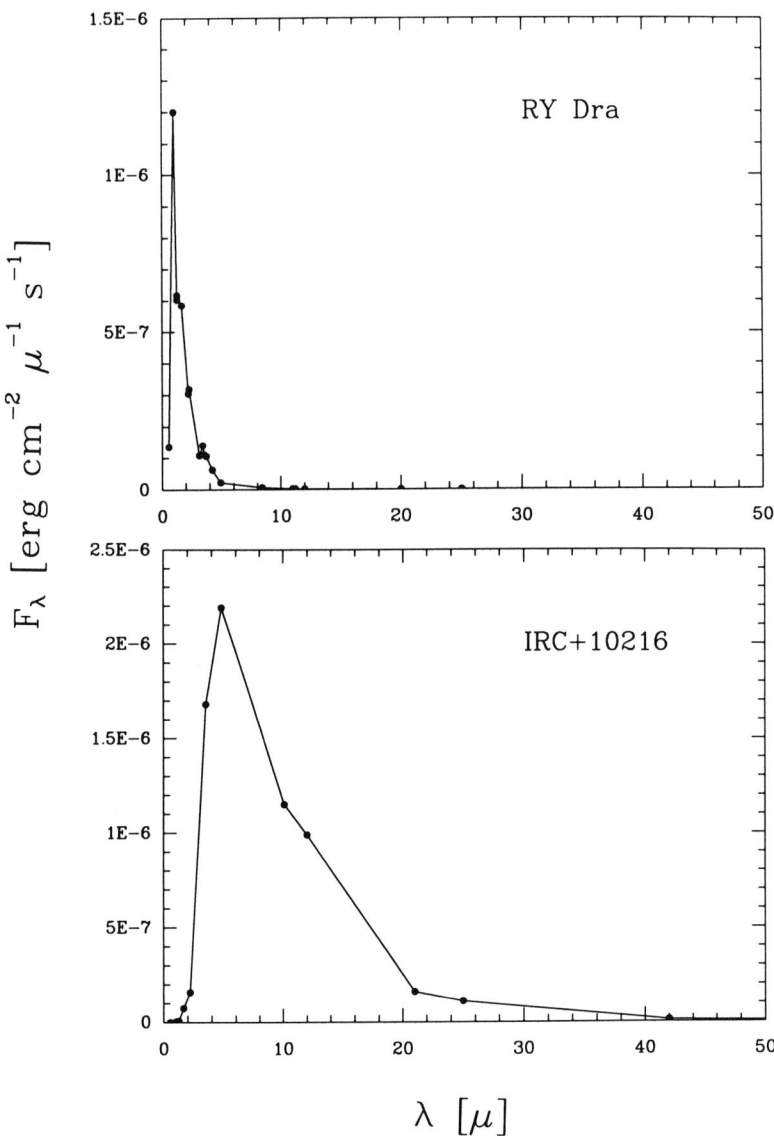

Fig. 4. Spectra of two carbon stars: RY Dra ($\dot{M} \sim 3 \times 10^{-7}$ $M_\odot$ yr^{-1}) and IRC+10216 ($\dot{M} \sim 6 \times 10^{-5}$ $M_\odot$ yr^{-1}).

THE TOTAL MASS LOSS RATE

Even with the much more extensive and complete catalog of evolved stars produced by the IRAS satellite, it is still not possible to measure the mass loss

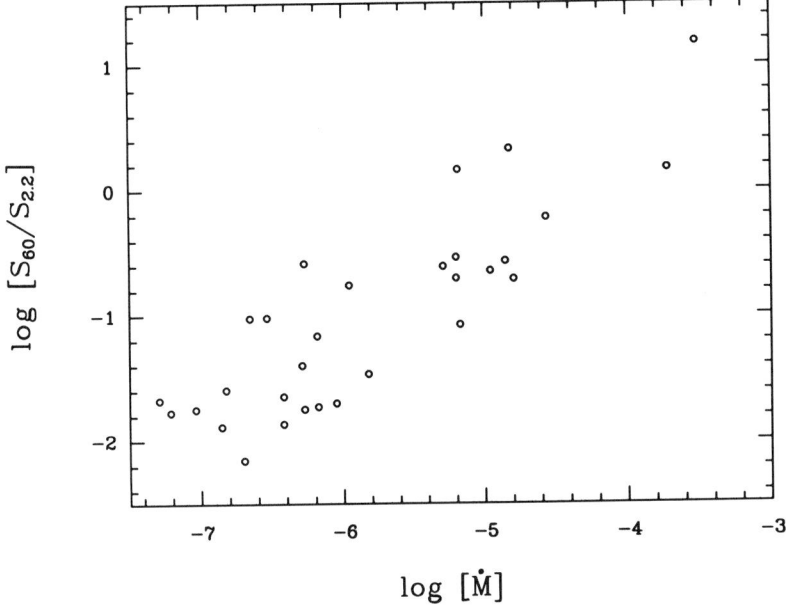

Fig. 5. Ratio of 60μ to 2.2μ flux density versus mass loss rate for a sample of oxygen stars.

rate of every star in the Galaxy, so all estimates to date have worked out the mass loss rate in a cylinder centered on the Sun, i.e. have found the mass loss rate $\dot{S}$ per unit area for the galactic plane. If it assumed that the mass-losing stars are distributed like the rest of the matter in the galactic disk, the total injection rate for the Galaxy, $\dot{M}_T$, is

$$\dot{M}_T = 2\pi \dot{S} \, R_o^2 \lambda^2 e^{1/\lambda} \qquad (8)$$

where R_o is the distance between the Sun and the galactic center and λ is the dimensionless scale length of the disk in units of R_o. With $\lambda = 0.4$ and $R_o = 8.5$ kpc, the effective area of the galactic disk is 9×10^8 pc^2.

The value of $\dot{S}$ is given by

$$\dot{S} = \frac{\Sigma \dot{M}}{\pi R^2} \qquad (9)$$

where R is the depth of the sample and is assumed to be much larger than the scale height of the galactic disk. However, if the mass loss rates are obtained from CO observations, stars in different mass loss rate ranges are sampled to very different depths as shown by Equation (5). This can be corrected for by calculating

$$\dot{S} = \sum \frac{\dot{M}}{\pi D_m^2} \qquad (10)$$

where D_m is the maximum distance at which the CO emission could have been detected and depends on the strength of the observed line and on the sensitivity of the observation. Fortuitously, the volume of the Galaxy sampled by an observation of given sensitivity depends on D^2 rather than D^3 (beyond a few hundred parsecs) since the Galaxy is flat, so that distance uncertainties roughly cancel out. A preliminary application of Equation (10) to CO observations of 50 stars by Knapp and Morris (1985) gave $\dot{S} = 2 - 4 \times 10^{-10}$ $M_\odot$ yr^{-1} pc^{-2} or $\dot{M}_T = 0.3$ $M_\odot$ yr^{-1}, with about half of the mass return coming from carbon stars and half from oxygen stars. These authors showed that roughly equal total contributions are made to $\dot{S}$ from stars in equal logarithmic intervals of mass loss rate.

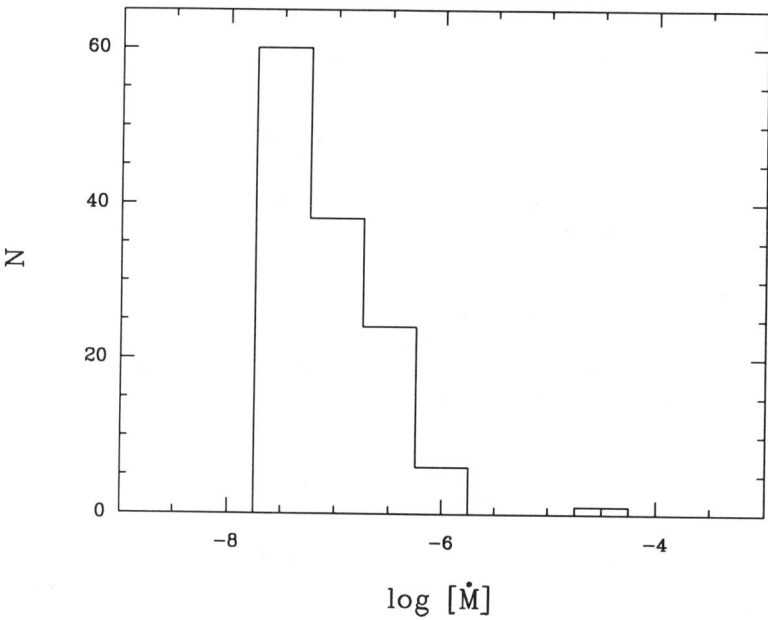

Fig. 6. Histogram of mass loss rates from IR colors for 129 oxygen stars at $|b| = 60°$ to $90°$.

CO emission has now been detected from over 200 evolved stars. Analysis of this sample suggests that $\dot{M}_T \sim 0.65$ $M_\odot$ yr^{-1}, with about 50% of the returned mass from oxygen and 50% from carbon stars. Thronson et al. (1987) use IRAS colors to find $\dot{M}_T \sim 0.4$ $M_\odot$ yr^{-1} with only 10% from carbon stars. The difference between this result and that above is that Thronson et al. find roughly similar values for the mean mass loss rate per oxygen star as per carbon star, but carbon stars comprise less than 10% of their sample. The CO observations, on the other hand, suggest that the average carbon star loses mass at a rate higher than the average oxygen star (cf. Figure 7). This discrepancy remains to be understood.

Jura and Kleinmann (1989) also estimate the total local rate of mass return to the ISM using IR flux densities (and Equation (7)) from a distance-limited sample of stars defined by the bolometric fluxes, and find $\dot{S} = 3-6 \times 10^{-10}$ $M_\odot$ yr^{-1} pc^{-2}, ($\dot{M}_T \sim 0.27$ to 0.54 $M_\odot$ yr^{-1}), about half from oxygen and half from carbon stars. Knapp et al. (1989) select a distance limited sample from the IRAS catalog by choosing stars near the galactic poles; the sensitivity of the IRAS observations allows the detection of an evolved star to a distance of about 1 kpc, so that the finite thickness of the galactic disk ensures completeness of the sample. The distribution of mass loss rates in this sample is shown in Figure 6.

Correction of M_p (the total mass loss rate) to $\dot{S}$ requires knowing the scale height of the sample; if the vertical distribution of the stars is described by

$$N(z) = N_o \text{sech}^2(z/z_o) \tag{11}$$

then

$$\dot{S}/\dot{M}_p = \frac{12}{\pi^3(z_o \tan(30°))^2} \tag{12}$$

For an isothermal distribution of vertical velocity dispersion σ_w, z_o is given by

$$z_o = \frac{\sigma_w}{(2\pi G \rho_o)^{1/2}} \tag{13}$$

where ρ_o is the local midplane mass density. The stellar radial velocities measured by the CO line give $\sigma_w = 20$ km s^{-1} (suggesting that the progenitor masses of the bulk of mass losing AGB stars are 1 - 2 $M_\odot$). This sample gives $\dot{S} \sim 6 \times 10^{-10}$ $M_\odot$pc^{-2} yr^{-1} or $\dot{M}_T = 0.55$ $M_\odot$ yr^{-1}, of which about 20 % comes from carbon stars.

There remains a lot of work to be done to improve the reliability of these estimates. But overall the observations suggest that the red giant and AGB stars are returning about 0.5 $M_\odot$ yr^{-1} to the Galaxy. The total mass of the interstellar medium is about 5×10^9 $M_\odot$; thus the above value for $\dot{M}_T$ suggests that the dust replenishment time is about once per Hubble time.

PLANETARY NEBULAE AND WHITE DWARFS

We now turn to the post-AGB phases of evolution, the planetary nebulae (PN) and white dwarfs. PN are bright emission line objects and are easily detectable to fairly large distances. The ionized gas is observed to be expanding away from the central stars, which are also cooling, and PN thus have a finite lifetime. The estimation of the total mass injection rate from planetaries is therefore made as follows: (1) the distances to as large a number of planetaries as possible are measured: (2) using these distances, a volume limited sample is defined: (3) the mass injection rate is then the number density of PN times the typical PN mass, divided by the observable lifetime.

Distances to PN, as for red giants, are very uncertain. There are several different methods for measuring these distances, which include (1) assumption of a constant total mass (2) assumption of a constant radius (3) statistical parallaxes (4) measurement of the extinction towards the nebula and the use of an extinction per unit distance law for the ISM (5) measurement of the radial expansion speed from optical emission line shapes and the angular expansion rate (most readily done at radio wavelengths - Masson 1986) and (6) observations of HI absorption to derive kinematic distances. These various distance estimators are reviewed by Pottasch (1984) and Lutz (1989).

Local complete samples have been compiled by Maciel (1981a,b), Pottasch (1984) and Phillips (1989); these authors find a local density of 50 kpc^{-3} or a column density of 25 kpc^{-2}. The typical mass of a PN is about 0.3 M$_\odot$ (though observed values range from 0.01 to 1 M$_\odot$) and the time scale for the nebula to be observable (before its surface brightness decreases below the detection threshold) is about 3×10^4 years. The local rate of mass injection from PN is then $\dot{S} = 2 - 4 \times 10^{-10}$ M$_\odot$ pc^{-2}, giving $\dot{M}_T = 0.2$ to 0.4 M$_\odot$ yr^{-1}, comparable to the rate derived for PN precursors.

Finally, we turn to the birthrate of white dwarfs. This can be estimated from the local space density of white dwarfs and the computed cooling time (Schönberner 1981). A census of local white dwarfs and their cooling times gives a birthrate of $1 - 2 \times 10^{-12}$ pc^{-2} yr^{-1} (Köster 1978; Weidemann 1978; Wesemael 1981). Assuming a typical white dwarf mass of 0.6 M$_\odot$, a typical progenitor mass of 1 M$_\odot$ and a scale height like that of the AGB stars, we find that this formation rate corresponds to a local mass injection rate of $\dot{S} = 3 - 5 \times 10^{-10}$ M$_\odot$ pc^{-2} yr^{-1}, or $\dot{M}_T = 0.3$ M$_\odot$, in good agreement with the above estimates.

CONCLUSIONS

Various methods for accounting for the mass injection rate from evolving red giants into the ISM give values of 0.3 to 0.6 M$_\odot$ yr^{-1} Galaxy wide, or 3 to 7×10^{-10} M$_\odot$ yr^{-1} pc^{-2} locally. It is instructive to compare this with the expected rate for mass return from stellar evolution considerations, $\dot{E} = R\psi$, where R is the returned fraction for a given IMF and ψ is the star formation rate. Miller and Scalo (1979) derive $\psi = 3$ to 7×10^{-9} M$_\odot$ yr^{-1} pc^{-2}. A comparison of the observed star formation and mass injection rates suggests that a large fraction of the returned mass is ejected while the stars are red giants, and also shows encouraging agreement among the expectations of stellar evolution theory, the local form of the IMF and the assumed (approximately) constant local star formation rate.

ACKNOWLEDGEMENTS

This work is supported by NASA under the IRAS Extended Mission Project (contract Number 957693) and by the National Science Foundation via grant AST87-02945 to Princeton University.

REFERENCES

Bieging, J.H. 1989, this symposium.
Bowers, P.F., Johnston, K.J. and Spencer, J.H. 1983, *Ap.J.* **274**, 733.
Bowers, P.F., and Knapp, G.R. 1988, *Ap.J.* **332**, 299.
Chevalier, R.P. 1989, this symposium.
Chu, Y.-H. 1989, in I.A.U. Symposium 131, 'Planetary Nebulae', ed. S. Torres-Piembert, Kluwer Scientific, p105.
Davis, L.E., Seaquist, E.R. and Purton, C.R. 1979, *Ap.J.* **230**, 434.
Diamond, P.J., Norris, R.P., Rowland, P.R., Booth, R.S. and Nyman, L.-Å. 1985, *M.N.R.A.S.* **212**, 1.
Engels, D. 1979, *Astron. Astrophys. Suppl.* **36**, 337.
Faber, S.M. and Gallagher, J.S. 1976, *Ap.J.* **204**, 365.
Gammie, C.F., Knapp, G.R., Young, K., Phillips, T.G. and Falgarone, E. 1989, *Ap.J. (Letters)* (in press).
Gehrz, R.D., and Woolf, N.J. 1971, *Ap.J.* **165**, 285.
Gezari, D.Y., Schmitz, M. and Mead, J.M. 1987, NASA Reference Publication No. 1196.
Hacking, P. *et al.* 1985, *P.A.S.P.* **97**, 616.
Hekkert, P. te Lintel, and Habing, H.J. 1989, in preparation.
Herman, J. 1983, Ph.D. Thesis, Leiden University.
Huggins, P.J., and Glassgold, A.E. 1982, *Ap.J.* **252**, 201.
Huggins, P.J., and Healy, A.P. 1989, *Ap.J.* (in press).
Iben, I. 1987, in 'Late Stages of Stellar Evolution' ed. S. Kwok and S.R. Pottasch, *Astrophys. Space Sci. Lib.* **132**, p175.
Impey, C.D., Wynn-Williams, C.G. and Becklin, E.E. 1986, *Ap. J.* **309**, 572.
IRAS Science Team 1986, *Astron. Astrophys. Suppl.* **65**, 607.
de Jager, C., Nieuwenhuijzen, H. and van der Hucht, K.A. 1988, *Astron. Astrophys. Suppl.* **72**, 259.
Jewell, P.R., Webber, J.C., and Snyder, L.E. 1981, *Ap.J.* **249**, 118.
Johnson, H.R., Beebe, R.F. and Sneden, C. 1975, *Ap.J.* **280**, 29.
Jura, M. 1987, *Ap.J.* **313**, 743.
Jura, M. 1988, in 'Millimetre and Submillimetre Astronomy' ed. R.D. Wolstencroft and W.B. Burton, *Astrophys. Space. Sci. Lib.* **147** (Kluwer Scientific), p189.
Jura, M., Kim, D.-W., Knapp, G.R. and Guhathakurta, P. 1987, *Ap.J. (Letters)* **312**, L11.
Jura, M. and Kleinmann, S.G. 1989, *Ap.J.* **341**, 359.
Kawabe, R., Ishiguro, M., Kasuga, T., Morita, K.-I., Ukita, N., Kobayashi, H., Okumura, S., Fomalont, E. and Kaifu, N. 1987, *Ap.J.* **314**, 322.
Knapp, G.R. 1985, *Ap.J.* **293**, 273.
Knapp, G.R. 1986, *Ap.J.* **311**, 731.
Knapp, G.R., and Morris, M. 1985, *Ap. J.* **292**, 640.
Knapp, G.R., Rauch, K.P. and Wilcots, E.M. 1989, in preparation.
Köster, D. 1978, *Astron. Astrophys.* **65**, 449.
Kwok, S. 1989, in I.A.U. Symposium 131 'Planetary Nebulae', ed. S. Torres-Piembert, Kluwer Scientific, p401.
Lockman, F.J. 1989, this conference.
Lutz, J.H. 1989, in 'I.A.U. Symposium 131: Planetary Nebulae' ed. S. Torres-Piembert, (Kluwer Scientific), p65.

Maciel, W. 1981a, *Astron. Astrophys.* **98**, 402.
Maciel, W. 1981b, *Astron. Astrophys. Suppl.* **44**, 123.
Mamon, G.A., Glassgold, A.E. and Huggins, P.J. 1988, *Ap.J.* **328**, 797.
Masson, C.R. 1986, *Ap.J. (Letters)* **302**, L27.
Miller, G.E., and Scalo, J.M. 1979, *Ap.J. (Suppl.)* **41**, 513.
Morris, M. 1975, *Ap.J.* **197**, 603.
Morris, M. 1980, *Ap.J.* **236**, 823.
Morris, M., Guilloteau, S., Lucas, R. and Omont, A. 1987, *Ap.J.* **321**, 888.
Morris, M., Redman, R., Reid, M.J. and Dickinson, D.F. 1979, *Ap.J.* **229**, 257.
Mufson, S.L., Lyon, J. and Marionni, P.A. 1975, *Ap. J. (Letters)* **201**, L85.
Neugebauer, G., and Leighton, R.B. 1969, NASA SP-3047.
Neugebauer, G., et al. 1984, *Ap.J. (Letters)* **278**, L1.
Netzer, N., and Knapp, G.R. 1987, *Ap.J.* **323**, 734.
Phillips, J.P. 1989, in 'I.A.U. Symposium 131: Planetary Nebulae' ed. S. Torres-Piembert, (Kluwer Scientific), p. 425.
Phillips, T.G., Young, K., Gammie, C.F. and Knapp, G.R. 1989, (in preparation).
Pottasch, S.R. 1984, 'Planetary Nebulae', *Astrophys. Space Sci. Lib.* **107** (D. Reidel Co.).
Rodriguez, L.F. 1989, this symposium.
Rowan-Robinson, M. and Harris, S. 1983a, *M.N.R.A.S.* **202**, 767.
Rowan-Robinson, M. and Harris, S. 1983b, *M.N.R.A.S.* **202**, 797.
Sahai, R. 1987, *Ap.J.* **318**, 809.
Schönberner, D. 1981, *Astron. Astrophys.* **103**, 119.
Schönberner, D. 1983, *Ap.J.* **272**, 708.
Skinner, C.J., and Whitmore, B. 1988a, *M.N.R.A.S.* **231**, 169.
Skinner, C.J., and Whitmore, B. 1988b, *M.N.R.A.S.* **234**, 79p.
Soifer, B.T., Rice, W.L., Mould, J.R., Gillet, F.C., Rowan-Robinson, M. and Habing, H.J. 1986, *Ap.J.* **304**, 651.
Spergel, D.N., Giuliani, J.L. and Knapp, G.R. 1983, *Ap.J.* **275**, 330.
Thronson, H.A., Latter, W.B., Black, J.H., Bally, J. and Hacking, P. 1987, *Ap.J.* **322**, 770.
Tinsley, B.M. 1975, *P.A.S.P.* **87**, 837.
Tinsley, B.M. 1980, *Fund. Cos. Phys.* **5**, 287.
Tsuji, T. 1976, *P.A.S. Japan* **28**, 567.
Tsuji, T., Unno, W., Kaifu, N., Izumiura, H., Ukita, N., Cho, S. and Koyama, K. 1988, *Ap.J. (Letters)* **327**, L23.
van der Veen, W.E.C.J., Habing, H.J. and Geballe, T. 1987, in 'Planetary and Protoplanetary Nebulae: from IRAS to ISO' ed. A. Preite Martinez, Kluwer Scientific, p69.
Weidemann, V. 1978, in 'I.A.U. Symposium 76: Planetary Nebulae', ed. Y. Terzian, D. Reidel Co.), p353.
Wesemael, F. 1981, *Ap.J.* **243**, 228.
Zuckerman, B., and Aller, L.H. 1986, *Ap.J.* **301**, 772.

PHOTODISSOCIATION REGIONS

DAVID J. HOLLENBACH
NASA Ames Research Center, Moffett Field, CA 94035

ABSTRACT   Photodissociation regions are interstellar regions of predominantly neutral gas where the FUV radiation field plays a significant role in the chemistry and/or the heating. Photodissociation, grain attenuation of the FUV flux and grain photoelectric heating lead to copious emission of CII(158μm), OI(63μm), SiII(35μm), CI(370,609μm), H$_2$ vibrational and CO rotational transitions, and IR continuum. Theoretical models compared with observations diagnose such physical parameters as the density and temperature structure, the elemental abundances, and the FUV radiation field. Applications are made to Orion, M17 and galactic nuclei. Theoretical photodissociation models can explain the correlation in the CII(158μm) and CO J=1-0 emission and the correlation of the CO J=1-0 luminosity with the molecular mass. Theoretical models also point to feedback mechanisms which may control the rate of star formation in galaxies and which may regulate the column density through giant molecular clouds.

INTRODUCTION

FUV(6eV-13.6eV) radiation plays an essential role in determining the structure and evolution of the interstellar medium and in affecting the process of star formation in galaxies. About half the bolometric luminosity of O and B stars is emitted in this band, and its interaction with the neutral gas surrounding these young stars illuminates the environment of massive star formation, and reveals the physical conditions including the density, temperature, clumping, morphology and gas phase elemental abundances in these regions.

Photodissociation regions (PDRs) occur where the FUV radiation dominates either the heating or a significant component of the chemical composition of the gas. Recent reviews on PDRs include Genzel *et al.* (1989) and Jaffe and Howe (1989). Figure 1, from Tielens and Hollenbach (1985a), schematically shows a 1D representation of a PDR. FUV flux from, for example, nearby hot stars or the ambient interstellar radiation field (ISRF) is incident from the left on a predominantly neutral cloud. Photons more energetic than 13.6eV are absorbed in HII plasma and at the HII/HI interface of the cloud. By the above definition, the PDR begins after that relatively sharp interface, whose thickness measured in units of extinction is $\Delta A_v < 0.1$. Typically, the FUV flux is sufficient to photodissociate molecules and to ionize those atoms

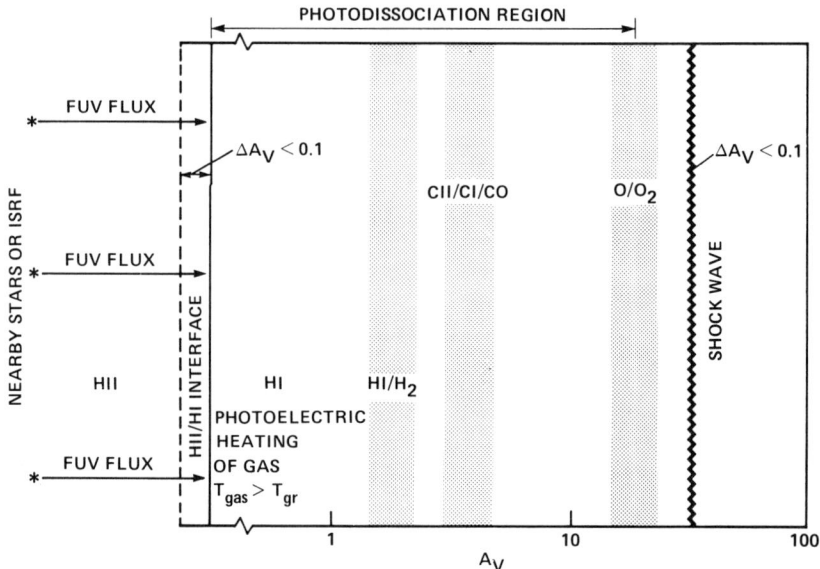

Figure 1. Schematic of a Photodissociation Region (PDR)

(e.g., C, S, Fe and Si) with ionization potentials less than 13.6eV to $A_v \gtrsim 1\text{-}2$, where significant dust attenuation of the FUV is obtained. Since any shock waves present have column densities of heated gas corresponding to $\Delta A_v < 0.1$ (McKee and Hollenbach 1980), the PDR region, by virtue of its higher column density of warm gas, has the potential to dominate the emission from the fine structure transitions of atoms and single ions. The hydrogen becomes molecular closest to the surface, typically at $A_v \sim 1-2$, and the C^+ typically becomes CO at $A_v \sim 2-4$. Because of their lower ionization potentials and the greater dust-penetrating power of less energetic photons, silicon and other trace species are kept ionized to $A_v \sim 3-5$. Any oxygen not combined in CO is kept photodissociated to $A_v \sim 5-10$, a result due in part to the slow formation rate of O_2 and the low photodissociation threshold. Of course, these approximate depths depend on the incident FUV flux G_o, but for high G_o the dependence is only logarithmic because of the exponential attenuation by dust.

Since interstellar FUV significantly affects the oxygen chemistry to $A_v \sim 5-10$, PDRs extend to these depths and therefore include regions where the hydrogen is H_2 and the carbon is locked in CO. Consequently, the emission from a PDR not only includes the grain IR continuum and the atomic and ionic fine structure lines, but also vibrational and rotational emission from H_2 and rotational transitions of CO. To be more precise, at $A_v \lesssim 1\text{-}2$ gas is heated by the grain photoelectric heating mechanism (to be discussed below) to temperatures $\sim 10^2 - 10^3$ K, significantly warmer than the grain temperatures $\sim 15-75$ K. Nevertheless, the emission from this region is largely FIR grain continuum, but with substantial production of CII(158μm), OI(63μm), and SiII(35μm), carbon recombination radiation (both radio and CI(9850)), and rovibrational transitions of H_2. Somewhat deeper into the cloud, in the region of the C^+ to CO transition, significant emission in the CI(370,309μm) and the low J ^{12}CO lines originates.

Theoretical models of PDRs have been constructed over the past decade by a number of research groups. Glassgold and Langer (1974, 1975, 1976), Black and Dalgarno (1977), Clavel et al. (1978), de Jong et al. (1980) and van Dishoeck and Black (1987) generally studied clouds with the ISRF ($G_o \sim 1$ since we express G_o in units of a "Habing"(1968) flux of 1.6×10^{-3} erg cm^{-2} s^{-1}) incident on neutral gas of hydrogen nucleus density $n \lesssim 10^{3-4}$ cm^{-3}. Black and van Dishoeck (1987) and Sternberg (1988) have done detailed analysis of the expected H$_2$ fluorescence spectra from these PDRs. Sternberg and Dalgarno (1989) have done a very thorough study of the structure and emission from PDRs over a wide range of G_o and n parameter space. However, this review is based primarily on our own studies of PDRs with fluxes $G_o = 1 - 10^6$ incident on gas with density $n = 10^2 - 10^7$ cm^{-3} (Tielens and Hollenbach 1985a,b,c; Wolfire et al. 1989a,b, 1990; Burton et al. 1989a,b; Hollenbach et al. 1989)

Theoretical models of PDRs are essential in understanding the interstellar medium (ISM) since most of the mass of the ISM resides in PDRs. By definition, all of the atomic gas in the ISM is PDR, but, in addition, the vast majority of molecular gas is PDR as well. This is true because the FUV from the ISRF significantly affects the chemistry to $A_v \sim 4$ from the surface of molecular clouds, and the typical half-thickness through a molecular cloud in the Galaxy is ~ 4 (Solomon et al. 1987). We shall discuss a possible explanation for this seeming coincidence in a later section.

PDRs are also important to understand since most of the grain radiation (IR emission) from the Galaxy (and, presumably, external galaxies) originates in PDRs. In addition, they are the origin of most of the CII(158μm), OI(63,145 μm), SiII(35μm), CI(370,609μm) and rotational ^{12}CO emission. The models also show that these FUV-excited regions can produce copious H$_2$ emission which sometimes mimics the spectrum from shock waves.

PDR models can be compared with this observed emission to derive the physical conditions in diverse regions such as the neutral gas associated with HII regions, reflection nebulae, planetary nebulae, the Galactic center, and the central (~ 1 kpc) regions of external galaxies. On a global scale, the effect of FUV on the neutral gas in a Galaxy may regulate the rate of star formation and the column density (A_v) through typical Giant Molecular Clouds (GMCs).

This paper will briefly review the physical processes incorporated into our theoretical models of PDRs and will then make several applications of these models to regions in our Galaxy and in external galaxies.

PHYSICAL PROCESSES IN THEORETICAL MODELS

Our theoretical models of PDRs (Tielens and Hollenbach 1985a, Burton et al. 1989b, Hollenbach et al. 1989) are 1D steady state models that self-consistently include chemistry, thermal balance and radiative transfer. The chemical matrix includes 165 reactions involving 41 species, mainly atoms, single ions and simple molecules. The FUV photodissociation keeps the abundances of most complex molecules very low. This chemical network has been checked for several of the PDR models against a much larger network taken from Prasad and Huntress (1980); the tight correspondence in these results indicates that the dominant reactions have been included in the smaller network. Perhaps the most important ingredient in the chemistry is the proper treatment of the FUV self-shielding of H$_2$ and CO. The PDR model has recently been updated (Burton et al. 1989) to include the new CO self-shielding rates given by van Dishoeck and Black (1988). As shown in Burton et al. (1989), the relative importance of self-shielding to grain shielding of the FUV is determined by the ratio G_o/n. Grains control the shielding and H_2 and CO become dominant at $A_v \gtrsim 1$-2 when $G_o/n \gtrsim 10^{-2}$ cm^3. H_2 self-shielding dominates and H_2 and CO form at $A_v \lesssim 1$ when $G_o/n \lesssim 10^{-2}$ cm^3. The new CO dissociation rates decrease the importance of CO self-shielding, which only becomes important at very low values of G_o/n seldom found in the ISM.

The thermal balance in PDRs is determined by equating the gas heating by photoelectric ejection of "hot" electrons from grains, collisional deexcitation of FUV-pumped H_2, C photoionization, H_2 formation and dissociation, cosmic rays, collisions with warm grains, and collisional deexcitation of IR-pumped OI to the gas cooling by fine structure emission of atoms and ions (mainly CII(158μm), OI(63μm) and SiII(35μm)), rotational transitions of molecules (mainly CO), and collisions with cooler grains. The first two mechanisms are sufficiently important to warrant further discussion.

<u>Grain Photoelectric Heating</u>.

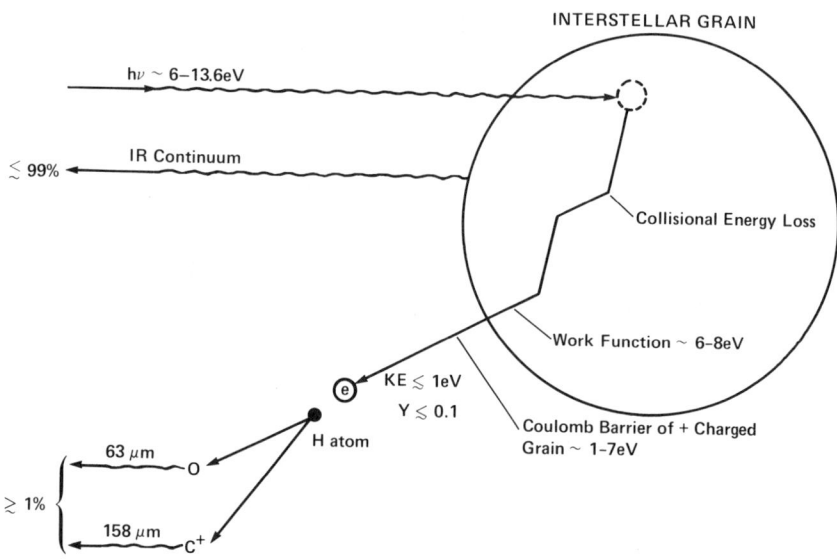

Figure 2. Grain photoelectric heating mechanism.

Grain photoelectric heating in PDRs is initiated by the grain absorption of FUV photons (see Figure 2). Electrons are freed inside the grain and roughly one in ten (i.e., a yield of $Y \sim 0.1$) is ejected from the grain. The escaping electron loses kinetic energy ($\lesssim$ 1eV) to collisions in the grain, to overcoming the work function of the grain material ($\sim 6-10$eV), and to the Coulomb potential of the grain (the photoelectric emission naturally leads to positively-charged grains with potentials of order 1-7 V). Thus, the kinetic energy of the escaping electron is of order $\lesssim 1$ eV, and this is delivered to the gas as heat via elastic collisions. A heating efficiency ϵ is defined as the ratio of the heat delivered to the gas to the total FUV energy absorbed by the grain. Assuming a typical FUV photon energy of about 10 eV and a yield $Y = 0.1$, the typical heating efficiency is therefore $\epsilon \lesssim (1\text{eV}/10\text{eV})\, Y \lesssim 10^{-2}$. The efficiency decreases as the positive charge on the grain increases. The charge is controlled by the photoelectric ejection rate and the eletron recombination rate onto grains, and depends on the ratio G_o/n_e, where n_e is the electron density. In PDRs $G_o/n_e = G_o/(x_C n)$, where x_C is the gas phase abundance of carbon and we have used the fact that most of the carbon is C^+ in the photoelectric heated region. For typical values of $G_o/n \lesssim 1$ cm^3 in interstellar clouds, $\epsilon \sim 10^{-2}-10^{-3}$. Approximately 0.1-1% of the FUV flux is converted to gas heating and emerges in the dominant gas cooling transitions. The other 99-99.9% of the FUV flux is converted to grain heating and emerges as IR continuum. Therefore, when observing neutral, opaque, interstellar clouds which are illuminated by radiation fields in which the FUV is dominant, we expect the luminosity in the OI(63μm) and CII(158μm) lines, $L_{CII} + L_{OI}$, to be about $10^{-3} - 10^{-2}$ of the luminosity of the grain continuum, L_{IR}. Figure 3 is a plot of this correlation taken from Hollenbach (1989) which shows some of the current observational data on reflection nebulae (lower left), HII regions and the central few parsecs of the Galaxy (middle), and starburst galaxies (upper right). The expected correlation for grain photoelectric heating is seen over 7 orders of magnitude, which demonstrates the global importance of this mechanism and of the PDR "phase".

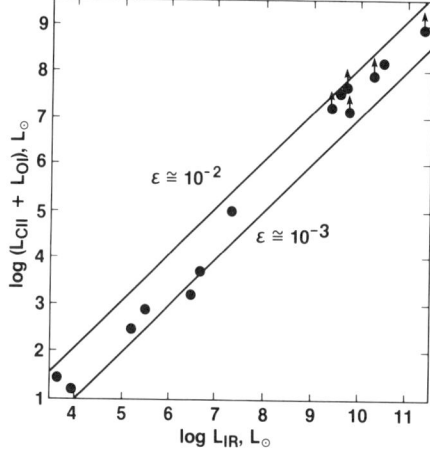

Figure 3. Correlation of $L_{CII} + L_{OI}$ with L_{IR}

<u>FUV Pumped H$_2$ Heating</u>
Molecular hydrogen absorbs photons in the range 11eV-13.6eV and becomes electronically excited; the subsequent decay back to the ground electronic state leads to the vibrational continuum (dissociation) 10% of the time (this is the H$_2$ photodissociation mechanism in PDRs) and to bound but vibrationally-excited states (typically $\sim$ 2eV above the ground state) 90% of the time. If the atomic hydrogen density is low, $n < n_{cr} \sim 6 \times 10^5 T^{-0.5} e^{(400/T)^2}$ (see Hollenbach 1988, but note that the critical density is uncertain because of uncertainties in the collisional deexcitation rates), the vibrationally excited H$_2$ *radiatively* cascades to the ground state, fluorescently emitting IR quanta. The resultant spectra have been investigated most recently by Black and van Dishoeck (1987)

and Sternberg (1988). One prime characteristic of this fluorescence is a 2-1S(1)/1-0S(1) ratio of ~ 0.5. However, if $n > n_{cr}$, then collisional deexcitation by H atoms proceeds prior to radiative decay and much of the vibrotational energy is transformed to translational energy (heat). The collisions greatly modify the spectra, pushing it towards LTE, and, for example, the 2-1S(1)/1-0S(1) ratio becomes << 0.5. The spectra in this case have been discussed by Hollenbach (1988), Sternberg (1988, 1989), Sternberg and Dalgarno (1989), and Burton et al. (1989a,b).

APPLICATIONS

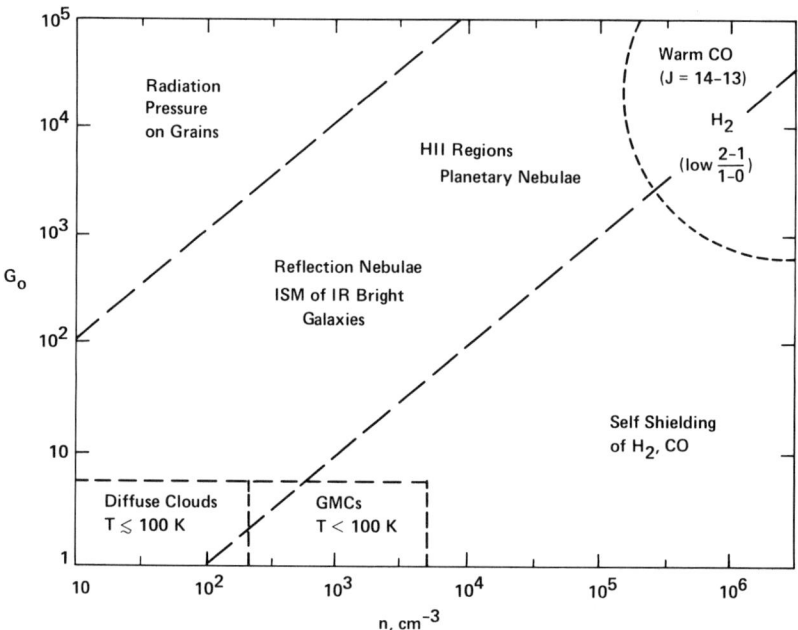

Figure 4. The n, G_o parameter space of PDRs

Figure 4 shows the parameter space G_o, n important in interstellar clouds. The parameter space breaks into three regions separated by the long dashed lines; radiation pressure on grains drives them supersonically through the gas in the upper left region with $G_o/n \gtrsim 10$ cm^3, the middle region includes those PDRs where grain attenuation of the FUV dominates over self-shielding, and the lower left region ($G_o/n \lesssim 10^{-2}$) is where self-shielding by H$_2$ brings the HI/H$_2$ and C$^+$/CO transitions to $A_v < 1$. The neutral PDR gas associated with various astrophysical phenomena are labelled on Figure 4. The upper right hand region of high density and G_o (short dashed line) is the range of parameter space where strong heating, self-shielding, and collisional deexcitation of H$_2$ combine to give intense emission from relatively high J CO (e.g., J=14-13) and low ratios of H$_2$ 2-1S(1)/1-0S(1).

HII Regions: Orion

Figure 5 shows an overlay of the ^{12}CO J=1-0 integrated brightness contours (dashed lines, Schloerb and Loren 1982) with the integrated CII(158μm) intensity contours (solid lines, Stacey *et al.* 1989) in the PDR region produced by the Trapezium stars shining on the Orion A molecular cloud. In Figure 5 Θ^2 A is the star to the lower left, Θ^1 C is the star in the center, and BN-KL is the box. The main point of the figure is to show the large extent of the PDR and the general spatial correlation of the ^{12}CO J=1-0 with the CII(158μm) over the 1-2 pc region. There are, however, some differences in detail due to the greater sensitivity of CII to geometrical and temperature effects.

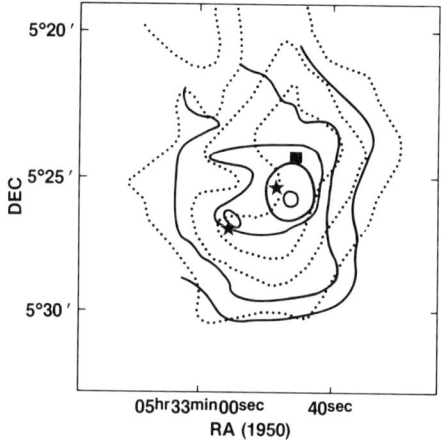

Figure 5. ^{12}CO J=1-0 (dotted contours) vs. CII(158μm) (solid contours) in Orion.

The PDR theoretical model of Tielens and Hollenbach (1985b) has been applied the the observed emission behind Θ^1 C and an excellent fit to the OI(63,145μm), CII(158(μm), CI(370,609μm), SiII(35μm), CI(9850), H$_2$ 1-0S(1), CO J=(1-0,2-1,3-2,7-6), and CII radio recombination lines has been obtained (differences between theory and observation $\sim$ 20%). This fit is extraordinary when one considers the paucity of free parameters in the theoretical model. The main parameters varied in the PDR modeling are n and G_o. However, in the case of Orion, where the PDR is seen face-on, G_o is tightly constrained to a value $\sim 10^5$ by the observations of the FIR continuum flux. Since the pressure in the PDR is probably equal to the pressure in the neighboring HII region and the electron densities there are of order 10^3 to 10^4 cm^{-3}, we would expect $n \sim 10^5$ cm^{-3}. Therefore, it is very satisfying that this best PDR fit to the line data is with $G_o = 10^5$ and $n = 10^5$ cm^{-3}. The standard model with these parameters produced, however, somewhat too little CII and CI emission. The best fit was obtained with a relatively high gas phase carbon abundance $x_C = 3 \times 10^{-4}$ and with a column of PDR gas along the line of sight about 2 times the column from a standard face-on PDR. The latter effect could be caused by geometrical effects, such as a corrugated PDR surface or a clumpy PDR, or by having non-standard grain properties that allow FUV to penetrate a factor of 2 further. However, it is important to stress that within the context of PDR modeling, a fit to the CII(158μm) emission naturally produces a fit to the CI(370,609μm) emission in Orion. We have argued (cf., Tielens and Hollenbach 1985c, Hollenbach et al. 1989) that all the observed CI fine structure emission originates in FUV-illuminated PDRs. A good summary of the origin of CI is given in Genzel et al. (1989).

The observed SiII(35μm) emission is fit with a gas phase silicon abundance of 3×10^{-6}, or ~ 0.1 solar (Haas et al. 1986). This result is another interesting aspect of the PDR modeling, since it suggests that the gas phase silicon abundance in this dense cloud is similar to that observed in diffuse clouds, and is not more heavily depleted due to accretion onto grains as one might have expected.

The model also predicts that for $A_v \lesssim 2$, the gas temperature is of order 500 - 1000 K, compared to the grain temperature of 75 - 100 K. Photoelectric heating has raised the gas temperature substantially above the grain temperature.

The PDR model can also be applied to the Orion Bar, where Hayashi et al. (1985) has noted that the H_2 2-1S(1)/1-0S(1) ratio varies from 0.6 on the Bar to 0.3 into the molecular cloud. Hayashi et al. suggest a combination of fluorescence (Bar) and shocks to explain this variation. Burton et al. (1989) propose that all the emission arises from PDRs, but that the lower ratio arises because of the presence of high density clumps in which collisional deexcitation of H_2 lowers the 2-1S(1)/1-0S(1) ratio. Figure 6 shows results from Burton et al. which demonstrates the variation of the 2-1S(1)/1-0S(1) ratio as a function of the PDR gas density. Above a density of $\sim 10^5$ cm^{-3} (again, recall that this density is uncertain because of uncertainties in the collisional deexcitation rates), the H_2 emission transitions from a "fluorescence" spectrum to a "thermal" spectrum.

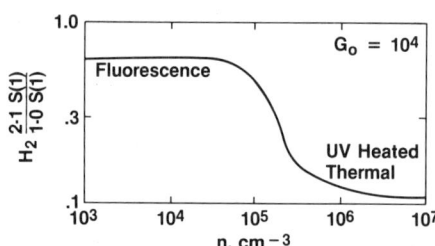

Figure 6. The H_2 2-1S(1)/1-0S(1) ratio vs n in PDRs with $G_o = 10^4$.

HII Regions: M17 and Clumping

Felli et al. (1984) have made a high spatial resolution VLA map of the M17 HII region and have uncovered numerous dense clumps inside the HII region. Stutzki et al. (1988), Genzel et al. (1989), and Jaffe and Howe (1989) have discussed some evidence for clumping in the PDR to the southwest of the M17 HII region. Snell et al. (1986), Evans et al. (1987) and Massi et al. (1988) have shown that the molecular gas just outside the PDR is clumped with densities reaching $\sim (3-10) \times 10^6$ cm^{-3}. We shall only briefly summarize the situation and make a few comments on the clumping in the PDR. Figure 7 (modified

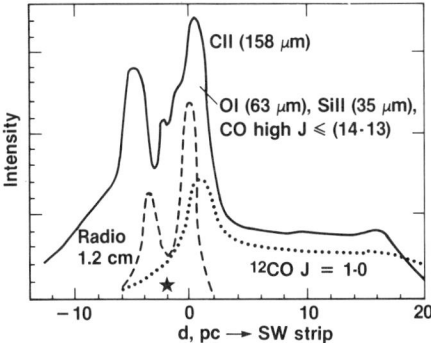

Figure 7. Strip maps of M17, an edge-on PDR.

from Stutzki et al. 1988) shows how the flux in CII(158µm) and ^{12}CO J=1-0 varies as a function of position along a strip map made from the northeast to the southwest through the infrared peak in M17. The radio free-free emission at 1.2 cm is also plotted, to demonstrate the relative positions of the HII, PDR and molecular gas in this edge-on geometry. The position of the exciting star is noted by the star in the figure. We have also indicated the general area where OI(63µm), SiII(35µm) (Haas et al. 1990) and CO high J ($\lesssim$ 14-13, Stutzki et al. 1988) emission originates. The basic argument for clumps or a dense core is that densities $\lesssim 10^3$ cm^{-3} are indicated in order for the CII peak to extend $\sim$ 2 pc into the cloud to the southwest of the HII gas. On the other hand, the OI, SiII and high J CO emission requires much higher densities in the PDR model. The CO J=14-13 intensity requires $n \gtrsim 10^6$ cm^{-3} in the context of PDR modeling (Burton et al. 1988a,b). Genzel et al. (1989) argue that the density in the PDR clumps cannot exceed $\sim 3 \times 10^4$ cm^{-3}, because the observed H$_2$ 2-1S(1)/1-0S(1) ratio has a "fluorescent" value $\sim$ 0.5 (Tanaka et al. 1989). A confirmation of the H$_2$ observations is suggested to settle this point, especially a map of the region around the CO J=14-13 peak.

The Correlation of CII(158µm) and ^{12}CO J=1-0

Crawford et al. (1985) plotted observations of CII(158µm) and ^{12}CO J=1-0 and noted that here was a tight linear correlation in the integrated intensities, I_{CII} and I_{CO}, of these lines when observing sources with relatively high G_o incident upon molecular gas. The observed linear correlation is plotted as the bold solid and dashed line on Figure 8. Wolfire et al. (1989a) show how this correlation can be understood in the context of PDR modeling. As discussed earlier, the PDR model assumes that, in molecular clouds illuminated by external FUV fluxes, the CII(158µm) emission arises from the warm, $A_v \lesssim 1$-2, outer regions whereas the ^{12}CO J=1-0 originates from the cooler gas somewhat deeper into the cloud. Figure 8, from Wolfire et al. , shows the result of theoretical PDR models for a range of n and G_o. The light solid lines connect runs of constant density, with the density in cm^{-3} given at the bottom. The points along these curves denoted with boxes, circles, etc denote factor of ten increases in G_o going up the curves. We have labelled with "2" the point $G_o = 10^2$. Note that for high G_o, the theoretical points cluster on the observed correlation line. We further note that the theoretical points are calculated assuming a beam filling factor of unity; variations in the filling factor would slide points parallel to the observed correlation line (i.e., points on the line would stay on the line).

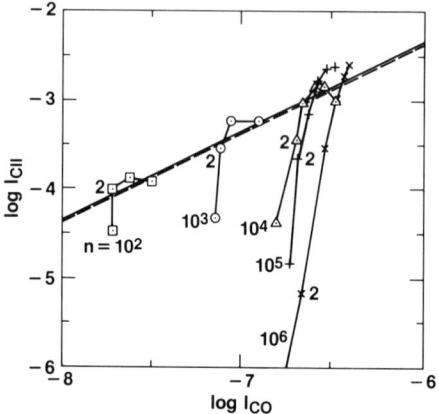

Figure 8. PDR modeling of the observed correlation of CII(158μm) with ^{12}CO J=1-0 for regions with high FUV fields.

The theoretical results can be best understood by first noting that for high densities, $n \gtrsim 3 \times 10^3$ cm^{-3}, and high incident FUV fluxes, $G_o \gtrsim 3 \times 10^3$, I_{CII} and I_{CO} are relatively constant. The CII intensity is constant because at high G_o the temperature of the gas exceeds the excitation energy, $\Delta E/k = 92$ K, of the CII(158μm) transition. In addition, the high density ensures LTE. With $T > \Delta E/k$, the CII(158μm) intensity is not temperature (G_o) sensitive. With $n > n_{cr}$ or LTE, the CII(158μm) intensity is not density sensitive, but rather only sensitive to the column density of C$^+$. At high G_o, the column of C$^+$ is roughly constant, varying only logarithmically with G_o because of dust attenuation. Therefore, I_{CII} is roughly constant at high n and G_o. The CO J=1-0 flux is constant because it is an optically thick transition which, at high densities or LTE, is mainly sensitive to the temperature of the emitting gas. The ^{12}CO J=1-0 transition originates from a deeper, cooler region where the gas temperature varies very slowly with changes in G_o. CII(158μm) and CO J=1-0 have similar critical densities so that the linear correlation is maintained at low densities, as the intensities drop because of non-LTE effects (see Figure 8).

The ISM of IR-Bright Galaxies

Wolfire et al. (1989b) have shown how theoretical PDR models can be compared with infrared and ^{12}CO observations to derive numerous interesting average physical parameters which describe the ISM in the central ~ 1 kpc of relatively nearby, IR-bright galaxies. The required observations for this modeling procedure are the CII(158μm), OI(63μm), SiII(35μm), and ^{12}CO J=1-0 lines and the IR continuum. OI(145μm) is optional, though desirable. Typically, the beam size for these observations is of order 1', which corresponds to 1.4 kpc at a distance of 5 Mpc. The model assumes that there exist N molecular clouds in the beam, all having characteristic radius R and gas density n and illuminated by an FUV flux G_o. The above observations compared with the PDR modeling can not only determine these four parameters, but also the mass M_a of warm atomic PDR gas in the beam, the average temperature T_a of this gas, the mass M_m of molecular gas, the average gas phase abundance of silicon x_{Si}, and the area and volume filling factors of these clouds, ϕ_a and ϕ_v. The reason that so many parameters can be derived from so few observations is that they are not independent.

The prescription for deriving the parameters, detailed in Wolfire et al. (1989b), proceeds as follows. Certain intensity ratios, e.g CII/OI and (CII+OI)/IR continuum, provide n and G_o. Given these parameters, the PDR model predicts T_a. The observed absolute CII intensity compared with the model CII intensity gives ϕ_a. The luminosity in CII, coupled with a knowledge of n and T_a, provides M_a. The luminosity in CO determines M_m, using the observed correlation in our Galaxy (an exception is M82, a galaxy with a peculiar CO J=2-1/J=1-0 ratio which indicates that the J=1-0 transition is thin). M_a, M_m, n and the size of the beam diameter give ϕ_v. Finally, ϕ_v, ϕ_a, and n give R and N.

Results have been obtained for M82 by Wolfire et al. (1989b) (see also Crawford et al. 1985 for a similar derivation of parameters) and for Arp 299, IC342, and NGC 6946 (in preliminary form) by Lord et al. (1990). The average gas densities obtained range from 3×10^3 cm^{-3} in Arp 299 (where we sample a bigger volume) to 5×10^4 cm^{-3} in M82. The incident FUV flux ranges from $G_o = 10^2$ in IC342 to 8×10^3 in M82 (recall these are very bright central regions with starburst activity). The atomic (CII) temperatures range from 100 K in IC342 to 400 K in M82. The warm atomic component is very significant, accounting for 1% of the gas mass in NGC 6946, 10% in M82, 30% in IC342 and 40% in Arp 299. The gas phase silicon abundances are high. For example, $x_{Si} \sim 2 \times 10^{-5}$ (nearly solar) in M82. This may result from the starburst activity which nucleosynthesizes additional silicon and which produces shock waves which vaporize interstellar grains and raise the *gas phase* abundances of refractories like silicon. The derived number of clouds and their sizes are surprising; there are numerous small clouds present. N ranges from 10^5 in M82 to 6×10^8 in Arp 299 and the sizes range from $R= 0.1$ pc in IC 342 to $R = 1$ pc in NGC 6946. These "clouds" are individual entities in the sense that they cannot shadow each other from the FUV flux. Nevertheless, they may be clustered together in GMC-like configurations or in sheets or filaments.

The Correlation of ^{12}CO J=1-0 with Molecular Mass

There has been great discussion over the last decade concerning the correlation of the luminosity of ^{12}CO J=1-0 with the molecular mass of a cloud (recent observational data and discussion is in Solomon et al. 1987). Since we believe that the observed ^{12}CO J=1-0 originates from FUV heated PDRs on the "surfaces" of opaque molecular clouds, we have applied the PDR models to try to understand this correlation (Wolfire et al. 1990). The observations indicate that there is roughly a constant column density, or A_v, through molecular clouds ($A_v \sim 7.5$, see Solomon et al. 1987 or McKee 1989). This means that for a cloud of given mass, the density and radius are specified. Wolfire et al. (1990) have subjected these clouds to an external FUV flux G_o (~ 1) and used the PDR code (which self-consistently calculates the temperature and CO abundance as a function of position in each cloud) to predict the ^{12}CO J=1-0 luminosity and line profiles. The preliminary results show that the theoretical models do match the observed correlation and that the results are insensitive to G_o

in the range expected for typical molecular clouds in the ISRF of the Galaxy ($G_o \sim 1-30$). However, clouds with $G_o=0$ (cosmic ray heated) do not produce the observed brightness temperatures in ^{12}CO J=1-0. The most interesting result, already suggested by Solomon *et al.* (1987), is that the line profiles in the standard models, which assume microturbulence, do not match the observed peaked profiles. The model profiles are flat-topped or self-absorbed. However, a macroturbulent model (i.e., clumps with a velocity spread and which are thin to each other) can fit the observed profiles. Hence, the observed ^{12}CO J=1-0 line profiles may provide yet further evidence for the clumping of the PDR regions.

Low Mass Star Formation and Constancy of A_v in Molecular Clouds

McKee (1989) proposes an explanation for the observed constancy of A_v (~ 7.5) in molecular clouds based on a PDR model. The basic idea derives from the assumption that the rate of low-mass star formation is governed by ambipolar diffusion (cf., Shu, Adams and Lizano 1987) and that newly-formed stars inject energy into the cloud which supports the cloud against gravitational collapse. Consider then a gravitationally-bound molecular cloud of given mass and suppose that $A_v \ll 7.5$ through the cloud. The ISRF ($G_o \sim 1$) incident on the cloud would then maintain a high ionization fraction through the cloud; the high ionization fraction means a low ambipolar diffusion rate; the low ambipolar diffusion rate means a low star formation rate; the cloud has therefore little support and it collapses, raising A_v. On the other hand, if the cloud began with $A_v \gg 7.5$, then the external FUV is completely shielded from the interior; the interior ionization fraction is low; ambipolar diffusion and star formation proceed rapidly; and the rapid injection of energy expands the cloud and lowers A_v. Using standard physical parameters, McKee shows that equilibrium is achieved when $A_v \sim 7.5$, and that therefore the external FUV flux regulates the cloud column density and the low-mass star formation rate.

Regulation of High Mass Star Formation

Parravano (1988) proposes that a feedback mechanism exists in galaxies whereby the rate of high-mass star formation is regulated by the interaction of the FUV with the neutral gas. He notes that non-gravitationally bound neutral gas may exist in two phases in a galaxy (cf., Field *et al.* 1968). Figure 9 shows

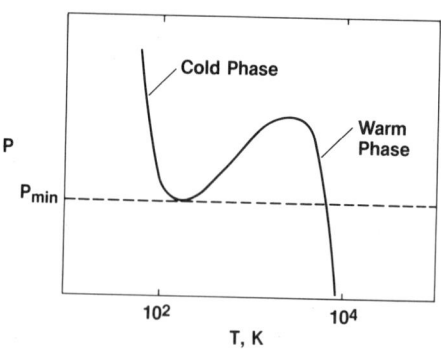

Figure 9. Schematic two phase diagram of a galactic ISM.

a schematic representation of a curve plotting the equilibrium between heating and cooling in pressure P, temperature T parameter space of neutral atomic gas. Stable phases exist where there is negative slope in this diagram, and Figure 9 indicates the stable "cold phase" (~ 100 K) and the "warm phase" ($\sim 10^4$ K). Both phases are PDRs. There is a minimum pressure P_{min} required for the cold phase to exist (see Figure 9). Parravano makes two assumptions in his modeling: (i) grain photoelectric heating dominates so that P_{min} monotonically increases with G_o, and (ii) molecular, star-forming clouds grow out of the cold phase (e.g., out of the coalescence of cold phase clouds). The feedback mechanism proceeds as follows. Suppose that the pressure P of the ISM is greater than P_{min} so that the cold phase exists. Then, molecular clouds grow, OB stars form, G_o increases and P_{min} rises. If $P_{min} > P$, the cold phase no longer exists, star formation drops, G_o drops, and P_{min} drops. Thus, the OB star formation rate is regulated so that $P_{min} \sim P$ in galaxies. Parravano (1988) offers some observational support of this prediction in external galaxies.

SUMMARY

PDRs emit much of the IR radiation (line and continuum) in galaxies. Most of the mass of the gas and dust in the Galaxy resides in PDRs and is significantly affected, either via chemistry or heating, by the FUV flux. Much of the gas is heated by the grain photoelectric heating mechanism. The spectra from PDRs is characterized by ratios $(L_{CII}+L_{OI})/L_{IR} \sim 10^{-3}-10^{-2}$. PDRs are the origin of most of the observed CII($158 m u$m), OI($63,145 \mu$m), SiII(35μm), CI($370,609 \mu$m), low J CO, and CII recombination radiation. PDRs also emit significant H_2 vibrational quanta; the 2-1S(1)/1-0S(1) ratios range from 0.5 (fluorescent) to $<<0.5$ when $n \gtrsim 10^5$ cm^{-3}. The spectra diagnose physical conditions such as the gas density, the FUV radiation field, and the elemental abundances. As examples, physical conditions were derived in Orion, M17 and galactic nuclei. The PDR models explain the correlation of CII(158μm) emission with ^{12}CO J=1-0 emission and of ^{12}CO J=1-0 with H_2 mass. The FUV flux in PDRs may regulate low and high mass star formation in galaxies, and may regulate the column density of gravitationally-bound, star-forming molecular clouds.

I would like to acknowledge that much of the research reported here was done in collaboration with A. Tielens, M. Burton, T. Takahashi, and M. Wolfire. This research was supported in part by NASA RTOP 188-44-53 and by a special NASA Theory Program which funds the Center for Star Formation Studies, a consortium of researchers from NASA Ames, UC Berkeley, and UC Santa Cruz.

REFERENCES

Black, J. H. and Dalgarno, A. 1977, *Ap. J. Suppl.*, **34**, 405.

Black, J. H. and van Dishoeck, E. F. 1987, *Ap. J.*, **322**, 412.

Burton, M., Hollenbach, D., and Tielens, A. G. G. M. 1989a, in *Infrared Spectroscopy in Astronomy*, ed. M. Kessler, in press.

Burton, M., Hollenbach, D., and Tielens, A. G. G. M. 1989b, *Ap. J.*, submitted.

Clavel, J., Viala, Y. P. and Bel, N. 1978, *Astr. Ap.*, **65**, 435.

Crawford, M. K., Genzel, R., Townes, C. H. and Watson, D. M. 1985, *Ap. J.*, **291**, 755.

deJong, T., Dalgarno, A. and Boland, W. 1980, *Astr. Ap.*, **91**, 68.

Evans, N. J., Mundy, L. G., Davis, J. H., Vanden Bout. P. 1987, *Ap. J.*, **312**, 344.

Felli, M., Churchwell, E. and Massi, M. 1984, *Astr. Ap.*, **136**, 53.

Field, G. B., Goldsmith, D. W. and Habing, H. J. 1969, *Ap. J. (Letters)*, **155**, L149.

Genzel, R., Harris, A. I., and Stutzki, J. 1989, in *Infrared Spectroscopy in Astronomy*, ed M. Kessler (ESA SP series), in press.

Glassgold, A. E. and Langer, W. D. 1974, *Ap. J.*, **193**, 73.

Glassgold, A. E. and Langer, W. D. 1975, *Ap. J.*, **197**, 347.

Glassgold, A. E. and Langer, W. D. 1976, *Ap. J.*, **206**, 85.

Haas, M., Meixner, M. M. and Tielens, A. G. G. M. 1990, in preparation.

Haas, M., Hollenbach, D. J. and Erickson, E. F. 1986, *Ap. J. (Letters)*, **301**, L57.

Habing, H. J. 1968, *Bull. Astr. Inst. Netherlands*, **19**, 421.

Hayashi, M., Hasegawa, T., Gatley, I., Garden, R. and Kaifu, N. 1985, *M.N.R.A.S.*, **215**, 31P.

Hollenbach, D. J. 1988, *Astr. Lett. and Communications*, **26**, 191.

Hollenbach, D. J. 1989, in *IAU Symposium 135, Interstellar Dust*, ed. A. G. G. M. Tielens and L. Allamandola, (Kluwer: Dordrecht), in press.

Hollenbach, D. J., Takahashi, T. and Tielens, A. G. G. M. 1989, in preparation.

Jaffe, D. T. and Howe, J. E. 1989, *Rev. Mex. de Astron. y Astr.*, in press.

Lord, S., Hollenbach, D. J., Lamb, S., Wolfire, M., and Erickson, E. F. 1990, in preparation.

Massi, M., Churchwell, E. and Felli, M. 1988, *Astr. Ap.*, **194**, 116.

McKee, C. F. 1989, *Ap. J.*, in press.

McKee, C.F. and Hollenbach, D. J. 1980, *Ann. Rev. Astron. Astrophys.*, **18**, 219.

Parravano, A. 1988 *Astr. Ap.*, **205**, 71.

Prasad, S. S. and Huntress, W. T. 1980, *Ap. J. Suppl.*, **43**, 1.

Shu, F. H., Adams, F. C., and Lizano, S. 1987, *Ann. Rev. Astron. Astrophys.*, **25**, 23.

Snell, R. C., Erickson, N. R., Goldsmith, P. F., Ulich, B. L., Lada, C. J., Martin, R. N. and Shulz, A. 1986, *Ap. J.*, **304**, 780.

Solomon, P. M., Rivolo, A. R., Barrett, J. W., and Yahil, A. 1987, *Ap. J.*, **319**, 730.

Sternberg, A. 1988, *Ap. J.*, **322**, 400.

Sternberg, A. 1989, in *Infrared Spectroscopy in Astronomy*, ed M. Kessler, in press.

Sternberg, A. and Dalgarno, A. 1989, *Ap. J.*, **338**, 197.

Stutzki, J., Stacey, G., Genzel, R., Harris, A., Jaffe, D., and Lugten, J. 1988, *Ap. J.*, **332**, 379.

Tanaka, M., Hasegawa, T., Hayashi, S., Brand, P., and Gatley, I. 1989, *Ap. J.*, **336**, 207.

Tielens, A. G. G. M. and Hollenbach, D. J. 1985a, *Ap. J.*, **291**, 722.

Tielens, A. G. G. M. and Hollenbach, D. J. 1985b, *Ap. J.*, **291**, 747.

Tielens, A. G. G. M. and Hollenbach, D. J. 1985c, *ICARUS*, **61**, 40.

van Dishoeck, E. F. and Black, J. H. 1987, *Ap. J. Suppl.*, **62**, 109.

van Dishoeck, E. F. and Black, J. H. 1988, *Ap. J.*, **334**, 711.

Wolfire, M. G., Hollenbach, D. J. and Tielens, A. G. G. M. 1989a, *Ap. J.*, in press.

Wolfire, M. G., Tielens, A. G. G. M. and Hollenbach, D. J. 1989b, *Ap. J.*, submitted.

Wolfire, M. G., Hollenbach, D. J., and Tielens, A. G. G. M. 1990, in preparation.

BIPOLAR OUTFLOWS: EVOLUTIONARY AND GLOBAL CONSIDERATIONS

LUIS F. RODRIGUEZ
Instituto de Astronomía, UNAM, Apdo. Postal 70-264, 04510 México, D. F., México

ABSTRACT The recently published surveys for bipolar outflows in molecular clouds allow for the first time a reliable discussion of the evolutionary and global implications of the outflow phenomenon. In this review I discuss three important questions related to these issues: 1) Do all stars go through the outflow phase? 2) Is the outflow phase the earliest observed stage in stellar evolution? and 3) Can the outflows sustain the turbulence observed in molecular clouds? The observational and theoretical information available at present is consistent with an affirmative answer for all three questions.

INTRODUCTION

Perhaps the most important recent advance in our understanding of the early stages of star formation is that they are characterized by the presence of powerful outflows. This paradigm contrasts strongly with that held 15 years ago, when infall motions were believed to dominate the star formation process. The discovery of high velocity molecular gas in Orion (Kwan and Scoville 1976; Zuckerman, Kuiper, and Rodríguez-Kuiper 1976) was followed by the finding that the geometry of the high velocity gas is bipolar (Snell, Loren, and Plambeck 1980; Rodríguez, Ho, and Moran 1980). Since then, a large number of molecular outflows have been detected and studied. In a recent review, Fukui (1989) catalogs 144 molecular outflows, with about 80 percent of them showing bipolarity and almost all of them associated with infrared sources.

Most of the intensive research activity around the bipolar outflow phenomenon has concentrated on the study of the characteristics of individual sources. Relatively little effort has been put into studying the evolutionary and global implications of the outflows. A very important observational development that will allow a better understanding of these aspects has been the availability of unbiased searches for molecular outflows in single molecular clouds, an approach pioneered by Fukui et al. (1986) and Margulis and Lada (1986).

In this paper I will discuss, in a broad context, three questions related to the outflow phenomenon: 1) Do all stars go through the outflow phase?, 2) Is the outflow phase the earliest observed stage in stellar evolution?, and 3) Can the outflows sustain the turbulence observed in molecular clouds?

DO ALL STARS GO THROUGH THE OUTFLOW PHASE?

It is fairly well established that stars of all luminosities can be associated with molecular outflows, since outflows have been detected in sources with luminosities from 0.6 $L_\odot$ (L1535; Terebey, Vogel, and Myers 1989) to $\sim 10^6$ $L_\odot$ (W49; Scoville et al. 1986). However, it is not clear what fraction of the stars of a given luminosity go through the outflow phase.

A key parameter for the discussion of these issues is the duration of the outflow phase. The accurate determination of this parameter is difficult. With more data it may be possible to find that the outflow duration actually depends on other parameters, such as the stellar luminosity. For the moment, I will try to estimate a typical duration that will be adopted for all outflows. A fairly solid upper limit for this duration can be obtained as follows. Mass outflows of order 10^{-6} solar masses per year have been found for the exciting stars of some of the best studied outflows (Lizano et al. 1988). Since the stars involved have masses of a few solar masses, the process cannot go on for more that $\sim 10^6$ years. Another way to estimate the duration of the outflow phase involves the fact that the exciting stars are usually embedded near the center of a dense core (Torrelles et al. 1983; Myers et al. 1988; Anglada et al. 1989). Assuming a radius of about 0.1 pc for these cores and a drift velocity for the stars of about 1 km s^{-1}, we obtain a value of $\sim 10^5$ years for the duration. Finally, a more direct way of estimating the duration of the outflow phase is to evaluate the kinematic age from dividing the outflow size over the outflow velocity. For example, for a compact outflow with typical size of about 0.1 pc and typical outflow velocity of about 10 km s^{-1}, we obtain a time scale of $\sim 10^4$ years. In what follows, I will adopt a characteristic duration of $\sim 10^5$ years for the outflow phase. As we will see, this duration is consistent with what we know of star formation in molecular clouds.

How many outflows are there, let us say, within 1 kpc from the Sun? By the time of the analysis of Rodríguez et al. (1982), about ten outflow sources had been detected within this region. The number had increased to ~ 50 by the time Lada (1985) reviewed the phenomenon. The most recent counting (Fukui 1989) gives about 100 outflow sources within 1 kpc from the Sun. This number is, of course, incomplete by an unknown factor. From the results obtained in the three clouds that have been surveyed for outflows (Mon OB1; Margulis et al. al. 1988, Orion South; Fukui et al. 1989, and S287; Iwata et al. 1989) it can be estimated that there is, on the average, one outflow for every $\sim 3 \times 10^3$ $M_\odot$ of molecular gas in these star-forming clouds. From Scoville and Sanders (1987) I estimate that there are about $\sim 3 \times 10^7$ $M_\odot$ of gas (H I + H$_2$ + H II) within 1 kpc from the Sun. If we assume that one tenth of this gas ($\sim 3 \times 10^6$ $M_\odot$) is in molecular clouds with active star formation, we expect of the order of 1000 detectable (detectable meaning that the mechanical luminosity of the outflow exceeds about 0.001 solar luminosities; Fukui 1989) molecular outflows within 1 kpc from the Sun. Since the duration of the outflow is $\sim 10^5$ years, we derive an outflow frequency of $\sim 3 \times 10^{-3}$ kpc^{-2} yr^{-1}. This value is similar to the formation rate in the solar vicinity of stars with mass greater than about 0.7 $M_\odot$ (Serrano 1978).

This is a rather remarkable conclusion because it seems to imply, in combination with the argument that follows, that a significant fraction, perhaps even all stars go through the outflow phase. The mechanical luminosity of an outflow is typically 0.003 of the bolometric luminosity of the exciting source (Rodríguez et al. 1982; Lada 1985). Then, since at present we can detect only the outflows with mechanical luminosities of 0.001 solar luminosities or more (Fukui 1989), we conclude that we can detect outflows only in stars with bolometric luminosities of about 0.3 $L_\odot$ or more. This rough estimate is consistent with the fact that the lowest bolometric luminosities in detected outflow sources are ~ 0.6 $L_\odot$ (Terebey, Vogel, and Myers 1989; Mizuno 1989). For a star in the main sequence this luminosity corresponds to a mass of about 0.7 $M_\odot$. This result would imply that all stars above this mass go through the outflow phase. Furthermore, since we are not sensitive enough to detect outflows in stars of lower luminosity, it is tempting to conclude, given the coincidence in outflow frequency and star formation rate for stars with $M_* \geq 0.7$ $M_\odot$, that all stars go through the outflow phase. There is, however, a correction that has to be made. The stars powering outflows are in the pre-main-sequence and thus overluminous with respect to the values in the main sequence. Adopting an approximate value of 3 for the overluminosity factor in low mass stars (Berrilli et al. 1989), we find that we can detect outflows in stars with masses in excess of ~ 0.5 $M_\odot$. Using again the star formation rates of Serrano (1978), this result implies that about one half of the stars go through the outflow phase. Given the uncertainties, it seems not unreasonable to conclude that all stars experiment an outflow phase during their early life.

IS THE OUTFLOW PHASE THE EARLIEST OBSERVED STAGE IN STELLAR EVOLUTION?

There is no doubt that the outflow phase is associated with young stars. The problem is to establish how young are these stars in comparison with other young objects like the T Tauri stars. The argument of the close spatial association of the exciting stars of outflows with dense cores, given above, suggests that the stars formed from these cores in the recent past ($\leq 10^5$ years ago). This time scale is about an order of magnitude smaller than the time scale of $\sim 10^6$ years usually attributed to the T Tauri phase (Cohen and Kuhi 1979). This implies that, in a given molecular cloud, one expects to find about 10 T Tauri stars per molecular outflow. This is close to the ratio of 1 outflow for every 7 T Tauri stars derived for L1641 by Fukui (1989) combining his CO radio data with the optical and infrared data of Strom et al. (1989).

It has also been argued that the outflow phase precedes the T Tauri phase because the outflow detection rates in embedded objects are several times larger than those in visible, T Tauri-like objects (Lada 1985). A drawback of this argument is that it is not yet clear if the relative paucity of outflows in visible objects is due to the lack of a powerful stellar wind or to the lack of sufficient circumstellar gas. A more definitive evidence may come from the suggestion of Myers et al. (1988) and Fukui (1989) that the outflow sources are characterized by significantly greater bolometric luminosity than the T Tauri stars. The excess luminosity is attributed to the release of accretion energy. Even when more data is required to test this result and be certain that it is due to evolutionary changes and not to sensitivity limitations, it can be concluded that the evidence is consistent with the outflow phase being a shorter, earlier episode than the T Tauri phase.

Do we expect earlier phases than the outflow stage in the evolution of a star? The current theoretical scenario for the evolution of young, low mass stars has been summarized by Shu, Adams, and Lizano (1987). The earliest phase of star formation is expected to be characterized by a pure infall phase. At a typical accretion rate of $\sim 10^{-5}$ solar masses per year, it will take about 10^4 years to build a 0.1 $M_\odot$ core. This brief phase will be followed by the outflow phase (with accretion still being present) with a duration of $\sim 10^5$ years. The star will later become visible during the T Tauri phase, that will last for $\sim 10^6$ years. The pure infall phase has not been detected unambiguously yet. One reason for this may be that pure infall objects are, in this scenario, rather rare: in a given molecular cloud we expect to find one pure infall source for every 10 outflow sources (or for every $\sim$100 T Tauri stars). Since about 10 outflow sources is the population of active star-forming molecular cloud complexes like Monoceros OB1 (Margulis, Lada, and Snell 1988), we expect only one pure infall source in a rather large piece of the sky. Even if we were able to locate the needle in the haystack, Anglada et al. (1987) have shown that the detection of the characteristic spectral emission from an infalling protostellar envelope is a rather difficult observational task. Since the pure infall phase has not been detected, we can conclude that the outflow phase is indeed the earliest observed stage in star formation.

CAN THE OUTFLOWS SUSTAIN THE TURBULENCE OBSERVED IN MOLECULAR CLOUDS?

The subject of turbulence in molecular clouds has been reviewed recently by Scalo (1987) and Falgarone and Puget (1988). In brief, the nature of the problem is as follows. Molecular clouds have kinetic temperatures in the order of 10 K. This temperature will produce a thermal width of $\sim$0.1 km s^{-1} in the rotational lines of CO, the tool usually employed to study molecular clouds. However, widths an order of magnitude larger ($\sim$1 km s^{-1}) are observed. Taking 10 pc as a typical dimension for a molecular cloud, the supersonic turbulence should dissipate in about 10^7 years, the crossing time for the cloud. Since the lifetime of molecular clouds is $\geq 10^8$ years (Scoville and Sanders 1987), a mechanism is required to pump energy.

Can outflows be this energy source? I already noted that from the results obtained in Mon OB1 (Margulis et al. al. 1988), Orion South (Fukui et al. 1989), and S287 (Iwata et al. 1989) it can be estimated that there is, on the average, one outflow for every $\sim 3 \times 10^3$ $M_\odot$ of molecular gas in these star-forming clouds. Furthermore, each outflow injects typically a momentum of $\sim$30 $M_\odot$ km s^{-1} into the cloud. Then, assuming momentum conservation we need 100 outflows per $\sim 3 \times 10^3$ $M_\odot$ of molecular gas to stir the cloud to the observed turbulence. Since at a given time one outflow is observed for this amount of mass, we require 100 generations of outflows (with a total duration of 100×10^5 years=10^7 years) to explain the observed turbulence. Since 10^7 years is the dissipation time of the turbulence, we conclude that to sustain a molecular cloud turbulent, a constant star formation rate (similar to the one derived for regions of active star formation) must be maintained.

This scheme has a problem, first stressed by Levreault (1988) in a somewhat different form. If we assume that each outflow requires, on the average, a star with one solar mass, we would have used up all of the initial $\sim 3 \times 10^5$ M$_\odot$ of molecular gas in about 3000 generations of outflows, that is, in a time of order $\sim 3 \times 10^8$ years. Since this time is of the order of the lifetime of the cloud, this scheme would then imply the existence of molecular clouds with star formation efficiency approaching unity, while the largest values observed are a few percent. Furthermore, for a molecular cloud to be turbulent, we would require it to be undergoing star formation <u>always</u> at rates similar to those observed in Mon OB1 or Orion, but it is known that even clouds with modest or even no clear evidence of star formation are turbulent.

One possible way out of this dilemma is to speculate that the dissipation time of turbulence is much larger that the value of 10^7 years discussed above. It is possible that molecular clouds have, in general, magnetic fields large enough for the clouds to be supersonic but subalfvénic (Shu <u>et al.</u> 1987). If this is the case, turbulent dissipation will be much less efficient and dissipation times will be much larger. A cloud does not have then to maintain continuous, active star formation to be turbulent. Instead, it may be possible to maintain the turbulence by going through a period of $\sim 10^7$ years of active star formation followed by a longer period (for example, $\sim 10^8$ years) of more modest star formation activity or even quiescence.

A POSSIBLE EVOLUTIONARY SCHEME

With what has been discussed, I will try to present a picture of stellar evolution that incorporates in an important manner the outflow phenomenon. Star formation probably starts with a brief ($\sim 10^4$ years) period of pure infall. This first stage is followed by the outflow phase, lasting $\sim 10^5$ years. Simultaneously, accretion continues during this period, probably with the intervention of a disk. The star eventually becomes visible in the T Tauri phase, which will last $\sim 10^6$ years. This active star formation period may persist for $\sim 10^7$ years and at its end the cloud would have experienced 100 generations of outflows, sufficient to feed the observed turbulence. About 3 percent of the mass of the cloud is transformed into stars in this episode. However, we see as T Tauri stars only the objects formed during the last $\sim 10^6$ years, that is, only 0.3 percent of the mass of the cloud. The gas in the cloud goes then into a period of lower or null star formation activity (as a quiescent molecular cloud, with some of the gas probably going into atomic or ionized form as a result of the star formation activity) that will last for $\sim 10^8$ years. Turbulence, being in a subalfvénic regime, persists during times of this order. After these $\sim 10^8$ years the cloud goes into another burst of active star formation that will feed turbulence once again. The timescale to transform gas available in the interstellar medium into stars will then be of the order of the time between outbursts ($\sim 10^8$ years) divided over the fraction of gas transformed into stars in each episode (~ 0.03). This gives a timescale of $\sim 3 \times 10^9$ years, reasonably close to the age of the Galaxy ($\sim 10^{10}$ years).

CONCLUSIONS

Since its detection, the molecular outflows have been proved to be a spectacular phenomenon on an individual basis. The data now available from the systematic surveys discussed here suggest that the outflow phenomenon also has profound implications in our understanding of stellar and cloud evolution. Even taking into account the considerable uncertainties present, it appears reasonable to conclude that the outflow phenomenon is a stage through which most, possibly all, stars pass. At present it seems to be as well the earliest observed phase in stellar evolution, although the long awaited detection of a pure infall source may displace outflows to a second place in the time history of star formation. Finally, the injection of momentum in the clouds by the outflows may be sufficient to account for the observed turbulence if the dissipation time for turbulence is about ten times longer that usually assumed.

ACKLOWLEDGEMENTS

I gratefully acknowledge support from a Fellowship from the Guggenheim Foundation. I am also indebted to G. Anglada, J. Cantó, A. Raga, F. H. Shu, and J. M. Torrelles for valuable comments.

REFERENCES

Anglada, G., Rodríguez, L. F., Cantó, J., Estalella, R., and López, R. 1987, Astron. Astrophys., 186, 280.
Anglada, G., Rodríguez, L. F., Torrelles, J. M., Estalella, R., Ho, P. T. P., Cantó, J., López, R., and Verdes-Montenegro, L. 1989, Ap. J., 341, 208.
Berrilli, F., Ceccarelli, C., Liseau, R., Lorenzetti, D., Saraceno, P., and Spinoglio, L. 1989, Mon. Not. R. ast. Soc., 237, 1.
Cohen, M., and Kuhi, L. V. 1979, Ap. J. Suppl., 41, 743.
Falgarone, E., and Puget, J. L. 1987, in Galactic and Extragalactic Star Formation, ed. R. E. Pudritz and M. Fich, NATO ASI Series, p. 195.
Fukui, Y., Sugitani, K., Takaba, H., Iwata, T., Mizuno, A., Ogawa, H., and Kawabata, K. 1986, Ap. J. (Letters), 311, L85.
Fukui, Y. 1989, in Low Mass Star Formation and Pre-Main Sequence Objects, ed. B. Reipurth, ESO Workshop, in press.
Fukui, Y. et al. 1989, in preparation.
Iwata, T. et al. 1989, in preparation.
Kwan, J., and Scoville, N. 1976, Ap. J. (Letters), 210, L39.
Lada, C. J. 1985, Ann. Rev. Astron. Astrophys., 23, 267.
Levreault, R. M. 1988, Ap. J., 330, 897.
Lizano, S., Heiles, C., Rodríguez, L. F., Koo, B. C., Shu, F. H., Hasegawa, T., Hayashi, S. S., and Mirabel, I. F. 1988, Ap. J. 328, 763.
Margulis, M., and Lada, C. J. 1986, Ap. J. (Letters), 309, L87.
Margulis, M., Lada, C. J., and Snell, R. L. 1988, Ap. J., 333, 316.
Mizuno, A. 1989, in preparation.

Myers, P. C., Heyer, M., Snell, R. L., and Goldsmith, P. F. 1988, Ap. J., 324, 907.
Rodríguez, L. F., Ho, P. T. P., and Moran, J. M. 1980, Ap. J. (Letters), 240, L149.
Rodríguez, L. F., Carral, P., Ho, P. T. P., and Moran, J. M. 1982, Ap. J., 260, 635.
Scalo, J. M. 1987, in Interstellar Processes, ed. D. J. Hollenbach and H. A. Thronson, Reidel, p. 349.
Scoville, N. Z., Sargent, A. I., Sanders, D. B., Claussen, M. J., Masson, C. R., Lo, K. Y., and Phillips, T. G. 1986, Ap. J., 303, 416.
Scoville, N. Z., and Sanders, D. B. 1987, in Interstellar Processes, ed. D. J. Hollenbach and H. A. Thronson, Reidel, p. 21.
Serrano, A. 1978, Ph. D. Thesis, Sussex University.
Shu, F. H., Adams, F. C., and Lizano, S. 1987, Ann. Rev. Astron. Astrophys., 25, 23.
Snell, R. L., Loren, R. B., and Plambeck, R. L. 1980, Ap. J. (Letters), 239, L17.
Strom, K. M., Newton, G., Strom, S. E., Seaman, R. L., Carrasco, L., Cruz-González, I., Serrano, A., and Grasdalen, G. L. 1989, Ap. J. Suppl., 71, 183.
Terebey, S., Vogel, S. N., and Myers, P. C. 1989, Ap. J., 340, 472.
Torrelles, J. M., Rodríguez, L. F., Cantó, J., Carral, P., Marcaide, J., Moran, J. M., and Ho, P. T. P. 1983, Ap. J., 274, 214.
Zuckerman, B., Kuiper, T. B. H., and Rodríguez-Kuiper, E. N. 1976, Ap. J. (Letters), 209, L137.

Section IV

Evolution of Dust, Gas, and Chemistry

EVOLUTION OF INTERSTELLAR DUST

BRUCE T. DRAINE

Princeton University Observatory, Princeton NJ 08544, USA

ABSTRACT Observational evidence for the evolution of interstellar dust is reviewed. Important physical processes are identified. The problem of interstellar depletions is emphasized: with our best estimates for how rapidly grains are eroded, it is difficult to understand why elements such as Fe, Ca, and Si are so strongly depleted from the gas phase. Some other puzzles are mentioned.

I. INTRODUCTION

Interstellar dust plays a pivotal role in the evolution of galaxies. Consider, for example, the very process of star formation in molecular clouds: Dust grains catalyze the formation of H_2, and, indirectly, other molecules whose formation pathways involve reactions with H_2. Dust grains provide shielding of molecular regions from starlight, thereby allowing ionization levels to drop to low enough values that ambipolar diffusion can deprive cloud cores of magnetic support (McKee 1989b). Under some circumstances dust grains play an additional role in the ionization balance by providing a nonradiative pathway for electron-ion recombination. Finally, the continuum infrared opacity provided by dust plays an important role in protostellar collapse.

Unfortunately, as this review will make evident, many aspects of the evolution of interstellar dust remain unclear! I will attempt to indicate the relatively few things which we feel we know, but mainly I will try to emphasize the major problems and puzzles.

In §II I summarize the observational evidence for active dust grain evolution. The overall framework for discussion of dust grain evolution is outlined in §III, and the major physical processes are described in §IV. The observed depletions present some serious puzzles; the "depletion problem" is the subject of §V. Several other difficulties are described in §VI. The final section summarizes the major points.

II. EVIDENCE FOR EVOLUTION OF INTERSTELLAR DUST

a) Extinction and Polarization Variations
The most direct evidence concerning interstellar dust comes from observations of interstellar extinction, scattering, and emission of photons by dust (see Mathis

1990). Many aspects of interstellar extinction are known to vary from one line-of-sight to another: *(i)* The ratio $R \equiv A_V/E(B-V) \approx 3.1$ in diffuse clouds, but $R \gtrsim 4$ for a number of dense clouds (Vrba and Rydgren 1984), presumably due to an increase in the mean grain size due to some combination of accretion and coagulation (cf. Jura 1980). *(ii)* The wavelength of peak linear polarization, λ_p, varies along with R, with $\lambda_p \approx 0.175R\,\mu$m (Serkowski *et al.* 1975; Vrba *et al.* 1981). *(iii)* The far-ultraviolet rise in the extinction curve exhibits considerable variation, with $E(1200\,\text{Å}-V)/E(B-V)$ ranging between 2.3 and 12.3 in the sample of Fitzpatrick and Massa (1988). *(iv)* The 2175 Å "bump" shows strong variation in both its strength (relative to $E(B-V)$) and width, although its central wavelength is nearly invariant (Fitzpatrick and Massa 1986). *(v)* The profile of the 10 μm "silicate" feature appears to be broader in dense clouds than in diffuse clouds (Roche and Aitken 1984, 1985). *(vi)* The 3.1 μm "ice" feature appears to be absent in diffuse clouds (with $\Delta A_{3.1\,\mu\text{m}}/A_V < 0.002$; Gillett *et al.* 1975) but present in many dense clouds, with strengths as large as $\Delta A_{3.1\,\mu\text{m}}/A_V = 0.06$ (Harris *et al.* 1978). *(vii)* The strengths (relative to $E(B-V)$) of at least some of the "diffuse bands" differ among different lines of sight (Krelowski and Westerlund 1988; Somerville 1988).

b) Depletions

Studies of elemental depletions provide important indirect evidence bearing on interstellar dust, since it is assumed that depletion results from the incorporation of atoms into solid grains (see the recent reviews by Jenkins 1987, 1989). Depletions are observed to vary from element to element – for example, Ca often shows large depletions while S shows little.

Evidence for *active evolution* of interstellar dust is provided by *(i)* the observation that dense clouds exhibit greater depletions than do low density regions, and *(ii)* the observation that high velocity gas tends to show more modest depletions than low velocity gas (Routly and Spitzer 1956; Siluk and Silk 1974; Cowie 1978).

III. THE SCENARIO

We observe stellar sources (cool giants and supergiants, planetary nebulae, and novae) which inject newly-formed dust into the interstellar medium. Whether supernova explosions form dust remains problematic. It is clear, however, that the depletions observed in the interstellar medium require that *significant depletion (i.e., grain growth) must take place in the interstellar medium*. There are two independent arguments leading to this conclusion: (1) Jura (1987) has pointed out that in the absence of depletion processes in the ISM, grain-free stellar winds from hot stars, mixed into the ISM, would result in unacceptable gas phase abundances of species, such as Fe, which are observed to be highly depleted. (2) Furthermore, workers (Barlow 1978; Draine and Salpeter 1979b; Dwek and Scalo 1980; McKee *et al.* 1987; McKee 1989a) who have tried to estimate destruction rates for interstellar grains find grain lifetimes which are short compared to the age of the Galaxy – unless processes in the ISM can recondense this material, this would result in unacceptably high gas phase abundances of depleted species such as Fe.

What is less clear is: *(i)* exactly what grain destruction rates apply in the different "phases" of the interstellar medium; *(ii)* where and how grain growth occurs in the ISM; and *(iii)* the rate of exchange of material between different phases of the ISM. To some degree these issues are unresolved because there is no agreement on what interstellar grains *are*; in addition, the structure and dynamics of the interstellar medium is by no means a solved problem.

IV. PHYSICAL PROCESSES

a) Dust Nucleation and Growth in Stellar Outflows

Dust is often present in outflows from cool stars. Stellar envelopes divide into two quite distinct classes: those with O/C> 1 (like the sun and the interstellar medium) and those where nucleosynthesis has raised the carbon abundance to the point that O/C< 1. In cooling oxygen-rich gas, the principal refractory condensates are expected to be silicate minerals. In cooling carbon-rich gas, carbonaceous solids (graphite or amorphous carbon) are expected to form. It is gratifying that precisely this dichotomy is observed: cool oxygen-rich stars with dusty outflows show a 10 μm emission feature which is securely identified as the Si-O "stretch" in silicates, while cool carbon-rich stars with dust have infrared spectra which are continuous with only weak spectral features (such as the 11.3 μm SiC feature) if any.

In the warm circumstellar environment, very small grains are unstable to evaporation or sublimation (i.e., more likely to become smaller by losing one or more atoms than to grow by accreting one or more atoms): there is therefore a barrier to grain formation which must be overcome by the process of "nucleation" (cf. Draine 1981). Because the circumstellar environment is far from thermodynamic equilibrium (e.g., the gas temperature, grain temperature, and radiation color temperature are all different), and because relatively complex and poorly understood chemistry may be involved (e.g., in the formation of small "silicate" clusters) the problem of nucleation is a difficult one. The discussions in the literature have generally been forced to make unrealistic simplifying assumptions, often resorting to so-called "homogeneous nucleation theory". Recent work has attempted to replace some of these simplifying assumptions with increasingly complex models (e.g., Gail and Sedlmayr 1987a,b; Frenklach and Feigelson 1989). It seems likely that in addition to the local microphysics of these nonequilibrium, chemically complex systems, realistic modelling will have to involve hydrodynamic complications: the temperature-density history of the outflowing gas is typically assumed to be that appropriate to steady, spherically-symmetric flow, whereas real stellar outflows are probably neither.

b) Accretion

The probability per unit time that a gas atom or ion with charge Ze will collide with a grain is given by

$$\tau_{accr}^{-1} = n_{\rm H} \left(\frac{8kT}{\pi m}\right)^{1/2} \alpha F \Sigma = \frac{1}{5.4 \times 10^7 \, \rm yr} \left(\frac{n_{\rm H}}{20 \, \rm cm^{-3}}\right) T_2^{1/2} \left(\frac{25 \, \rm amu}{m}\right)^{1/2} \alpha F \Sigma_{21} \quad (4.1)$$

where Σ is the grain geometric cross section per H nucleus, $\Sigma_{21} \equiv \Sigma/10^{-21} \, \rm cm^2$, $T_2 \equiv T/100 \, \rm K$, $\alpha \leq 1$ is a "sticking efficiency", and the dimensionless "focussing factor" $F(Z,T,...)$ allows for changes in collision rates due to electrostatic interactions between the grain and the approaching ion.

The "MRN" graphite-silicate model, with a lower radius cutoff of 50 Å, has $\Sigma_{21} = 1.1$ (Mathis, Rumpl, and Nordsieck 1977; Draine and Lee 1984); this can be regarded as a conservative estimate for Σ_{21} for regions with "normal" extinction.

If the grains are large and charged to a potential U, then F depends only on $\phi \equiv ZeU/kT$ (Spitzer 1978):

$$F = \exp(-\phi) \quad \text{for } \phi > 0 \quad ; \quad F = 1 - \phi \quad \text{for } \phi < 0 \quad (4.2)$$

In diffuse clouds the $a \gtrsim 100$ Å grains responsible for the optical and ultraviolet extinction are expected to be positively charged: Draine (1978) estimated $U \approx 0.5$ V for standard diffuse cloud conditions ($n_e \approx 0.02$ cm^{-3}, $T \approx 100$ K, and average starlight background). Thus for positive ions (e.g. Si$^+$) $F \approx e^{-58} \ll 1$, and accretion of Si would be limited to Si^0; since $n(\text{Si}^0)/n(\text{Si}^+) \ll 1$, accretion of Si onto dust would be strongly suppressed in diffuse clouds.

The implications which ultrasmall grains may have on the value of $F\Sigma_{21}$ are discussed in §V.

Another timescale of physical interest is τ_{mono}, the time required to accrete one monolayer of atoms. If n_{cond} is the number density of condensible atoms, then the time scale to add one monolayer of these atoms to a surface is just

$$\tau_{mono}^{-1} = \left(\frac{kT}{2\pi m}\right)^{1/2} \Omega^{2/3} \alpha F$$

$$= \frac{1}{4.7 \times 10^6 \text{ yr}} \left(\frac{n_H}{20 \text{ cm}^{-3}}\right) T_2^{1/2} \left(\frac{25 \text{amu}}{m}\right)^{1/2} \left(\frac{n_{cond}/n_H}{10^{-4}}\right) \left(\frac{\Omega}{10^{-23} \text{ cm}^3}\right)^{2/3} \alpha F \quad (4.3)$$

where Ω is the volume per atom.

c) Grain-Grain Collisions

The dust population will in general have a spread in velocities, as the result of Brownian motion, shocking of the gas in which they are entrained, acceleration of the gas by turbulence, "betatron" acceleration of charged grains when magnetic field compression occurs, drifts driven by radiation pressure, or "superthermal" linear motion due to photoelectric emission, hydrogen recombination, or simple variations of geometry or accomodation coefficient on aligned, spinning grains (Purcell 1979). Because of this velocity dispersion, grain-grain collisions are inevitable, with a variety of possible outcomes: coagulation, rebound, shattering, and vaporization. The outcome depends on the collision speed, and on the composition and sizes of the collision partners (Draine 1985; Tielens 1989). Tielens et al. (1987) have proposed that carbon grains could undergo a phase transition to diamond in grain-grain collisions of appropriate relative velocity.

d) Erosion by Atoms or Ions

Atoms or ions impinging on the grain surface can remove atoms or molecules from the surface. When the mechanism by which this happens is based on the transfer of kinetic energy from projectile to target, it is referred to as "physical sputtering". For physical sputtering to be be effective, the impinging atoms and ions must have sufficient kinetic energy: either the gas must be hot or the grain must be moving through the gas with a large velocity. Sputtering yields have been estimated by Barlow (1978) and Draine and Salpeter (1979a).

Chemical reactions with impinging particles might be responsible for removal of atoms from the target: this is referred to as "chemisputtering". Chemisputtering "yields" for H and O projectiles impinging on carbon targets remain controversial (cf. Draine 1979; Bar-Nun, Litman, and Rappaport 1980), and it is unclear whether chemisputtering is of importance in the interstellar medium.

e) Irradiation

Interstellar grains are subject to irradiation by both starlight, X-rays, and cosmic rays. Photodesorption is the process whereby electronic excitation of an adsorbed atom or molecule, or excitation of the substrate, results in desorption of the adsorbed atom or molecule. Photodesorption yields are poorly known (cf. Draine and Salpeter 1979b); Bourdon, Prince and Duley (1982) have recently reported very small yields in the mid-UV, but yields are expected to be larger at shorter wavelengths. UV fluxes in diffuse clouds are large enough that photodesorption may play a major role in determining the composition of grain surfaces.

Ultraviolet photolysis of ice mixtures ($H_2O:NH_3:CH_4$) can produce a nonvolative organic residue (e.g., Greenberg 1982). X-rays should have similar effects. Cosmic rays will produce both chemical alteration (e.g., Strazzula *et al.* 1988) and radiation damage; Fe nuclei can deposit enough energy in very small grains to desorb volatile molecules (Leger, Jura, and Omont 1985).

V. THE PUZZLES OF INTERSTELLAR DEPLETIONS

a) Observed Depletions

It is well established that observed depletions are correlated with density (see, e.g., Jenkins 1987). Let δ_X denote the ratio of the abundance in the gas phase to the total abundance (gas + grains) of some element X. (Henceforth, by "large depletion" we shall refer to $\delta \ll 1$.) Consider Si as a well-studied example: in low density atomic clouds $\delta_{Si} \approx 1/4$, while in dense regions $\delta_{Si} \approx 1/40$. These depletions are consistent with the observed strength of the $10\,\mu m$ extinction feature, which appears to require that the bulk of the Si be incorporated in a silicate mineral, perhaps amorphous olivine (Draine and Lee 1984). Some other elements show much more extreme depletions: $\delta_{Ca} \approx 1/15$ in diffuse regions, and 1/4000 in dense regions!

b) Depletion: The Problems

There are several problems here: (1) the element-to-element variations in depletion; (2) the region-to-region variations; and (3) the overall level of depletion.

Field (1974) noted that the depletions of elements tended to vary monotonically with the position of the element in a theoretical condensation sequence for solar abundance material: the elements which would be first to condense (e.g., Ca) were also those which were observed to be most depleted. Snow (1975), on the other hand, called attention to the apparent correlation between depletion and ionization potential. It seems clear that chemical selectivity is involved, and since chemical properties are correlated with both position in the condensation sequence and ionization potential, it is not surprising that the correlations noted by Field and Snow are present, although the actual mechanics of depletion remain somewhat mysterious. Recent work by Joseph (1988) concludes that *relative* depletions (*i.e.*, δ_X/δ_Y) of depletable species in diffuse clouds do not vary even while the absolute depletions do, a surprising result since some species would be expected to deplete more rapidly than others. This problem will not be discussed further here, but certainly deserves more attention.

The rest of this section will concentrate on the interrelated problems of region-to-region variations and overall depletion level. The central difficulty is that *observed depletions are larger than we would have expected!* As will be seen, it will be necessary to invoke large rates of mass exchange in and out of molecular clouds to explain the observations.

c) Overall Balance Between Grain Growth and Destruction

The question of the absolute level of depletion, and the region-to-region variations, involve the overall balance between grain growth and destruction. Most grain destruction probably takes place in interstellar shock waves; see the recent reviews by Seab (1988) and McKee (1989a).

TABLE I   Sites of Grain Growth and Destruction

phase	$n_H(\text{cm}^{-3})$	$T(K)$	f_V	f_M	δ_{Si}
HIM	.005	5×10^5	0.5	0.0013	0.5 ?
WNM	0.3	6000	0.5	0.08	0.25
CNM	20	100	0.04	0.4	0.025
MC	300	30	0.003	0.5	0.01 ?

In Table I I list my current favorite values of parameters for four different "phases" of the ISM: the "hot ionized medium" (HIM), "warm neutral medium" (WNM) and "cool neutral medium" (CNM) of the "three-phase" model (McKee and Ostriker 1977), plus a "molecular cloud" (MC) phase. The density $n_H = 300\,\text{cm}^{-3}$ adopted for the MC is somewhat arbitrary, but is approximately the median density for molecular gas in the Solar neighborhood. The quantity f_V is the volume filling factor, and f_M is the fraction of the mass of the ISM contributed by each component.

TABLE II   Rates of Grain Growth and Destruction ($10^{-10}\,\text{yr}^{-1}$)

phase	τ_{eros}^{-1}	τ_{accr}^{-1}	$f_M(1-\delta_{Si})\tau_{eros}^{-1}$	$f_M \delta_{Si} \tau_{accr}^{-1}$
HIM	100	$3F\Sigma_{21}$	.07	$.002\alpha F\Sigma_{21}$
WNM	300	$20F\Sigma_{21}$	18.	$0.4\alpha F\Sigma_{21}$
CNM	10	$180F\Sigma_{21}$	4	$2\alpha F\Sigma_{21}$
MC	1	$1500F\Sigma_{21}$	0.5	$7\alpha F\Sigma_{21}$
Total			~ 20	$\sim 10\overline{\alpha F\Sigma_{21}}$

Table II shows growth and destruction rates for each phase. The accretion rate τ_{accr}^{-1} is given by eq. 4.1. The quantity τ_{eros} is the grain lifetime against erosion, defined so that the rate of conversion of solid mass to gas mass by destructive processes is $M_{dust}\tau_{eros}^{-1}$, where M_{dust} is the mass in dust. McKee (1989a) has recently reviewed grain destruction, using the results of a detailed study of time-dependent interstellar shocks (McKee *et al.* 1987). Adopting an *effective* rate of supernovae (for purposes of grain destruction) of 0.008 yr^{-1} [35% of van den Bergh's (1983) rate of .022 yr^{-1} for the Galaxy] McKee estimates the overall grain lifetime to be $\tau_{eros} = 4.3 \times 10^8$ yr^{-1}, with the bulk of the grain destruction taking place in the WNM. With the parameters in Table 1, the WNM accounts for only .08 of the total mass; hence the lifetime of grains in the WNM is only $\tau_{eros,WNM} = 3.5 \times 10^7$ yr. McKee notes that for the cold phase, SN-driven shocks would give a lifetime $\tau_{eros,CNM} \approx (n_{H,CNM}/n_{H,WNM}\tau_{eros,WNM} = 2 \times 10^9$ yr; we reduce this lifetime by a factor 2 to allow for other destruction processes (e.g., cloud-cloud collisions). In the molecular clouds, supernova shocks are relatively unimportant, but there is still some grain destruction due to phenomena such as the shock waves resulting from outflows from newly-formed stars; we estimate a destruction rate $\tau_{eros,MC}^{-1} \approx 1 \times 10^{-10}$ yr^{-1}.

Consider now some specific element X (e.g., Si), and let $\delta_{X,j}$ denote the value of δ_X in phase j. One may readily derive the following equation for the rate of change of $\delta_{X,j}$ (McKee 1989; note that our definition of δ differs from his):

$$\dot\delta_{X,j} = \frac{(1-\delta_{X,j})}{\tau_{eros,j}} - \frac{\delta_{X,j}}{\tau_{accr,j}} + \sum_{k \neq j}\frac{\dot M_{k \to j}}{M_j}(\delta_{X,k} - \delta_{X,j}) \ . \tag{5.1}$$

The following simplifying assumptions have been made: *(i)* all phases have the same "cosmic abundance"; and *(ii)* time variation of this "cosmic abundance" may be neglected. The sum over $k \neq j$ includes the contribution from stellar winds to phase j. If we multiply eq.(5.1) by M_j, sum over j, and assume for simplicity that the mass of each phase is constant, we obtain the equation of global balance

$$\sum_j f_{M,j}\dot\delta_{X,j} = \sum_j \left[f_{M,j}(1-\delta_{X,j})\tau_{eros,j}^{-1} - f_{M,j}\delta_{X,j}\tau_{accr,j}^{-1}\right] \ , \tag{5.2}$$

where the fraction of the mass in a given phase is $f_{M,j} \equiv M_j/\Sigma M_j$. Thus it is evident that a steady-state solution must have

$$\sum_j f_{M,j}(1-\delta_{X,j})\tau_{eros,j}^{-1} = \sum_j f_{M,j}\delta_{X,j}\tau_{accr,j}^{-1} \ . \tag{5.3}$$

The individual terms $f_{M,j}(1-\delta_{X,j})\tau_{eros,j}^{-1}$ and $f_{M,j}\delta_{X,j}\tau_{accr,j}^{-1}$ are given for Si in Table 1. The following points are evident: (1) The HIM makes a negligible contribution to the sums in eq. (5.2) and (5.3), mainly because f_m is so small. (2) Grain destruction exceeds grain growth in the WNM by at least an order of magnitude (assuming $F\Sigma_{21} \lesssim 1$). (3) The global balance of eq. (5.3) *could* be achieved if $F\Sigma_{21} \approx 1$ in both the CNM and dense H$_2$ phases. However, since we have concluded above that $F << 1$ for Si$^+$ accretion in the CNM, the global balance (5.3) appears to be achievable only by doing the bulk of the grain growth in the dense phase, shielded from starlight, where the grains may be neutral or negatively charged. (4) If the bulk of grain growth takes place in the dense H_2 phase, then an overall steady state requires rapid replenishment of the WNM and CNM with freshly-depleted material from the dense H_2 phase. The turnover time for the WNM must be of order $\simeq 10^7$ yr.

There are many different ways to invoke mass exchange to help balance the four equations (5.1). To see this, let us altogether ignore the HIM, and consider two simple schemes for mass exchange, which we refer to as "Scheme A" and "Scheme B":

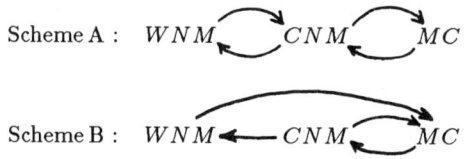

For these two schemes one may obtain the mass exchange rates by requiring mass balance and $\dot{\delta} = 0$ for each phase; the resulting mass exchange rates are shown in Table III ($M_{ISM} \approx 5 \times 10^9 M_\odot$ is the total ISM mass).

TABLE III   Mass Exchange Rates (10^{-10} yr^{-1})

$\dot{M}_{i \to j}/M_{ISM}$	Model A	Model B
$CNM \to WNM$	80	80
$WNM \to CNM$	80	0
$MC \to CNM$	1500	300
$CNM \to MC$	1500	200
$WNM \to MC$	0	80

The important point here is that very large mass exchange rates in and out of the molecular phase are required independent of which mass exchange scenario is posited, since grain destruction takes place predominantly in the WNM, but grain growth can only take place in the MC. In the case of Scheme A, the MC turnover time $M_{MC}/\dot{M}_{MC \to CNM} \approx 3 \times 10^6$ yr implies an extremely short lifetime for molecular material. In Scheme B the molecular cloud lifetime $\sim 2 \times 10^7$ yr is still short, but perhaps acceptable.

d) Role of Ultrasmall Grains

Recent observations of $1 - 25\,\mu$m infrared emission from interstellar clouds now makes it appear likely that the grain population includes very large numbers of very small grains with radii smaller than 10 Å (Sellgren, Werner and Dinerstein 1983), probably including polycyclic aromatic hydrocarbons (Leger and Puget 1984). These numerous small particles could potentially imply values of Σ_{21} as large as $\simeq 10$ (Draine and Sutin 1987). Because of induced polarization effects as well as charge fluctuations, collision rates of ions and atoms with small grains can differ appreciably from the "classical" result (3.1) (Draine and Sutin 1987);

in some cases *very* large values of F can result, particularly in regions which are effectively shielded from interstellar UV.

VI. FURTHER PUZZLES AND CONJECTURES

a) Origin of the grain size distribution

There is evidence that the grain size distribution is approximately given by a power law $dn \propto a^{-3.5} da$ for radii $.005 \lesssim a \lesssim 0.25\,\mu$m (Mathis, Rumpl and Nordsieck 1977), but the origin of this size distribution remains vague. The size distribution is presumably maintained by a combination of grain growth by accretion of atoms, coagulation (Draine 1985; Tielens 1989), grain fragmentation in grain-grain collisions (Biermann and Harwit 1980), and sputtering in shock waves (Seab and Shull 1983; McKee *et al.* 1987; Liffman and Clayton 1989).

b) The role of ultrasmall grains

As noted above, observations of $1 - 25\,\mu$m emission from diffuse clouds seem to require that large numbers of ultrasmall grains, containing as few as $30 - 50$ atoms, be present. These particles are estimated to contribute values of $\Sigma_{21} \gtrsim 5$ (the value of 5 is obtained by extrapolating the MRN power-law size distribution down to 3 Å, but this may be an underestimate).

In addition to potentially increasing the value of Σ by an order of magnitude, the small particles may substantially increase the effective value of F. There are several reasons for this. First of all, when the particle radius becomes smaller than the "photoelectron escape length", the rate per area of photoelectron emission will begin to drop, leading to a reduction in the "steady-state" potential to which the grain will be charged. Second, for small grains the polarization potential can lead to an increase in the rate of capture (per grain surface area) of free electrons (Draine and Sutin 1987). Third, for small grains the grain charge will be a fluctuating quantity, and the grain will spend some fraction of the time at smaller than average potentials (including zero or negative values) during which the rate of accretion of positive ions may be enormously enhanced.

A detailed study of these effects appears to be called for.

c) The persistence of ultrasmall grains

In regions which are not shielded from starlight, very small particles will be destroyed by thermal sublimation following absorption of UV photons: the minimum stable size for a silicate cluster is estimated to be $\simeq 37$ atoms (Guhathakurta and Draine 1989). Clusters smaller than this critical size are rapidly destroyed. Clusters somewhat larger than the critical size should also be short-lived, since they should accrete and grow quickly: accretion of one monolayer, in $\sim 10^7 (0.5/\alpha F)\,$yr, would double the grain mass. The apparent ubiquity of ultrasmall grains with $\lesssim 50$ atoms is therefore puzzling: is there something which suppresses grain growth (e.g. very small effective sticking coefficients, as might result if "filled shell" configurations were unusually stable). Or is there some process which resupplies ultrasmall grains on a $\sim 10^7$ yr timescale (e.g., shattering in grain-grain collisions?).

d) Could nucleation occur in the ISM?

It has generally been taken for granted that "nucleation" of grains could only take place in a dense region like a stellar envelope. Nucleation in diffuse regions in the presence of ultraviolet starlight still seems impossible, but we should perhaps reconsider the possibility of nucleation in dense, dark molecular clouds. In some moderately dense clouds (e.g. TMC-1 in Taurus), carbon chain molecules as large as $HC_{11}N$ have formed in detectable quantities, apparently via gas-phase ion-molecule reactions (Winnewisser and Herbst 1987). Collision of two such chain molecules should produce a highly stable carbonaceous cluster, whose fate (in a shielded region) would be to grow. If mass exchange between dense gas and diffuse gas is rapid (as has been argued above), perhaps this could be a significant source of ultrasmall particles.

e) Maintenance of chemically distinct grain species?

Most interstellar grain models postulate the existence of two or more distinct materials. For example, the MRN model (Mathis, Rumpl, and Nordsieck 1977, Draine and Lee 1984) assumes the existence of separate graphite and silicate particles. Greenberg (1982) invokes separate graphite particles, small bare silicate particles, and larger silicate particles (coated with the residue from photolysis of accreted ices). Duley (1987) postulates the existence of very small silicate particles and larger hydrogenated amorphous carbon particles. Mathis and Whiffen (1989) have proposed a model in which chemically distinct small particles (graphite, silicate, amorphous carbon) are agglomerated in a fluffy structure. Models which attribute the 2175 Å feature to graphite appear to require that the graphite be relatively "pure" (and, indeed, crystalline) in order to obtain a good match with the observed profile. Similarly, the absorption and polarization profiles of the $10\,\mu m$ feature appear to require that the grains responsible have dielectric properties close to that expected for pure amorphous silicates (although this point has not been fully explored).

If chemically distinct grains exist (and make up a significant fraction of the total grain mass) then we have a puzzle: if it is true that an appreciable fraction of solid grain material is formed in the interstellar medium, and if atoms from the interstellar medium collide indiscriminately with different grain types, why doesn't the interstellar grain population (or at least that fraction of it which was grown in the ISM) consist of a single chemical composition?

One possibility is that grain growth may be selective: perhaps a C atom alighting on a silicate surface (or a Si atom on a carbon surface) will not "bond" to the substrate and will be removed by some "cleaning" mechanism such as either photodesorption or chemisputtering by impinging atomic oxygen. Such selectivity could conceivably maintain the "purity" of different grain components. Similar processes could work for conglomerate grains: an adsorbed atom may explore the grain surface until either finding a chemically inviting region or being removed by the "cleaning" process.

f) Origin and Survival of isotopically anomalous grains

In recent years careful studies of primitive chondritic meteorites have disclosed the presence of appreciable numbers of inclusions with anomalous isotopic compositions. Some of these inclusions, while retaining (perhaps diluted) isotopic anomalies (e.g., Ca-Al rich inclusions with ^{26}Mg) were evidently melted and chemically reprocessed during during the formation of the solar system.

Some others (diamond and SiC) appear to actually be well-preserved interstellar grains (Anders et al. 1989). Approximately 2% of the carbon in primitive meteorites is in diamond form (Tang and Anders 1988). Based on their isotopic composition, the SiC grains appear to have been formed in red giant atmospheres or novae (Anders et al. 1989). The diamond grains appear to have been formed around one or more red giants; subsequent ion-implantation is thought to be responsible for the presence of r-process Xe in the microdiamonds (Lewis et al. 1989).

If the grain destruction rates in Table I are correct, how can small particles have survived their exposure in the ISM between the time when they were formed in an outflow from a cool star and the time of formation of the solar nebula? The answer seems to be that the stochastic nature of the grain destruction means that there was some period of time τ between the last grain destruction episode (e.g., SN shockwave) and incorporation of the gas into the solar nebula. During this time τ, any newly-formed grains injected by red giants, planetary nebulae, or novae would not be destroyed, and would have retained their mineralogical structure and peculiar isotopic composition. During their stay in the ISM these grains may have acquired coatings, and may have coagulated with other grains; subsequent processing in the solar nebula or terrestrial test tube could have obliterated such associations. In Table I we have estimated the destruction timescale to be of order 3×10^7 yr in the WNM and 1×10^9 yr in the CNM; hence it would not be surprising if τ were of order $\simeq 10^{8\pm1}$ yr. The typical well-preserved grain would then have an exposure of $\tau/2$ in the ISM – this seems not inconsistent with the estimated presolar cosmic-ray exposure time of 1.5×10^7 yr estimated for the interstellar SiC grains (Amari and Lewis 1989). In fact the relatively short age determined for the SiC grains seems to support the age estimates in Table I.

VII. SUMMARY

The interstellar medium is actively processing dust, but the details of this processing remain unclear. Various enigmas remain before us: perhaps the most conspicuous of these is the problem of understanding the observed interstellar depletions, which seem to require that *most of the interstellar dust volume was condensed in molecular clouds, not in stellar outflows!* A corollary is that there must be rapid turnover of material between dense molecular clouds (where depletion can take place) and the less dense regions where depletions are observed, with implied molecular cloud lifetimes $\lesssim 2 \times 10^7$ yr!

Progress on understanding the depletion question, and other puzzles, will require advances in: *(i)* understanding the structure and dynamics of the ISM (the arena); *(ii)* characterization of the interstellar grain population (the players); *(iii)* surface physics and chemistry (the rules).

ACKNOWLEGEMENTS

I thank C. L. Joseph for helpful discussions. This research was supported in part by NSF grant AST86-12013.

REFERENCES

Amari, S., and Anders, E. 1989, preprint.
Anders, E., Lewis, R. S., Ming, T., and Zinner, E. 1989, in *IAU Symposium 135: Interstellar Dust*, ed. L. J. Allamandola and A. G. G. M. Tielens (Dordrecht: Reidel), in press.
Barlow, M. J. 1978, *M.N.R.A.S.*, **183**, 367.
Bar-Nun, A., Litman, M., and Rappaport, M. L. 1980, *Astr. Ap.*, **85**, 197.
Biermann, P., and Harwit, M. 1980, *Ap. J. (Letters)*, **241**, L105.
Bourdon, E. B., Prince, R. H., and Duley, W. W. 1982, *Ap. J.*, **260**, 909.
Cowie, L. L. 1978, *Ap. J.*, **225**, 887.
Draine, B. T. 1978, *Ap. J. Suppl.*, **36**, 595.
Draine, B. T. 1979, *Ap. J.*, **230**, 106.
Draine, B. T. 1981, in *Physical Processes in Red Giants*, ed. I. Iben and A. Renzini (Dordrecht: Reidel), p. 317.
Draine, B. T. 1985, in *Protostars and Planets II*, ed. D. C. Black and M. S. Matthews (Tucson: Univ. of Arizona Press), p. 621.
Draine, B. T., and Lee, H. M. 1984, *Ap. J.*, **285**, 89.
Draine, B. T., and Salpeter, E. E. 1979a, *Ap. J.*, **231**, 77.
Draine, B. T., and Salpeter, E. E. 1979b, *Ap. J.*, **231**, 438.
Draine, B. T., and Sutin, B. 1987, *Ap. J.*, **320**, 803.
Duley, W. W. 1987, *M.N.R.A.S.*, **229**, 203.
Field, G. B. 1974, *Ap. J.*, **187**, 453.
Fitzpatrick, E. L, and Massa, D. 1986, *Ap. J.*, **307**, 286.
Fitzpatrick, E. L, and Massa, D. 1988, *Ap. J.*, **328**, 734.
Frenklach, M., and Feigelson, E. D. 1989, *Ap. J.*, **341**, 372.
Gail, H.-P., and Sedlmayr, E. 1987a, in *Physical Processes in Interstellar Clouds*, ed. G. E. Morfill and M. Scholer, (Dordrecht: Reidel), p. 275.
Gail, H.-P., and Sedlmayr, E. 1987b, *Astr. Ap.*, **171**, 97.
Gillett, F. C., Jones, T. W., Merrill, K. M., and Stein, W. A. 1975, *Astr. Ap.* **45**, 77.
Greenberg, J. M. 1982, in *Submillimetre wave astronomy*, ed. J. E. Beckman and J. P. Phillips (Cambridge: Cambridge Univ. Press, p. 261.
Guhathakurta, P., and Draine, B. T. 1989, *Ap. J.*, **345**, in press.
Harris, D. H., Woolf, N. J., and Rieke, G. H. 1978, *Ap. J.*, **226**, 829.
Jenkins, E. B. 1987, in *Interstellar Processes*, ed. D. J. Hollenbach and H. A. Thronson (Dordrecht: Reidel), p. 533.
Jenkins, E. B. 1989, in *IAU Symposium 135: Interstellar Dust*, ed. L. J. Allamandola and A. G. G. M. Tielens (Dordrecht: Reidel), in press.
Joseph, C. L. 1988, *Ap. J.*, **335**, 157.
Jura, M. 1980, *Ap. J.*, **253**, 63.
Jura, M. 1987, in *Interstellar Processes*, ed. D. J. Hollenbach and H. A. Thronson (Dordrecht: Reidel), p. 3.
Krelowski, J., and Westerlund, B. E. 1988, *Astr. Ap.*, **190**, 339.
Leger, A., Jura, M., and Omont, A. 1985, *Astr. Ap.*, **144**, 147.
Leger, A., and Puget, J. L. 1984, *Astr. Ap.*, **137**, L5.
Lewis, R. S., Anders, E., and Draine, B. T. 1989, *Nature*, **339**, 117.
Liffman, K., and Clayton, D. D. 1989, *Ap. J.*, **340**, 853.
Mathis, J. S. 1990, this volume.
Mathis, J. S., Rumpl, W., and Nordsieck, K. H. 1977, *Ap. J.*, **217**, 425.

Mathis, J. S., and Whiffen, G. 1989, *Ap. J.*, **341**, 808.
McKee, C. F. 1989a, in *IAU Symposium 135: Interstellar Dust*, ed. L. J. Allamandola and A. G. G. M. Tielens (Dordrecht: Reidel), in press.
McKee, C. F. 1989b, *Ap. J.*, submitted.
McKee, C. F., Hollenbach, D. J., Seab, C. G., and Tielens, A. G. G. M. 1987, *Ap. J.*, **318**, 674.
McKee, C. F., and Ostriker, J. P. 1977, *Ap. J.*, **218**, 148.
Purcell, E. M. 1979, *Ap. J.*, **231**, 417.
Roche, P. F., and Aitken, D. K. 1984, *M.N.R.A.S.*, **208**, 481.
Roche, P. F., and Aitken, D. K. 1985, *M.N.R.A.S.*, **215**, 425.
Routly, P. M., and Spitzer, L., Jr. 1952, *Ap. J.*, **115**, 227.
Seab, C. G. 1988, in *Dust in the Universe*, ed. M. E. Bailey and D. A. Williams (Cambridge: Cambridge Univ. Press), p. 303.
Seab, C. G., and Shull, J. M. 1983, *Ap. J.*, **275**, 652.
Sellgren, K., Werner, M. W., and Dinerstein, H. L. 1983, *Ap. J. (Letters)*, **271**, L13.
Serkowski, K., Mathewson, D. S., and Ford, V. L. 1975, *Ap. J.*, **196**, 261.
Siluk, R. S., and Silk, J. 1974, *Ap. J.*, **192**, 51.
Snow, T. P. 1975, *Ap. J. (Letters)*, **202**, L87.
Somerville, W. B. 1988, *M.N.R.A.S.*, **234**, 655.
Strazzula, G., Massimino, P., Spinella, F., Calcagno, L., and Foti, A. M. 1988, *Infrared Phys.*, **28**, 183.
Tang, M., and Anders, E., *Geochim. Cosmochim. Acta*, **52**, 1245.
Tielens, A. G. G. M. 1989, in *IAU Symposium 135: Interstellar Dust*, ed. L. J. Allamandola and A. G. G. M. Tielens (Dordrecht: Reidel), in press.
Tielens, A. G. G. M., Seab, C. G., Hollenbach, D. J., and McKee, C. F. 1987, *Ap. J. (Letters)*, **319**, L109.
van den Bergh, S., 1983, *Pub. A.S.P.*, **95**, 388.
Vrba, F. J., Coyne, G. V., and Tapia, S. 1981, *Ap. J.*, **243**, 489.
Vrba, F. J., and Rydgren, A. E. 1984, *Ap. J.*, **283**, 123.
Winnewisser, G., and Herbst, E. 1987, *Topics in Current Chemistry*, **139**, 119.

THE CHEMISTRY OF THE DIFFUSE INTERSTELLAR GAS

EWINE F. VAN DISHOECK
Div. of Geological and Planetary Sciences
California Institute of Technology 170-25
Pasadena, CA 91125.

ABSTRACT. The chemical and physical structure of diffuse, translucent and high–latitude clouds is reviewed with reference to recent observational results. The observational data are compared with a variety of current theoretical models, and the strengths and weaknesses of each of the models are pointed out. New developments regarding the thermal balance and the small–scale structure of the clouds are discussed.

1. HISTORICAL PERSPECTIVE

Diffuse interstellar clouds have been studied since the beginning of this century by the absorption lines of atoms and molecules superposed on the spectra of bright background stars. The first interstellar lines due to Na and Ca^+ were discovered by Hartmann (1904) and Heger (1919). Nearly thirty years would pass, however, before the first interstellar molecule was identified. In fact, it was at a meeting of the Astronomical Society of the Pacific in Berkeley on June 19, 1934 that Merrill announced his discovery of four new interstellar lines. In contrast with the Na and Ca^+ lines, the new lines were not sharp but diffuse, and were therefore attributed to molecules. Although the origin of these diffuse bands has still not been firmly identified nearly 55 years later, it is indeed plausible that some molecular carrier is responsible for them (Greenberg 1987). Three years after Merrill's announcement, Dunham and Adams discovered two sharp interstellar lines at 4300 and 4232 Å, which were again announced in the *Publications of the Astronomical Society of the Pacific*. Swings and Rosenfeld (1937) and Douglas and Herzberg (1941) subsequently correctly assigned these lines to CH and CH^+, respectively. At the same time, McKellar (1940) identified the interstellar CN molecule through its strong absorption lines around 3874 Å. It is clear that West Coast observatories played a pivotal role in the early studies of the chemistry of the diffuse interstellar gas.

Nearly thirty years later, Carruthers (1970) detected the most important interstellar molecule, H_2, at ultraviolet wavelengths through a rocket experiment. The *Copernicus* satellite subsequently added HD, CO and OH to the list. Finally, in 1977 diatomic carbon, C_2, was detected by its absorption lines around 8750 Å (Souza and Lutz 1977).

The chemistry of the diffuse gas has been reviewed previously by Dalgarno and Black (1976), Watson (1978), Crutcher and Watson (1985), Black (1987*a,b*), Dalgarno (1988), van Dishoeck (1989*a,b*), and in great detail by van Dishoeck and Black (1988*a*; hereafter vDB). The present review will be mostly an abbreviated version and update of the latter paper. The reader is also referred to the upcoming review article by Black and van Dishoeck (1991).

2. DIFFUSE CLOUDS

2.1. Motivation

A detailed understanding of the physical and chemical structure of diffuse interstellar clouds (i.e., clouds with a total visual extinction A_V^{tot} <2 mag) is considered a prerequisite for having any confidence in the modeling of the more complex dense clouds. Diffuse clouds contain only the simplest diatomic molecules, so that the least number of reactions (with often unknown reaction rates!) are involved in describing their abundances. Stringent tests of the basic framework of interstellar chemistry are therefore possible. Consequently, much higher standards are applied to the modeling of diffuse clouds compared with dense clouds: whereas dense cloud modelers often consider an order of magnitude difference with observations satisfactory, the goal of diffuse cloud models is to reproduce the measured abundances to a factor of two or better.

The key questions regarding diffuse clouds are: (a) What are the physical conditions in diffuse clouds? In particular, what is their temperature and density structure? (b) Can the abundances of most species indeed be described mainly by quiescent steady–state reactions, and to what extent do shock–heated chemistry and grain–surface chemistry play a role? (c) What is the origin and evolution of diffuse clouds? Are they the evolutionary products or precursors of dense clouds?

The aim of the diffuse cloud studies is to understand the chemistry well enough that molecular abundances can be used as diagnostics to provide insight into the above questions. For example, some molecules such as CH^+ may be sensitive indicators of shock chemistry, whereas the hope is that the abundances of others could be used as "clocks", to determine the evolutionary state of the cloud. Further information about the relation between diffuse and dense clouds may come from studies of so–called translucent clouds ($A_V^{tot} \approx 2 - 5$ mag), as well as the clouds recently discovered at high–galactic latitudes (Magnani, Blitz and Mundy 1985). These clouds will be discussed in more detail in §3.

2.2. Observations

The modeling of diffuse cloud chemistry is hampered by the fact that only a few molecules have been detected, as mentioned in §1. However, in contrast with dense clouds, direct information on the atomic abundances is available from visible and ultraviolet (UV) absorption line observations. Because the UV radiation penetrates diffuse clouds, most molecules are readily photodissociated, so that all species (with the exception of hydrogen) exist primarily in atomic form. The atomic lines can thus indicate the amount of depletion of the element from the gas–phase onto grains, and put limits on the ionization balance. In practice, the analysis is complicated by the fact that the UV lines measured by *Copernicus* are heavily saturated and unresolved (Jenkins 1987).

An overview of the molecular lines searched for in diffuse clouds has been given by vDB. Although many molecules have not been detected, the upper limits

Table 1. Available Diagnostics of Physical Conditions

Species	Phys. Condition Probed[a]	Diffuse Clouds	High–latitude Clouds	Translucent Clouds
H_2 low J	T	+	-	-
H_2 high J	I_{UV}	+	-	-
H_2/H	$n_H y_f / I_{UV}$	+	-	-
C_2 low J	T	+	+	+
C_2 high J	I_R/n_H	+	+	+
C, C^+, O J	n_H, T	+	-	-
CO low J	n_H, T	+[b]	+[c]	+[c]

[a] I_{UV} =scaling factor for radiation field at ultraviolet wavelengths λ=912-1100 Å; I_R=scaling factor for radiation field at far–red wavelengths $\lambda \approx 1 \mu m$; y_f=grain formation efficiency (see vDB).
[b] From UV absorption line observations.
[c] From millimeter emission line observations.

on their column densities can often provide useful constraints on the chemistry as well. As millimeter receivers become more sensitive, it becomes possible to search for some molecules by their millimeter emission lines. An example of this approach is provided by the CS molecule, for which a very tentative detection of the J=2–1 line at 98 GHz has been reported by Drdla, Knapp and van Dishoeck (1989) in the ζ Oph cloud. The derivation of column densities from these millimeter observations is highly uncertain, and has to take account of excitation not only by neutral species, but also by electrons.

2.3. Models

2.3.1. Physical conditions. The various approaches to the modeling of diffuse clouds have been discussed in detail by vDB. If a specific line of sight, such as that toward ζ Oph, is modeled, the first step in the procedure is to constrain the physical conditions as closely as possible. Table 1 lists the various diagnostics that have been used to determine the density $n_H=n(H)+2n(H_2)$, temperature T and strength of the ultraviolet radiation field I_{UV} in diffuse clouds. As the table shows, observations of the H_2 rotational excitation and the H_2/H abundance ratio alone are sufficient to constrain most of the physical parameters. Indeed, the early diffuse cloud models, such as those of Black and Dalgarno (1977) and Jura (1975), were based primarily on the H_2 observations. Subsequent analyses of the CO and C_2 rotational excitations (Crutcher and Watson 1981; van Dishoeck and Black 1982, 1986; Le Bourlot *et al.* 1987) have led to better estimates of the central densities, which in some cases (such as the ζ Oph cloud) had to be revised down by a factor of 2–4. The C_2 rotational excitation in diffuse clouds is remarkably uniform (Danks and Lambert 1983), and suggests that the central densities are typically of the order of a few hundred cm^{-3}.

The central temperature is constrained by both the C_2 low J population ratio and the H_2 $J=1/J=0$ ratio, and is typically 25–40 K. However, a cloud of uniform temperature $T \approx 40$ K cannot provide the large observed populations in the H_2 $J=2$ and 3 levels, so that higher temperatures, $T=100$–200 K, are required at the edge of the cloud. A crucial parameter in the models is the strength of the incident ultraviolet radiation field, since photodissociation is the primary destruction mechanism of the neutral species in the clouds. This strength is usually indicated with a scaling factor I_{UV} with respect to the standard radiation field by Draine (1978), and is constrained only by the observed high-J population of H_2 under the assumption that ultraviolet pumping is the dominant excitation mechanism. If some other mechanism, such as collisional excitation in a high-temperature gas, contributed as well, the inferred I_{UV} would only be an upper limit.

2.3.2. Steady–state chemistry. Once the physical structure of the cloud has been determined, the network of chemical reactions can be solved at each depth into a cloud. Integration over depth then gives the column densities which can be compared directly with observations. For many purposes, it is simply incorrect to assume that diffuse clouds are homogeneous, and that the observed ratios of column densities of species are equal to the ratios of their local densities in the center of the cloud. Owing to effects of radiative transfer, the abundances of many species are expected to vary significantly with depth. In a cloud of $A_V=1$ mag, the molecular fraction H_2/H can vary by 4–6 orders of magnitude from boundary to center.

The basic outline of this network was established more than 15 years ago (see e.g. Dalgarno and Black 1976; Watson 1978 for reviews), but many key aspects have still not been fully tested. This is partly due to the fact that a number of the crucial reaction rates, such as the $C^+ + H_2$ radiative association rate and the H_3^+ dissociative recombination rate, are still not well determined. The early models of Black and Dalgarno (1977) were quite successful in reproducing the observations using reasonable parameters for the unknown reaction rates. However, as more reaction rates become better determined through experimental or theoretical work, it becomes increasingly difficult to develop a model which reproduces the whole array of observational data.

Significant progress in our understanding of interstellar molecular processes has been made in the following areas: (1) Ion–molecule reaction rates are now measured at temperatures as low as 10 K, and significant enhancements compared with values measured at 80 K are found for reactions between ions and molecules with a permanent dipole moment (Rowe 1988); (2) Experiments on the dissociative recombination of molecular ions are focusing more on vibrationally cold ions. Of particular importance are the branching ratios to the various products: the first measurements of branching ratios for the dissociative recombination of H_3O^+ to H_2O and OH have just been reported (Herd, Adams and Smith 1989), and theories are being developed which could predict such branching ratios from first principles (see Mitchell and Guberman 1989); (3) Attempts are being made to provide better estimates of radiative association rates at low temperatures through measurements of the infrared radiative relaxation rates (Gerlich and Kaefer 1989); and (4) Photodissociation rates of a number of important species such as CO, CN, CH, NH and OH, have been accurately determined by experiments or theory (see van Dishoeck 1988 for

Table 2. Computed and Observed Column Densities (in cm^{-2})

Species	ζ Per vDB[a] Model F	Obs	ζ Oph vDB[a] Model G	Obs	vDB[b] Model T1
δ_C	0.50		0.52		0.30
δ_D	1.10		0.30		0.60
ζ_o (s^{-1})	6.0(-17)		8.0(-17)		4.0(-17)
k_{ra} (cm^3 s^{-1})	7.0(-16)		7.0(-16)		3.0(-16)
I_{UV}	3.5		3.5		1.0
C	3.1(15)	(3.3±0.4)(15)	3.1(15)	(3.2±0.6)(15)	2.9(15)
C$^+$	3.8(17)	(3.0±1.0)(17)	3.2(17)	(1.0±0.5)(17):	1.0(17)
CH	2.1(13)	(2.0±0.3)(13)	2.3(13)	(2.5±0.3)(13)	2.0(13)
C$_2$	2.0(13)	(1.9±0.3)(13)	2.2(13)	(2.4±0.3)(13)	2.4(13)
OH	5.0(13)	(4.2±0.5)(13)	4.9(13)	(4.8±0.5)(13)	5.4(13)
H$_2$O	8.6(11)	...	8.1(11)	≤2.2(13)	1.9(12)
HD	5.9(15)	(3.8±1.4)(15)	2.6(14)	(2.1±1.0)(14)	5.2(15)
CO	7.1(13)[c]	(6.1±3.0)(14)	7.0(13)[c]	(2.0±0.3)(15)	1.2(15)[d]
CH$^+$	2.6(11)	(3.5±0.4)(12)	3.0(11)	(2.9±0.3)(13)	7.6(10)
CN	4.8(11)[c]	(3.0±0.3)(12)	4.2(11)[c]	(2.9±0.3)(12)	2.3(12)[d]
NH	2.4(10)	<6.3(11)	2.3(10)	<7.5(12)	2.5(10)

[a] Models from van Dishoeck and Black (1986), but with updated reaction rates. The adopted branching ratio for dissociative recombination of H$_3$O$^+$ to form OH is 0.85.

[b] Model T1 from van Dishoeck and Black (1988b).

[c] Obtained using the standard Draine (1978) radiation field and grain model 2 of Roberge et al. (1981). For the ζ Per cloud, the CO column density increases to 8.0(14) and the CN column density to 1.5(12) cm^{-2} if the modified Draine radiation field is used (cf. van Dishoeck and Black 1988b) with the HD 204827 extinction curve (cf. van Dishoeck and Black 1989). For the ζ Oph cloud, they increase to 4.0(14) and 2.0(12), respectively, if the modified Draine field with the ζ Oph extinction curve is adopted.

[d] Obtained using the modified Draine radiation field with the HD 204827 extinction curve.

a summary).

Table 2 compares the observed and model column densities of a number of species in the ζ Per and ζ Oph clouds. It is seen that the measured abundances of molecules such as CH, C$_2$, OH and HD can be reproduced quite well in the models, *provided* that the C$^+$ + H$_2$ radiative association rate is of order $(5-7) \times 10^{-16}$ cm^3 s^{-1} and the cosmic ray ionization rate $\zeta_o \approx 6 \times 10^{-17}$ s^{-1}. The very uncertain abundance of CS in diffuse clouds can also be reasonably well understood by gas–phase ion–molecule reactions (Drdla *et al.* 1989). In contrast, the same models fail to reproduce the observed CO, CN and CH$^+$ column densities by 1–2 orders of magnitude. The problem with the CH$^+$ abundance in diffuse clouds is well known, and will be discussed further in §2.3.4 and §3.3.5. The large discrepancies for CO and CN compared with earlier models result from the fact that the photodissociation rates for these molecules have increased recently by an order of magnitude or more (van Dishoeck and Black 1988b; Viala et al. 1988a). The adopted CO photodissociation rate is based on the measured cross sections by Letzelter et al. (1987), and recent experiments at even higher spectral resolution by Stark et al. (1989) confirm their integrated values. The discrepancy for CO is therefore unlikely to be due to uncertainties in the adopted molecular parameters. The CO and CN problems could be removed if the scaling factor I_{UV} were reduced in the models. However, this would lead to larger column densities of CH, C_2, OH and HD as well, species whose abundances were already well reproduced in the models. A simultaneous reduction of both the C$^+$ + H_2 rate and ζ_o would be necessary to prevent the column densities of these molecules from becoming too large. The carbon depletion factor δ_C could also be lowered somewhat in that case, but it would have to remain large enough to reproduce the observed C and C$^+$ abundances. Another solution to the problem is a modification of the shape of the radiation field at the shortest wavelengths. Both CO and CN are photodissociated primarily by photons with wavelengths shorter than 1000 Å. This is exactly the region where neither the shape of the interstellar radiation field, nor the extinction curve are well determined (van Dishoeck and Black 1988b). If the intensity of the radiation field fell off more steeply at λ <1000 Å, the discrepancies could be reduced significantly. Also, the shape of the extinction curve may affect the predicted column densities, as demonstrated by van Dishoeck and Black (1989). If all these factors are taken to be most favorable, the models for ζ Per can reproduce the observed CO and CN column densities, but those for ζ Oph still predict too little CO by a factor of 2–3, indicating that it is likely that I_{UV} has been slightly overestimated for that cloud. Note that these changes in radiation field and extinction curve also affect the amount of UV pumping in the models (although to a lesser extent), so that it remains difficult to construct a self-consistent model which reproduces both the observed H_2 high-J population and the measured CO column density. For comparison, the abundances resulting from a model with I_{UV}=1 and with a lower gas-phase carbon abundance are included in the table. Although this model reproduces the observed molecular column densities equally well, it has, of course, much less UV pumping of H_2.

2.3.3. *Thermal balance.* The steady-state diffuse cloud models of Black and Dalgarno (1977), vDB and Viala, Roueff and Abgrall (1988b) do not consider the thermal balance explicitly, but use the above mentioned diagnostics to constrain the temperature structure. VDB do, however, calculate the heating and cooling rates at each depth into the cloud. The principal heating source is the photoelectric effect on dust grains, whereas the principal cooling agent is emission following the fine-structure excitation of C$^+$. At the center of the models, the heating and cooling rates approximately balance at a temperature of 20–30 K for the inferred carbon depletion factor. However, it has long been recognized (see e.g. Roberge 1981) that at the edge of the cloud, the known heating sources are insufficient to maintain the temperatures of 100–200 K inferred from the H_2 J=2 and 3 populations at densities $n_H \approx$100 cm^{-3}. d'Hendecourt and Léger (1987) have suggested that photoionization of large molecules such as PAH's may significantly increase the heating rates. Lepp and Dalgarno (1988) pointed out that heating due to photodetachment of negatively–charged PAH's may contribute as well, but showed that these additional heat sources are insufficient to maintain $T \approx 250$ K. However, the presence of PAH's also results in reduced gas-phase carbon abundances, which lowers the cooling rates considerably. If the heating due to PAH's as prescribed by Lepp and Dalgarno (1988) is included in the models of vDB with the PAH and carbon abundances inferred by Lepp et al. (1988), temperatures of 100–150 K can be maintained at the edges of the ζ Oph and ζ Per clouds. The same models would require, however, somewhat higher temperatures in the centers of the clouds, $T \approx 30$–40 K.

The detailed models of these "classical" diffuse clouds raise the question whether in general, the edges of interstellar clouds have temperatures in excess of 100 K when exposed to the normal interstellar radiation field. This question is particularly relevant to the prediction of intensities of the submillimeter fine-structure lines, such as the C^+ ($^2P_{3/2} \to {}^2P_{1/2}$) 158 μm and O ($^3P_1 \to {}^3P_2$) 63 μm lines. The excitation energies of these levels lie at 91 and 228 K, respectively, so that the presence of a warm edge at $T > 100$ K can result in orders of magnitude difference in predicted line strengths. For example, the predicted line intensities in the ζ Oph model G of vDB (not including PAH's) are $I(C^+ \ 158 \ \mu m) = 5 \times 10^{-5}$ and $I(O \ 63 \ \mu m) = 1 \times 10^{-6}$ ergs s^{-1} cm^{-2} sr^{-1}. In a model with a constant temperature of 40 K, they would be reduced to 5×10^{-6} and 2×10^{-8} ergs s^{-1} cm^{-2} sr^{-1}, respectively. These predictions include corrections for saturation. The models of depth–dependent abundances and excitation in diffuse clouds are relevant also to the detailed description of "photodissociation regions" in UV-illuminated dense molecular clouds. For example, the measured C^+ line intensity 25 pc away from the M17 ionization front is about 2×10^{-4} ergs s^{-1} cm^{-2} sr^{-1} (Matsuhara et al. 1989; Stutzki et al. 1988). The diffuse and translucent cloud results demonstrate that it is possible that most of this radiation is produced in the warm outer layer of the cloud, rather than deep inside, so that no "exotic" chemistries would be required to explain its abundance.

2.3.4. Beyond steady-state models. To what extent can time–dependent and evolutionary effects modify the chemistry in diffuse clouds? The time scales for most chemical processes are so short, only about 10^3 years, that the abundances of most species (including CO) are not affected. The only exception is formed by H and H_2, since the time scale for H_2 photodissociation decreases from about 10^3 years at the edge to about 5×10^6 years in the center of a diffuse cloud. The time–dependent treatment of the H_2 abundance and excitation has recently been considered by Wagenblast and Hartquist (1988; 1989) in models in which the physical conditions are kept fixed with time. The results depend on the history of the cloud, in particular on the initial H/H_2 ratio. If the hydrogen was initially all in molecular form, the observed H/H_2 ratio in a cloud like ζ Oph can be reproduced in models with densities as low as 50 cm^{-3} if the cloud is younger than 5×10^5 years. On the other hand, if the hydrogen was initially all atomic, the models at early times require greater densities than those in steady–state. The rotational excitation, and hence the inferred strength of the ultraviolet radiation field, is not a sensitive function of initial conditions or cloud age.

Evolutionary models, in which physical conditions are allowed to change with time, have been developed by Tarafdar et al. (1985) and Prasad et al. (1987). In this picture, interstellar clouds spend most of their lifetime as diffuse clouds and evolve into dense clouds on a time scale of $10^6 - 10^7$ years. Although this model results in greatly different abundances in dense clouds, the abundances during the diffuse life span are very similar to those found in the steady–state diffuse cloud models, because of the short time scales of the chemical processes. It will thus be very difficult to find any good diagnostic of the evolutionary state of a diffuse cloud. A good diagnostic of the evolutionary state of denser clouds may again be provided by the H/H_2 ratio in the centers. H I observations of dark clouds such as L 134 suggest incomplete $H \rightarrow H_2$ conversion in some cases (van der Werf, Goss and Vanden Bout 1988).

The presence of additional shocked layers in diffuse clouds has been suggested by Elitzur and Watson (1978; 1980) to explain the observed abundances of the CH^+ molecule. The higher temperatures in the shocked region result in greatly enhanced rates for the $C^+ + H_2 \rightarrow CH^+ + H$ reaction which is endothermic by about 0.4 eV. Subsequent, more sophisticated magnetohydrodynamic (MHD) shock models have been developed by Draine and Katz (1986) and Pineau des Forêts et al. (1986). The presence of the magnetic field causes the ions and neutral species to have different flow speeds, and this velocity difference can contribute part of the energy required for the $C^+ + H_2$ reaction to occur. The attractive feature of the MHD models is that they provide a natural explanation for the fact that CH^+ production is enhanced in shocks, but that of neutral species such as OH to a much smaller extent. Although the recent models have been quite successful in reproducing the measured CH^+ column densities with reasonable shock parameters, they tend to overestimate the OH abundance and the H_2 rotational excitation. Other problems with the current shock models have been summarized by vDB and Hartquist et al. (1989). Observational support for the formation of the CH^+ ion in high–temperature gas comes from the measured correlation of CH^+ column density with that of rotationally excited H_2 (Lambert and Danks 1986). Another strong argument in favor of the shock models was the observed velocity difference between the CH and CH^+ lines of typically a few km s^{-1} (Federman 1982; 1987), which is well reproduced in the models. However, this velocity difference has recently been remeasured by Hawkins and Craig (1989) to be less than 0.5 km s^{-1} for the ζ Oph cloud. Since there is independent evidence in the Ophiuchus region for a shock resulting from a supernova explosion about 10^6 years ago (de Geus 1988), the absence of this signature in the CH^+ data is surprising. In addition, Crawford (1989) has found no velocity differences greater than 1 km s^{-1} for about 10 lines of sight in the Scorpius OB1 association. Finally, Jenkins et al. (1989) note that the CH^+ absorption in the direction of π Sco coincides exactly with their central H_2 velocity component. Further observational tests of the shock vs. steady–state scenario using the Hubble Space Telescope have been suggested by vDB.

3. TRANSLUCENT AND HIGH–LATITUDE CLOUDS

3.1 Characteristics

Translucent clouds with $A_V^{tot} \approx$2–5 mag have properties which lie in between those of the classical diffuse clouds such as ζ Oph ($A_V^{tot} \approx$1 mag) and the classical dense clouds such as TMC–1 ($A_V^{tot} \gtrsim$ 10 mag). Detailed studies of them may provide further insight into the evolutionary state of the diffuse clouds. Another motivation for studying translucent clouds is the increasing amount of evidence that dense molecular clouds have a "clumpy" structure through which the UV radiation can penetrate (e.g. Stutzki *et al.* 1988); the individual clumps may well be similar in size to the translucent clouds, so that a detailed study of their chemistry may be of paramount importance to understanding the chemistry of larger cloud complexes.

The clouds are called translucent because they are thin enough to allow optical absorption line studies toward background stars. Thus, photoprocesses still play a role in the chemistry, even though their importance diminishes toward the center. On the other hand, translucent clouds are thick enough to permit millimeter emission line observations. As discussed by Crutcher (1985), the combination of the optical and millimeter techniques has several advantages over studies based on either technique alone. Briefly, the big advantage of the millimeter observations is that the lines are in emission, so that mapping of the surroundings is possible. Also, the high spectral resolution ($\geq 10^6$) ensures that the lines are resolved, in contrast with the optical lines. Finally, a larger variety of molecules, especially the more complex ones, have been observed at millimeter wavelengths. On the other hand, the optical observations have the advantage of very high angular resolution, so that one can be sure that the same material is sampled for all species. Also, the derivation of column densities from the optical lines is more straight-forward, if the lines are not too saturated. Some important molecules with great diagnostic value, such as H_2 and C_2, cannot be observed at radio wavelengths, but have strong visible and ultraviolet absorption lines. Lines of atoms in various stages of ionization can be measured at optical wavelengths, which provides direct information on the depletion of the elements and the electron abundances. Finally, the extinction and polarization can be measured, providing information on the properties of the dust. The big disadvantage is, of course, the requirement of a bright star located fortuitously behind the cloud.

The most comprehensive study of a translucent cloud is that of Crutcher (1985) of the line of sight toward HD 29647, a heavily reddened late B star which lies behind a substantial part of the Taurus clouds. Early searches for molecular absorption lines toward heavily reddened stars were by Münch (1964) and Cohen (1973), whereas the first attempts to perform millimeter observations of diffuse clouds were those of Knapp and Jura (1976) and Dickman *et al.* (1983). More recent studies include those of Hobbs, Black and van Dishoeck (1983), Lutz and Crutcher (1983), van Dishoeck and de Zeeuw (1984), Crutcher and Chu (1985), Gredel and Münch (1986), Cardelli and Wallerstein (1986; 1989), Federman and Lambert (1988), Jannuzi *et al.* (1988) and van Dishoeck and Black (1989). About 25 stars with strong molecular absorption lines have so far been identified. Some of these shine through the outer edges of dense molecular clouds (such as the Taurus clouds), whereas others appear to lie behind isolated translucent clouds. Finally, it should be mentioned that translucent clouds can also be observed by absorption line studies at millimeter wavelengths against bright (extra)galactic background sources (see Cox, Güsten and Henkel 1988 for a summary). An excellent example is provided by the line of sight toward 3C123, which lies behind a region with $A_V^{tot} \approx 2-3$ mag in the Taurus clouds. This line of sight should therefore be very similar in properties to that toward HD 29647.

A similar class of clouds is formed by the high-latitude molecular clouds,

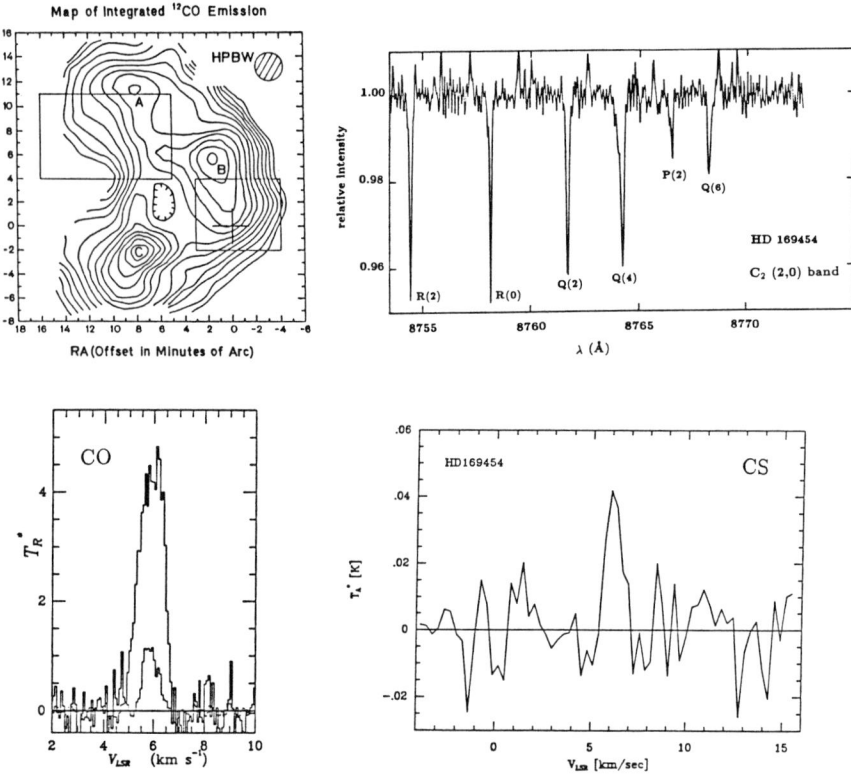

Figure 1. *Millimeter and optical observations toward HD 169454. (a) CO 1–0 map with the position of the star indicated by the cross (Jannuzi et al. 1988); (b) C_2 absorption lines (van Dishoeck and Black 1989); (c) ^{12}CO and ^{13}CO 1-0 emission at the position of the star; (d) CS 2–1 emission at the position of the star (Drdla et al. 1989).*

which were discovered through their millimeter CO emission by Magnani, Blitz and Mundy (1985) and Keto and Myers (1986). They are also associated with the more extended IRAS 100 μm cirrus emission (Weiland et al. 1986; Désert, Bazell and Boulanger 1988). Most of them are very nearby, with distances of 100 pc or less (Hobbs et al. 1986, 1988). Their averaged extinctions from star counts are found to be similar to those of the classical diffuse clouds, $A_V^{tot} \approx 1$ mag (Magnani and de Vries 1986), but it is likely that the extinctions are somewhat higher on small scales, especially in places where CO is detected. Thus, the high–latitude clouds of Magnani et al. may be more characteristic of the translucent clouds.

Based on large observed CO, OH and H_2CO abundances, Magnani, Blitz and Wouterloot (1988) suggested that the chemistry in high–latitude clouds differs significantly from that found in the classical diffuse clouds. In particular, they suggested that the clouds are very young, and have undergone frequent shocks during their formation from the atomic gas. This hypothesis can be tested through absorption line observations of species which are thought to be good shock diagnostics, such as the CH^+ molecule (see however §2.3.4).

Unfortunately, only few stars have so far been found behind high–latitude clouds which are bright enough to perform absorption line observations. De Vries and van Dishoeck (1988) have identified about 10 lines of sight with molecular absorption, whereas Welty et al. (1989) and Penprase and Blades (1989) have added another few.

In the following, discussion will focus on two well–studied lines of sight. The first is a good example of an isolated translucent cloud. Fig. 1 shows selected optical and millimeter observations toward HD 169454, a heavily reddened B1 supergiant (B1.5Ia, $V=6.6$, $A_V^{tot} \approx 3.3$ mag). As the figure shows, the cloud consists of 3 major condensations, and the line of sight toward the star passes fortuitously through one of the densest regions. The second example is the high–latitude cloud in the direction of HD 210121, a B5–6 V star of magnitude $V=7.5$ with $A_V^{tot} \approx 1$ mag (de Vries and van Dishoeck 1988).

3.2. Physical conditions. Translucent and high–latitude clouds can be analyzed in the same way as the classical diffuse clouds. The first step is again to try to constrain the physical conditions as good as possible. Table 1 includes the diagnostic species available for these clouds. The biggest difference with the diffuse clouds is that no observations of the H_2 rotational excitation are available; it is not yet certain whether the high–resolution UV spectrometer on the HST will be able to provide any data on H_2 for such clouds in the near future. Thus, other diagnostics which can be observed from Earth need to be used to constrain the temperatures and densities in the clouds. Note that, because of the absence of any H_2 data, no constraints are available on the scaling factor for the UV radiation field, I_{UV}. The most useful probe is the C_2 molecule, which has been observed in a number of translucent clouds (van Dishoeck and Black 1989). The central temperatures are typically found to be somewhat lower, $T \approx 15\text{--}25$ K, and the densities somewhat higher, $n_H \approx 700\text{--}1500$ cm^{-3}, than in the classical diffuse clouds. C_2 has very recently been detected in the high–latitude cloud toward HD 210121 (Gredel, van Dishoeck, de Vries and Black 1990a). A preliminary analysis suggests that the temperature is quite low, $T \lesssim 20$ K, and the density similar to that in the translucent clouds.

Another possible probe of the density is provided by the CO molecule. In contrast with high–dipole moment molecules such as CS, the excitation of CO in translucent clouds is dominated by collisions with neutral species, rather than with electrons. In diffuse clouds, the CO rotational excitation can be obtained from UV absorption lines in the A–X band around 1300–1500 Å. For translucent and high–latitude clouds, such UV data are not yet available, but the CO excitation can also be inferred from millimeter observations of the molecule. In particular, the CO $J=1\text{--}0/J=3\text{--}2$ antenna temperature ratio is a sensitive function of the density in the cloud, and the observed ratios result in densities similar to those found from the C_2 excitation (van Dishoeck et al. 1990).

In principle, the analysis of high–dipole moment molecules such as CN and CS could determine the electron densities in the cloud. In practice, their excitation is found to be very close to that of the cosmic background radiation, $T_{ex} \approx 2.7$ K (Meyer and Jura 1985; van Dishoeck and Black 1989), and no lines of sight with significant excess have yet been firmly identified.

3.3. Molecular abundances. Table 3 compares the observed column densities for

Table 3. Observed column densities (in cm^{-2})

Species	Diffuse		High–Latitude	Translucent	
	ζ Per	ζ Oph	HD 210121	HD 169454	3C123
H_2	(4.8±1.0)(20)	(4.2±0.3)(20)	(6–10)(20)	(1.5–2.2)(21)	(1–2)(21)
H	(6.5±0.7)(20)	(5.2±0.2)(20)	$\geq$1(20)	$\geq$2(20)	$\leq$1.7(21)
CH	(2.0±0.3)(13)	(2.5±0.3)(13)	(3.5±0.5)(13)	(5.8±0.8)(13)	7(13)
CH$^+$	(3.5±0.4)(12)	(2.9±0.3)(13)	(6±2)(12)	(3.0±1.0)(13)	...
C_2	(1.9±0.3)(13)	(2.4±0.3)(13)	(3±1)(13)	(7.0±1.4)(13)	...
CN	(3.0±0.3)(12)	(2.9±0.3)(12)	(1.4±0.6)(13)	(5.5±0.6)(13)	...
OH	(4.2±0.5)(13)	(4.8±0.5)(13)	...	...	2(14)
CO	(6.1±3.0)(14)	(2.0±0.3)(15)	(2–8)(15)	(1–5)(16)	2(17)
CS	...	(0.7–5)(12)	(0.4–8)(13)	(0.4–7)(13)	...
H_2CO	...	...	...	...	1(13)
C_3H_2	...	...	...	...	4(12)
A_V (mag)	1.0	0.8	1.1	3.3	2–3

^{a} See vDB, de Vries and van Dishoeck (1988), Jannuzi et al. (1988) and Cox et al. (1988) for references to the observations.

two classical diffuse clouds, the high–latitude cloud toward HD 210121, and the translucent cloud toward HD 169454. Also included are the results of millimeter absorption line observations of the translucent cloud toward 3C123. The main conclusion to be drawn from this table is that for many species, such as CH, C_2, CN, CO and probably CS, there is a gradual increase in molecular column densities from the diffuse clouds to the translucent clouds. The column densities in the high–latitude cloud fall in the middle, and do not appear anomalous at all. The CO column density increases more steeply than that of other species, which is well understood from the theoretical models (see below). The CH$^+$ column density is surprisingly small toward HD 210121, and suggests that if shocks are responsible for the formation of this molecule, they affect the chemistry to a smaller extent in this high–latitude cloud than in the classical diffuse clouds.

Table 3 illustrates that care has to be taken in the comparison of molecular abundances in diffuse and translucent clouds, especially with respect to claims of "normal" or "abnormal" abundances. Because the concentrations of many species change drastically with depth (e.g. the fraction of carbon in CO can change from less than 1% to 100%), the only correct method is to compare the total observed *column* densities. In contrast with dense clouds, column density ratios cannot in general be equated with local density ratios in the center of the clouds. For example, the abundances of some species like CH are large only in the outer zone of the cloud, where photoprocesses play a role. Thus, the column densities of these species do not continue to grow with cloud thickness, but will level off. Once this limit has been reached, the column density ratio of such a molecule with respect to that of molecules like H_2 or CO will start to decrease with increasing cloud thickness. This could lead to incorrect conclusions about the molecule being underabundant. The only proper way to analyze such data is through depth–dependent models. Tabulations of *average* abundances as functions of *average* extinctions determined from e.g. star counts are of little use for comparison with models, since the high–latitude and translucent clouds show structure on very small scales.

In the following, the individual species will be discussed in more detail.

3.3.1. H_2 and CH. As illustrated by van Dishoeck and Black (1989), there is in general not a close correlation between the column densities of molecules such as CH and the extinctions toward the stars. This is most likely due to the fact that many of the stars in the translucent cloud sample are distant supergiants, for which large fractions of the extinction are contributed by diffuse atomic clouds along the line of sight not associated with the molecular regions. Thus, the extinction toward the star provides only an upper limit to the H_2 column density. Because of the closer location of the stars to the clouds, this problem should be less for the high–latitude clouds. However, these clouds are thin enough that the amount of H associated with the cloud should be considered in deriving H_2 column densities. Also, assumptions need to be made about the ratio of total to selective extinction, $A_V/E(B-V)$, and the overall gas to dust ratio, A_V/N_H. These ratios have been calibrated for local diffuse clouds, but do not necessarily apply to the thicker translucent and high–latitude clouds.

A better indicator of the amount of molecular material is provided by the CH molecule, for which a linear relation with H_2 has been established by Danks *et al.* (1984) for classical diffuse clouds. The work of Mattila (1986, 1989) suggests that the correlation also holds for larger extinctions in dark clouds. The two proposed relations generally give similar H_2 column densities. It should be noted, however, that these relations are averaged over a number of clouds, and that for individual clouds, the CH/H_2 ratio may be different. For example, a higher than average radiation field, a lower gas–phase carbon abundance and/or a lower density will lower the CH/H_2 ratio. CH has also been detected in a number of high–latitude clouds by its 9 cm emission line (Magnani *et al.* 1989), and column densities similar to those found for HD 210121 have been inferred.

If the measured C_2 column densities are plotted as functions of the observed CH column densities, a very good correlation is found, indicating that the chemistries of these species are closely related (van Dishoeck and Black 1989; Federman and Huntress 1989).

3.3.2. CO. Models of the CO abundance in diffuse and translucent clouds, which take into account the latest molecular data regarding the photodissociation of the molecule, have been constructed by van Dishoeck and Black (1988b) and Viala *et al.* (1988a). It is now well established that the CO photodissociation occurs through a large number of line absorptions in the 912–1100 Å range. Van Dishoeck and Black (1988b) simulate the full absorption spectrum of CO and its isotopes at each depth into the cloud and consider not only self–shielding in the lines and continuum shielding by dust, but also mutual shielding of the isotopes by ^{12}CO and shielding by lines of H and H_2. Figure 2 illustrates the increase in CO column density with total H_2 column density in the cloud. For $N(H_2) \approx 5 \times 10^{20}$ cm^{-2} and $I_{UV} \approx 1$, the CO column density is typical of a diffuse cloud with $A_V^{tot} \approx 1$ mag, and the CO/H_2 column density ratio is about 10^{-6}. Most of the carbon is in the form of C^+ in these clouds. For $N(H_2) > 2 \times 10^{21}$

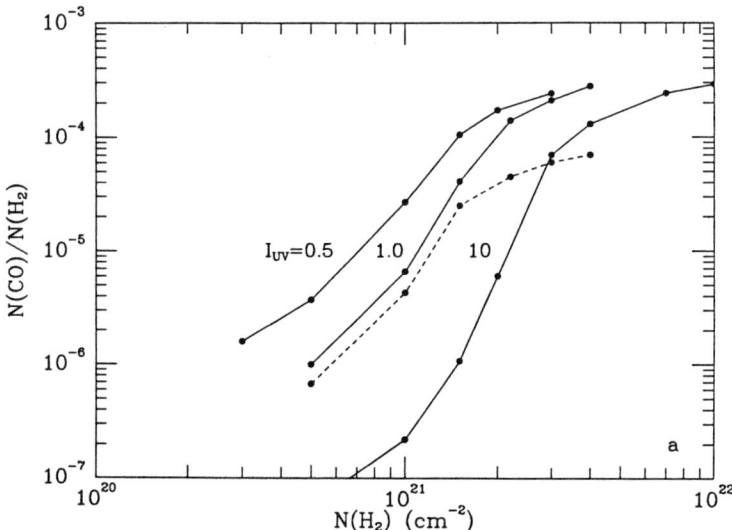

Figure 2. *Computed CO/H_2 column density ratio as a function of total H_2 column density for translucent clouds exposed to different strengths of the radiation field (from: van Dishoeck and Black 1988b).*

cm^{-2}, corresponding to $A_V^{tot} \geq 3$ mag, all the carbon in the center of the cloud is transformed into CO, and the CO/H_2 column density ratio reaches the dense cloud value of 10^{-4}. In the range $A_V^{tot} \approx 1-3$ mag, the CO column density increases very rapidly. The exact location and steepness of the transition region from diffuse to dense cloud abundances depend on the physical parameters of the cloud. If I_{UV} is decreased or the density increased, the transition region shifts to smaller A_V^{tot}. If I_{UV} is increased, the transition zone shifts to larger A_V^{tot} and is much steeper.

This steep increase in CO column density with H_2 column density is reproduced in the observations listed in Table 3. The HD 169454 cloud lies in the interesting region of parameter space where CO just starts to account for most of the gas–phase carbon. The CO column density in the HD 210121 high-latitude cloud can be well reproduced if the cloud is exposed to somewhat less UV radiation than the classical diffuse clouds, $I_{UV} \approx 0.5-1$, or if it has slightly larger densities. The CO abundance in this cloud is not "abnormal", and has not yet reached the dark cloud limit.

The CO column densities in classical diffuse clouds derived by Federman *et al.* (1980) from UV absorption line observations also show a steeper than linear relation with H_2 column density, even though these CO column densities are too small for the molecule to become self–shielding. This relation can be well reproduced in the current models, if the radiation field is not enhanced (see also §2.3.2.). The UV observations have measured CO column densities down to about 10^{12} cm^{-2}. Routine millimeter observations can detect CO down to an antenna temperature $T_A^* \approx 0.1$ K, corresponding to CO column densities of a few times 10^{14} cm^{-2}, more than 2 orders of magnitude larger! Lada and Blitz

(1988) compared CO column densities derived from millimeter emission with visual extinctions for a number of diffuse clouds, and concluded that there are two populations of diffuse clouds: CO rich and CO poor. This statement is somewhat misleading, since some of the CO emission in their "CO rich" clouds may come from material located behind the star. Also, the "CO poor" clouds have well determined CO column densities from UV observations, which are below the millimeter detection limit. Since the clouds studied by Lada and Blitz lie just in the transition region $A_V^{tot} \approx$ 1–3 mag, it is not surprising that they find larger CO column densities for some of the thicker translucent clouds. It merely reflects the transition zone illustrated in Figure 2, rather than the existence of two different "populations" of clouds.

3.3.3. CN. In contrast with the C_2–CH relation, the column densities of CN in translucent clouds show a much larger scatter when plotted as functions of CH column density. There appear to be lines of sight such as the HD 29647 and HD 147889 clouds which have very similar CH and C_2 column densities, but for which the CN column densities differ by more than an order of magnitude. Although the current chemical models can reproduce the *average* observed CN column densities if the radiation field is not enhanced (Federman *et al.* 1984), they cannot account for these large variations (van Dishoeck and Black 1989). Several possible solutions have been listed in the latter paper. One possible explanation lies in the observed variations of the shape of the UV extinction curve (Seab *et al.* 1981; Cardelli and Savage 1988). Since CN photodissociates primarily at the shortest wavelengths, λ <1000 Å, these variations can result in orders of magnitude differences in the central photorates in a cloud with $A_V^{tot} \approx 3$ mag. However, this effect is mitigated by the fact that at a depth of only 1 mag into a cloud, the removal of CN by atomic oxygen starts to compete with photodissociation. Unless O is substantially depleted in the center of translucent clouds, these variations in CN abundances remain poorly understood.

3.3.4. Other molecules. The OH radical has been seen in classical diffuse clouds by its UV absorption lines, and the measured column densities can be reproduced well with a cosmic ray ionization rate $\zeta_o \approx 6 \times 10^{-17}$ s^{-1}, if the H_3^+ dissociative recombination is slow. If a similar value of ζ_o is adopted in the models for the translucent and high–latitude clouds, the OH column densities are increased considerably. This is due mostly to the reduced radiation field in these clouds, which lowers not only the OH photodissociation rate, but also the abundance of C$^+$, the other principal destroyer of OH. OH column densities of a few times 10^{14} cm^{-2} have been found in high–latitude clouds and in the translucent cloud toward 3C 123, which can be well reproduced in the models.

The chemistry of CS in diffuse and translucent clouds has been discussed most recently by Drdla *et al.* (1989). The predicted CS abundance is sensitive to the rate of the reaction of S$^+$ with CH at low temperatures and the CS photodissociation rate, which is uncertain by a factor of five. Using reasonable values for these rates, the models reproduce well the observed column densities.

H_2CO has been detected in absorption in a number of high–latitude clouds (Magnani *et al.* 1988; Turner, Rickard and Lan–Ping 1989), and in absorption toward galactic and extragalactic radio continuum sources (Colgan, Salpeter and Terzian 1986; Güsten and Downes 1981). The inferred column densities are typically a few times 10^{12} cm^{-2}. Since H$_2$CO has not yet been seen in classical diffuse clouds, it is not known whether such a column density is "abnormal" or not. The H$_2$CO chemistry has been outlined e.g. by de Jong, Dalgarno and Boland (1980), although several reaction rates have changed since the appearance of that paper. The principal formation route appears to be through the reactions of CH$_3$ and CH$_5^+$ with O. The model column densities in translucent clouds are typically 10^{10} cm^{-2}, about 2-3 orders of magnitude smaller than observations.

C$_3$H$_2$ has been detected in emission in a few high–latitude cores by Turner et al. (1989) and in absorption toward radio continuum sources by Cox et al. (1988). The inferred C$_3$H$_2$ column densities are about $(3-5)\times 10^{12}$ cm^{-2} for the radio continuum lines of sight behind Taurus, and somewhat lower in the high–latitude cores. The C$_3$H$_2$ molecule is thought to be formed mostly from reactions of C$^+$ with CH$_4$ and C$_2$H$_2$, leading to the cyclic C$_3$H$_3^+$ ion, which subsequently recombines to form C$_3$H$_2$. The model predictions are typically 10^{10} cm^{-2} for a translucent cloud like HD 169454, about 2 orders of magnitude below observations. Both the H$_2$CO and C$_3$H$_2$ column densities would be increased substantially if some other source of CH$_4$ were available in the clouds, e.g. through grain surface formation or reactions with large molecules (Lepp et al. 1988). The widespread occurrence of C$_3$H$_2$ in diffuse clouds is surprising, since the molecule appears vulnerable to photodissociation (van Dishoeck et al. 1990b).

3.3.5. CH$^+$. The observations of CH$^+$ in high–latitude and translucent clouds add to the mystery of the origin of this molecule. As mentioned above, the CH$^+$ column densities in the high–latitude clouds are comparatively small (de Vries and van Dishoeck 1988) suggesting that the mechanism responsible for the formation of CH$^+$ does not play a more dominant role in the chemistry and structure of these clouds than in the classical diffuse clouds.

The measured CH$^+$ column densities in translucent clouds vary considerably from cloud to cloud (Cardelli and Wallerstein 1989; Gredel, van Dishoeck and Black 1990b). The CH$^+$ column density still continues to increase with total thickness of the cloud: lines of sight have been found with CH$^+$ column densities as large as 10^{14} cm^{-2}. Such an increase would not be expected on the basis of shock models, unless the number of shocked layers increases with total thickness of the cloud. On the other hand, there are also translucent clouds which, like the high–latitude clouds, show very little CH$^+$ absorption. Examples are HD 29647 (Crutcher 1985) and HD 62542 (Cardelli and Wallerstein 1989). The inferred densities for these clouds, and for the high–latitude clouds, are on the high side of the observed range ($n_H \gtrsim 1000$ cm^{-3}). Could the CH$^+$ formation be "quenched" by higher densities?

Of particular interest are also the high–resolution observations of Crawford (1989) of CH and CH$^+$ toward a number of highly reddened stars in the Sco OB1 association. Several velocity components are observed toward most stars, which can be divided into three distinct types: components present in both CH and CH$^+$, components present in CH$^+$ only, and components present in CH only. As mentioned in §2.3.4., the observed velocity differences between CH and CH$^+$ in components in which both are present are too small (<1 km s^{-1}) for significant shock production of CH$^+$, unless a very special geometry is adopted. The CH$^+$ column density in the "CH$^+$ only" components was found to be relatively constant, about $(5-10)\times 10^{12}$ cm^{-2}, and may arise in (shocked?) low density foreground clouds with only a small fractional abundance of H$_2$.

3.4. Small scale structure

There is growing evidence that diffuse and translucent clouds show significant variations in density, column density and/or velocity on very small scales (<30", corresponding to <0.02 pc at the distance of the nearest molecular clouds). Early observations of Liszt (1979) of the CO $J=1-0$ 2.6 mm and CH 9 cm emission toward ζ Oph suggested already complicated line profiles. Higher quality data of the CO line emission have been obtained by Langer, Glassgold and Wilson (1987), Crutcher and Federman (1987) and Le Bourlot, Gérin and Pérault (1989). The spectra of Langer et al. suggest that the gas toward ζ Oph is located in 4 distinct clumps, whose velocities differ by less than 1 km s^{-1}, and whose line widths ($\Delta V \leq 0.5$ km s^{-1}) indicate temperatures less than 100 K. Moreover, at least one of the components shows strong variations in the emission strength over scales as small as 12", corresponding to a linear size of only 0.006 pc, which has been interpreted in terms of large local density fluctuations (Le Bourlot et al. 1989). The small CO line widths exclude shock formation of the molecule.

Inhomogeneous structure is also found in even more diffuse atomic clouds, which can be studied only by absorption line observations. Jenkins et al. (1989) obtained high resolution (2 km s^{-1}) UV absorption spectra toward π Sco. This star has a large atomic hydrogen column density in front of it, $N(H) \approx 2 \times 10^{20}$ cm^{-2}, but an H$_2$ column density of only 2×10^{19} cm^{-2}, a factor of 20 below that of the ζ Oph cloud. The H$_2$ was found to be distributed over several components spaced 2.5 km s^{-1} apart, with the smallest components having H$_2$ column densities of only 10^{14} cm^{-2}.

The translucent cloud toward HD 169454 was mapped in CO emission by Jannuzi et al. (1988), and shows three major condensations with scales of several arcmin (see Figure 1), but with evidence for further structure on even smaller scales. Gredel, van Dishoeck and Black (1990b) are in the process of mapping small areas around the positions of each of the translucent and high-latitude background stars in CO $J=1-0$, and find small-scale structure in a significant fraction of the cases. Indeed, evidence for multiple velocity components and strong variation in emission strengths is now found in virtually all interstellar clouds, ranging from the diffuse clouds to the giant molecular clouds (Falgarone 1989). Broad, non-Gaussian line wings are found in many clouds not associated with star-forming activity, and the structure of the clouds appears self-similar on all scales (Blitz, Magnani and Wandel 1988; Falgarone and Phillips 1990; Falgarone et al. 1990).

If these small scale structures are so common in diffuse clouds, what is their origin and what will the effect be on the models and the chemistry? The clumpiness will certainly increase the surface area of the cloud, so that all photorates are effectively increased. Thus, the inferred scaling factors for the radiation field for each of the clumps will have to be lowered compared with the homogeneous models. It still remains to be investigated to what extent the various clouds shield each other; the radiative transfer in both continuum and lines through the various clumps will be non-trivial, and the results will depend strongly on the adopted geometry and the sizes of the individual clumps.

The observed broad line wings have recently been interpreted by Keto and Lattanzio (1989) as the result of clump–clump collisions. Using a three-dimensional smoothed-particle hydrodynamics code, they simulated the velocity and density fields due to two colliding clumps in various geometries. The computed CO emission profiles from such a source are remarkably similar to those observed in high–latitude clouds by Blitz et al. (1988). It will be of interest to incorporate the chemistry into a structure of colliding clumps, or into any other structure with localized supersonic turbulent motions. One particular question to be addressed is whether such turbulent motions can account for the chemical energy needed to drive the formation of the CH^+ ion. There is a tendency for lines of sight with large CH^+ column densities to show a more complicated velocity profile (Gredel et al. 1990b). The turbulent motions can also cause mixing of species located in the core of the cloud with those located in the envelope. Such chemical mixing was first considered by Boland and de Jong (1982), and more recently by Chièze and Pineau des Forêts (1989).

4. CONCLUDING REMARKS

To what extent can we answer the key questions raised in §2 regarding diffuse clouds?

(a): A number of good diagnostic tools are currently at our disposal to constrain the physical conditions in the classical diffuse clouds. The major problem lies in the appropriate strength of the incident ultraviolet radiation field. If some other mechanism such as shocks indeed contributes significantly to the high–J population of H_2, the strength may have been overestimated in a number of steady–state models. Observations of lines of H_2 arising in vibrationally excited levels using the HST should be able to distinguish between these two mechanisms (cf. vDB). Because also the photodissociation cross sections for molecules such as CO and CN have increased by an order of magnitude or more in recent years, the current steady–state diffuse cloud models produce column densities of these species that are an order of magnitude below observations. Some reduction of the radiation field in the models appears therefore needed. Whether the intensity has to be lowered over the full wavelength range or only specifically at the shortest wavelengths is still an open question. Also, deviations of the extinctions from the "standard" extinction curve may play a role. No constraints are currently available on the strength of the UV radiation field incident on the translucent and high–latitude clouds. The average densities of the diffuse clouds sampled by the absorption line observations appear to be of the order of a few hundred cm^{-3}, and may be somewhat higher in the translucent and high–latitude clouds. The temperatures in diffuse clouds appear to be of the order of 100–150 K at the edges, and 25–40 K in the centers. Such high temperatures at the edges could be maintained by the heating due to photoionization of large molecules such as PAH's, if they contained about 1–10% of the carbon in the cloud. The centers of translucent clouds, and possibly the high–latitude clouds, appear to be somewhat colder, $T \approx 15$–25 K.

(b): The abundances of many simple radicals such as CH and OH appear to be well described by steady–state gas–phase ion–molecule chemistry, although lingering uncertainties in crucial rate coefficients prevent firm conclusions. Even molecules such as CO and CN, whose abundances are currently more difficult to understand in the steady–state networks, are most likely produced by gas–phase low–temperature chemistry. In contrast, the ubiquitous presence of the CH^+ ion is still most easily explained by some high–temperature layer in the cloud, but current shock models have difficulties in reproducing the array of observational data. In particular, the recently measured absence of any significant velocity shift between CH and CH$^+$ will pose problems for the shock models. The most convincing argument against significant grain–surface production of the simplest molecules is still based on the observation by Crutcher and Watson (1976) that NH is much less abundant than CH and OH in diffuse clouds. This difference follows very naturally from the gas–phase low–temperature reactions, whereas the three species would be expected to have comparable abundances if grain chemistry dominated. Grains do, however, play an indirect role in the chemistry since they are responsible for the shape of the extinction curve. As discussed in §2.3.2 and 3.3.3, differences in shapes can have significant effects on the computed column densities. Also, grains may be important in the formation of more complex species.

(c): Unfortunately, few diagnostics of the evolutionary state of diffuse clouds can currently be identified. The most sensitive indicator may be the H/H$_2$ ratio in the centers of the clouds, since these species require the longest time scale for equilibration. However, it will be difficult to distinguish the evolutionary effects from other possible interpretations, such as different densities or different H$_2$ grain–surface formation rates. As a consequence, it is currently not clear whether diffuse clouds evolve into dense clouds or vice–versa. Systematic studies of the translucent and high–latitude clouds may provide further insight.

The study of diffuse clouds has suffered from a lack of new observational data over the last decade, especially from the fact that no sensitive high resolution spectrograph is available in space for searches at ultraviolet wavelengths, where most molecules have their strongest electronic transitions. The launch of the Hubble Space Telescope with its high–resolution spectrometer is therefore awaited anxiously. In the mean time, new impetus for the modeling of diffuse clouds has come from recent high spectral and spatial resolution millimeter observations, and it is anticipated that the small–scale structure of diffuse and translucent clouds will be an important topic of study in the coming years.

ACKNOWLEDGMENT. Much of the research reviewed in this paper has been performed in collaboration with J.H. Black. The author is grateful for his critical reading of the manuscript.

REFERENCES

Black, J.H. 1987a, in IAU Symposium **120**, *Astrochemistry*, eds. M.S. Vardya and S.P. Tarafdar (Reidel, Dordrecht), p. 217.
Black, J.H. 1987b, in *Spectroscopy of Astrophysical Plasmas*, eds. A. Dalgarno and D. Layzer (Cambridge University), p. 279.
Black, J.H. and Dalgarno, A. 1977, *Ap. J. Suppl.*, **34**, 405.
Black, J.H. and van Dishoeck, E.F. 1991, *Ann. Rev. Astr. Astrophys.*, in preparation.
Blitz, L., Magnani, M., and Wandel, A. 1988, *Ap. J.*, **331**, L127.
Boland, W. and de Jong, T. 1982, *Ap. J.*, **261**, 110.
Cardelli, J.A. and Savage, B.D. 1988, *Ap. J.*, **325**, 864.
Cardelli, J.A. and Wallerstein, G. 1986, *Ap. J.*, **302**, 492.
Cardelli, J.A. and Wallerstein, G. 1989, preprint.

Carruthers, G.R. 1970, *Ap. J. (Letters)*, **161**, L81.
Chièze, J.P. and Pineau des Forêts, G. 1989, *Astr. Ap.*, **221**, 89.
Cohen, J.G. 1973, *Ap. J.*, **186**, 149.
Colgan, S.W.J., Salpeter, E.E., and Terzian, Y. 1986, *Astron. J.*, **91**, 107.
Cox, P., Güsten, R., and Henkel, C. 1988, *Astr. Ap.*, **206**, 108.
Crawford, I.A. 1989, *M. N. R. A. S.*, in press.
Crutcher, R.M. 1985, *Ap. J.*, **288**, 604.
Crutcher, R.M. and Chu, Y.-H. 1985, *Ap. J.*, **290**, 251.
Crutcher, R.M. and Federman, S.R. 1987, *Ap. J. (Letters)*, **316**, L71.
Crutcher, R.M. and Watson, W.D. 1976, *Ap. J.*, **209**, 778.
Crutcher, R.M. and Watson, W.D. 1981, *Ap. J.*, **244**, 855.
Crutcher, R.M. and Watson, W.D. 1985, in *Molecular Astrophysics*, eds. G.H.F. Diercksen, W.F. Huebner and P.W. Langhoff, NATO ASI Series **157** (Reidel, Dordrecht), p. 255.
Dalgarno, A. 1988, *Astro. Lett. and Communications*, **26**, 153.
Dalgarno, A. and Black, J.H. 1976, *Rep. Prog. Phys.*, **39**, 573.
Danks, A.C., Federman, S.R., and Lambert, D.L. 1984, *Astr. Ap.*, **130**, 62.
Danks, A.C. and Lambert, D.L. 1983, *Astr. Ap.*, **124**, 188.
de Geus, E.J. 1988, Ph. D. thesis, Leiden University.
de Jong, T., Dalgarno, A., and Boland, W. 1980, *Astr. Ap.*, **91**, 68.
Désert, F.X., Bazell, D., and Boulanger, F. 1988, *Ap. J.*, **334**, 815.
de Vries, C.P. and van Dishoeck, E.F. 1988, *Astr. Ap.*, **203**, L23; *The Messenger*, **53**, 47; and in preparation.
d'Hendecourt, L.B. and Léger, A. 1987, *Astr. Ap.*, **180**, L9.
Dickman, R.L., Somerville, W.B., Whittet, D.C.B., McNally, D., and Blades, J.C. 1983, *Ap. J. Suppl.*, **53**, 55.
Douglas, A.E. and Herzberg, G. 1941, *Ap. J.*, **94**, 381.
Draine, B.T. 1978, *Ap. J. Suppl.*, **36**, 595.
Draine, B.T. and Katz, N.S. 1986, *Ap. J.*, **306**, 655; **310**, 392.
Drdla, K., Knapp, G.R., and van Dishoeck, E.F. 1989, *Ap. J.*, **345**, in press.
Dunham, T. 1937, *Pub. A.S.P.*, **49**, 26.
Elitzur, M. and Watson, W.D. 1978, *Ap. J.*, **222**, L141.
Elitzur, M. and Watson, W.D. 1980, *Ap. J.*, **236**, 172.
Falgarone, E. 1989, to appear in *Structure and Dynamics of the Interstellar Medium*, eds. G. Tenorio-Tagle, M. Moles, and J. Melnick (Kluwer, Dordrecht).
Falgarone, E. and Phillips, T.G. 1990, preprint.
Falgarone, E., Phillips, T.G., and Walker, C. 1990, in preparation.
Federman, S.R. 1982, *Ap. J.*, **257**, 125.
Federman, S.R. 1987, in IAU Symposium **120**, *Astrochemistry*, eds. M.S. Vardya and S.P. Tarafdar (Reidel, Dordrecht), p. 123.
Federman, S.R. and Huntress, W.T. 1989, *Ap. J.*, **338**, 140.
Federman, S.R. and Lambert, D.L. 1988, *Ap. J.*, **328**, 777.
Federman, S.R., Danks, A.C., and Lambert, D.L. 1984, *Ap. J.*, **287**, 219.
Federman, S.R., Glassgold, A.E., Jenkins, E.B., and Shaya, E.J. 1980, *Ap. J.*, **242**, 545.
Gerlich, D. and Kaefer, G. 1989, *Ap. J.*, in press.
Gredel, R. and Münch, G. 1986, *Astr. Ap.*, **154**, 336.
Gredel, R., van Dishoeck, E.F., de Vries, C.P., and Black, J.H. 1990a, in preparation.

Gredel, R., van Dishoeck, E.F., and Black, J.H. 1990*b*, in preparation.
Greenberg, J.M. 1987, in IAU Symposium **120**, *Astrochemistry*, eds. M.S. Vardya and S.P. Tarafdar (Reidel, Dordrecht), p. 501.
Güsten, R. and Downes, D. 1981, *Astr. Ap.*, **99**, 27.
Hartmann, J. 1904, *Ap. J.*, **19**, 268.
Hartquist, T.W., Flower, D.R., and Pineau des Forêts, G. 1989, to appear in *Molecular Astrophysics —A volume honoring Alexander Dalgarno*, ed. T.W. Hartquist (Cambridge University).
Hawkins, I. and Craig, N. 1989, private communication.
Heger, M.L. 1919, *Lick Obs. Bull.*, **10**, 59.
Herd, C.R., Adams, N.G., and Smith, D. 1989, *Ap. J.*, in press.
Hobbs, L.M., Black, J.H., and van Dishoeck, E.F. 1983, *Ap. J. (Letters)*, **271**, L95.
Hobbs, L.M., Blitz, L., and Magnani, L. 1986, *Ap. J. (Letters)*, **306**, L109.
Hobbs, L.M., Blitz, L., Penprase, B.E., Magnani, L., and Welty, D.E. 1988, *Ap. J.*, **327**, 356.
Jannuzi, B.T., Black, J.H., Lada, C.J., and van Dishoeck, E.F. 1988, *Ap. J.*, **332**, 995.
Jenkins, E.B. 1987, in *Interstellar Processes*, eds. D. Hollenbach and H.A. Thronson (Reidel, Dordrecht), p. 533.
Jenkins, E.B., Lees, J.F., van Dishoeck, E.F., and Wilcots, E.M. 1989, *Ap. J.*, **343**, 785.
Jura, M. 1975, *Ap. J.*, **197**, 581.
Keto, E.R. and Myers, P.C. 1986, *Ap. J.*, **304**, 466.
Keto, E.R. and Lattanzio, J.C. 1989, *Ap. J.*, in press.
Knapp, G.R. and Jura, M. 1976, *Ap. J.*, **209**, 782.
Lada, E.A. and Blitz, L. 1988, *Ap. J. (Letters)*, **326**, L69.
Lambert, D.L. and Danks, A.C. 1986, *Ap. J.*, **303**, 401.
Langer, W.D., Glassgold, A.E., and Wilson, R.W. 1987, *Ap. J.*, **322**, 450.
Le Bourlot, J., Roueff, E., and Viala, Y. 1987, *Astr. Ap.*, **188**, 137.
Le Bourlot, J., Gérin, M., and Pérault, M. 1989, *Astr. Ap.*, **219**, 279.
Lepp, S. and Dalgarno, A. 1988, *Ap. J.*, **335**, 769.
Lepp, S., Dalgarno, A., van Dishoeck, E.F., and Black, J.H. 1988, *Ap. J.*, **329**, 418.
Letzelter, C., Eidelsberg, M., Rostas, F., Breton, J., and Thieblemont, B. 1987, *Chem. Phys.*, **114**, 273.
Liszt, H.S. 1979, *Ap. J. (Letters)*, **233**, L147.
Lutz, B.L. and Crutcher, R.M. 1983, *Ap. J. (Letters)*, **271**, L101.
Magnani, L. and de Vries, C.P. 1986, *Astr. Ap.*, **168**, 271.
Magnani, L., Blitz, L., and Mundy, L. 1985, *Ap. J.*, **295**, 402.
Magnani, L., Blitz, L., and Wouterloot, J.G.A. 1988, *Ap. J.*, **326**, 909.
Magnani, L., Lada, E.A., Sandell, G., and Blitz, L. 1989, *Ap. J.*, **339**, 244.
Matsuhara, H. *et al.* 1989, *Ap. J. (Letters)*, **339**, L67.
Mattila, K. 1986, *Astr. Ap.*, **160**, 157.
Mattila, K. 1989, *Astr. Ap.*, **210**, 389.
McKellar, A. 1940, *Pub. A.S.P.*, **52**, 187; 312.
Merrill, P.W. 1934, *Pub. A.S.P.*, **46**, 206.
Meyer, D.M. and Jura, M. 1985, *Ap. J.*, **297**, 119.
Mitchell, J.B.A. and Guberman, S.L. 1989, *Dissociative recombination: theory, experiment and applications* (World Scientific, Singapore).

Münch, G. 1964, *Ap. J.*, **140**, 107.
Penprase, B.E. and Blades, J.C. 1989, private communication.
Pineau des Forêts, G., Flower, D.R., Hartquist, T.W., and Dalgarno, A. 1986, *M. N. R. A. S.*, **220**, 801.
Prasad, S.S., Tarafdar, S.P., Villere, K.R., and Huntress, W.T. 1987, in *Interstellar Processes*, eds. D. Hollenbach and H.A. Thronson (Reidel, Dordrecht), p. 631.
Roberge, W.G. 1981, Ph. D. thesis, Harvard University.
Roberge, W.G., Dalgarno, A., and Flannery, B.P. 1981, *Ap. J.*, **243**, 817.
Rowe, B.R. 1988, in *Rate Coefficients in Astrochemistry*, eds. T.J. Millar and D.A. Williams (Kluwer, Dordrecht), p. 135.
Seab, C.G., Snow, T.P., and Joseph, C.L. 1981, *Ap. J.*, **246**, 788.
Souza, S.P. and Lutz, B.L. 1977, *Ap. J. (Letters)*, **216**, L49.
Stark, G., Smith, P.L., Yoshino, K., Ito, K., and Parkinson, W.H. 1989, in preparation.
Stutzki, J., Stacey, G.J., Genzel, R., Harris, A.I., Jaffe, D.T., and Lugten, J.B. 1988, *Ap. J.*, **332**, 379.
Swings, P. and Rosenfeld, L. 1937, *Ap. J.*, **86**, 483.
Tarafdar, S.P., Prasad, S.S., Huntress, W.T., Villere, K.R., and Black, D.C. 1985, *Ap. J.*, **289**, 220.
Turner, B.E., Rickard, L.J., and Lan-ping, X. 1989, *Ap. J.*, **344**, 292.
van der Werf, P.P., Goss, W.M., and Vanden Bout, P.A. 1988, *Astr. Ap.*, **201**, 311.
van Dishoeck, E.F. 1988, in *Rate Coefficients in Astrochemistry*, eds. T.J. Millar and D.A. Williams (Kluwer, Dordrecht), p. 49.
van Dishoeck, E.F. 1989a, in *Highlights of Astronomy*, ed. D. McNally (Kluwer, Dordrecht), **8**, p. 323.
van Dishoeck, E.F. 1989b, to appear in *Molecular Astrophysics —A volume honoring Alexander Dalgarno*, ed. T.W. Hartquist (Cambridge University).
van Dishoeck, E.F. and Black, J.H. 1982, *Ap. J.*, **258**, 533.
van Dishoeck, E.F. and Black, J.H. 1986, *Ap. J. Suppl.*, **62**, 109.
van Dishoeck, E.F. and Black, J.H. 1988a, in *Rate Coefficients in Astrochemistry*, eds. T.J. Millar and D.A. Williams (Kluwer, Dordrecht), p. 209.
van Dishoeck, E.F. and Black, J.H. 1988b, *Ap. J.*, **334**, 771.
van Dishoeck, E.F. and Black, J.H. 1989, *Ap. J.*, **340**, 273.
van Dishoeck, E.F. and de Zeeuw, T. 1984, *M. N. R. A. S.*, **206**, 383.
van Dishoeck, E.F., Phillips, T.G., Black, J.H., and Gredel, R. 1990a, in preparation.
van Dishoeck, E.F., van Hemert, M.C., and Stehouwer, A. 1990b, in preparation.
Viala, Y.P., Letzelter, C., Eidelsberg, M., and Rostas, F. 1988a, *Astr. Ap.*, **193**, 265.
Viala, Y.P., Roueff, E., and Abgrall, H. 1988b, *Astr. Ap.*, **190**, 215.
Wagenblast, R. and Hartquist, T.W. 1988, *M. N. R. A. S.*, **230**, 363.
Wagenblast, R. and Hartquist, T.W. 1989, *M. N. R. A. S.*, **237**, 1019.
Watson, W.D. 1978, *Ann. Rev. Astr. Astrophys.*, **16**, 585.
Welty, D.E., Hobbs, L.M., Blitz, L., and Penprase, B.E. 1989, *Ap. J.*, in press.
Weiland, J.L., Blitz, L., Dwek, E., Hauser, M.G., Magnani, L., and Rickard, L.J. 1986, *Ap. J. (Letters)*, **306**, L101.

EVOLUTION OF THE CHEMISTRY IN DENSE CLOUDS

LUCY M. ZIURYS
Department of Chemistry, Arizona State University,
Tempe, AZ 85287-1604

ABSTRACT Molecule synthesis in dense interstellar clouds is examined from both theoretical and observational aspects. The chemistry of cold, quiescent clouds, which is based upon gas phase ion-molecule reactions, is reviewed. Changes to this chemistry resulting from star formation shock waves, and outflows are discussed.

INTRODUCTION

One of the most interesting ways in which the dense interstellar medium evolves is through its chemistry. This chemistry takes place in clouds with sufficiently large particle densities (10^3-10^7 cm^{-3} and higher) such that hydrogen is primarily in the form of H_2, and most other gaseous constituents are in the form of molecules. The high densities found in these objects also means that they are shielded from local UV radiation, as opposed to diffuse clouds. Thus, the interiors of these dense clouds are generally quite cold, with kinetic temperatures near 10 K, except for energetic regions where temperatures can approach 100-10,000 K (e.g. McKee and Hollenbach 1980).

Dense clouds are quite chemically rich, as is shown in Table I. To date over 65 different chemical species have been observed in these objects in the gas phase, ranging from simple diatomic hydrides such as CH and OH, to complex organic compounds like CH_3CH_2CN and CH_3OCH_3. As the table also illustrates, many molecules present in dense clouds are ions and free radicals, so called "transient species", which are quite chemically short-lived on earth. Also of note are the polymer-like, long-chain carbon compounds of the general formula, $H-(C\equiv C)_n CN$. Detection of these molecules has essentially established the field of astrochemistry, and is primarily the contribution of millimeter-wave astronomy.

Additionally present in dense clouds are dust grains, whose composition is less than certain. There is some

TABLE I   Molecules Detected in Dense Clouds

Species with General Abundances

H_2	CCH	HCO^+	H_2CS	CH_3CN
OH	HCN	N_2H^+	HNCO	CH_3CCH
CH	HNC	HCS^+	HC_3N	CH_2CN
CO	HCO	$H_2D^{+(t)}$	H_2CCO	CH_3OH
NO	OCS	$HOCO^+$	NH_2CN	CH_3NH_2
SO	SO_2	$HCNH^+$	CH_2NH	CH_2CHCN
CN	H_2S	$H_3O^{+(t)}$	C_3H_2	HC_5N
NS	HCl	H_2CO	$c-C_3H$	HC_2CHO
CS	H_2O	NH_3	CH_3CHO	

Species Abundant Only in Cold, Quiescent Clouds

C_3N	C_3H	C_4H	C_5H	HC_7N
C_6H	C_3O	C_2S	C_3S	HC_9N
CH_3C_2CN	CH_3C_4CN	CH_3C_4H		$HC_{11}N$

Species Abundant Only in Hot, Perturbed Gas

SiO	SiS	CH_3SH	$HCOOCH_3$	CH_3CH_2CN
PN	HCOOH	HNCS	CH_3CH_2OH	CH_3OCH_3
		NH_2CHO		

t) tentative detection

evidence for the grains to be silicate or graphite in nature
(e.g. Knacke 1988). Another suggestion is that they are
composed in part of "PAHs", i.e., large polycyclic aromatic
hydrocarbons (Léger and Puget 1984; Allamandola et al.
1985). In addition, there is somewhat stronger evidence for
grains possessing molecular mantles composed of H_2O, CO, and
perhaps NH_3 (Knacke et al. 1982) and H_2S (Geballe et al
1985). The exact composition of dust grains, however,
remains controversial, in contrast to gas phase molecules,
whose identification is securely based on high resolution
(i.e. rotational) spectroscopy.

Studying the evolution of the chemistry in dense clouds
is difficult. First of all, determination of cloud ages is
highly uncertain. Estimates are made from dynamical
timescales and times of free-fall collapse, but it is not

even clear how molecular clouds are formed in the first place (e.g. Kwan and Valdes 1983). However, it is evident from observations that many dense clouds are quite cold and quiescent, with no obvious indication of energetic activity (so-called "dark" clouds such as those in the Taurus complex, and many Lynd's objects). It is also observationally clear that other dense clouds such as Orion are highly perturbed, which is usually a result of star formation or supernova destruction. The presence of outflows, shocks, and young stars in these clouds can result in elevated temperatures and densities. Thus, examining the chemical evolution in dense clouds actually becomes a matter of formulating the chemistry in the quiescent regions, and then studying how it changes as a result of energetic activity.

CHEMISTRY IN COLD, QUIESCENT CLOUDS

The Ion-Molecule Scheme of Interstellar Synthesis

It is currently accepted that molecules in cold dense clouds are primarily formed by gas phase, positive ion-molecule reactions of the form:

$$A^+ + B \rightarrow C^+ + D.$$

Synthesis by such reactions, originally postulated by Herbst and Klemperer (1973) and Watson (1973), was suggested because about 50% of such processes are thought to be exothermic. Also, the exothermic channels usually have negligible activation energy barriers, and proceed at the Langevin rate, which is fast ($k \sim 10^{-9}$ $cm^3 s^{-1}$). Additionally, ion-molecule reactions concern only two-body collisions. Given such essential characteristics, these reactions can readily occur in the low temperature ($T \sim 10$ K), and low density ($n \sim 10^4 - 10^5$ cm^{-3}) environment of dark "dense" clouds. This is in contrast to other types of chemical reactions, in particular those between two neutral species, which usually have activation barriers and often may involve three-body collisions (requiring $n \geq 10^{11}$ cm^{-3}). Consequently, these reactions cannot occur quickly enough to efficiently produce molecules during the lifetime of a dense cloud, presumably $10^7 - 10^8$ years.

The basic network of ion-molecule chemistry (Herbst and Klemperer 1976, Watson 1976) relies only on gas phase processes; grains are ignored, except for the production of H_2 (see the last section in this article). The chemistry is initiated by cosmic rays which leads to the formation of H_3^+, a "key ion" in the network:

$$H_2 + c.r. \rightarrow H_2^+ + e^-$$

$$H_2^+ + H_2 \rightarrow H_3^+ + H.$$

H_3^+ then reacts with other species to begin chemical chains which lead to a variety of compounds, for example (e.g. Millar 1989):

$$H_3^+ + C \rightarrow CH^+ \rightarrow CH_2^+ \rightarrow CH_3^+$$

$$H_3^+ + O \rightarrow OH^+ \rightarrow H_2O^+ \rightarrow H_3O^+$$

$$H_3^+ + CO \rightarrow HCO^+ \rightarrow H_3CO^+.$$

Also included in the chemical network are certain dissociative recombination reactions, which are important as they lead to neutral products:

$$AH_2^+ + e^- \rightarrow AH + H.$$

Radiative association processes, of the general form

$$A^+ + B \rightarrow AB^+ + h\nu$$

are incorporated as well, since they are a likely route to large molecules (Herbst and Klemperer 1976). Additionally, neutral-neutral reactions which have reasonably fast rates at low temperatures are used.

In order to calculate abundances, this reaction network is incorporated into a coupled system of non-linear equations. The early models of ion-molecule chemistry (i.e. Herbst and Klemperer 1976; Mitchell, Ginsburg and Kuntz 1978) determined molecular abundances by allowing the chemistry to reach steady-state, i.e., for all species $d[A]/dt = 0$. It was found in general that steady-state could be reached in about 10^6-10^7 years, or well within the expected lifetime of a dense cloud (Herbst and Klemperer 1976). Later models (e.g. Prasad and Huntress 1980, Millar and Nejad 1985; Herbst and Leung 1986) adopted a "pseudo" time-dependent approach, i.e., abundances were followed as a function of time. These models showed that abundances of many species "peaked" before steady-state, at a so-called "early time" of ~ 3×10^5 years (Millar 1989).

Comparison of Theory and Observation
Molecules have been detected toward numerous cold, quiescent clouds. However, only two such objects, TMC-1 and L13N, have been observed in a sufficient number of species to serve as tests of ion-molecule theory. Observations towards these

sources have primarily been the work of Irvine and
collaborators (e.g. Irvine et al. 1989), Swade (1987), and
Winnewisser, Walmsley, and collaborators (e.g. Walmsley,
Winnewisser and Toelle 1980). These studies have shown that
almost all interstellar molecules that have been detected in
dense clouds have been observed in TMC-1 and L134N, with the
exception of the largest organic species that contain oxygen,
i.e. CH_3OCHO, CH_3OCH_3, CH_3CH_2OH and a few containing
refractory elements (see Table I). Molecular abundances
derived from observations of TMC-1 and L134N have been found
to be in good agreement with the predictions of ion-molecule
models. In fact, ion-molecule chemistry has been quite
successful in accounting for the concentrations of dark cloud
species, especially if "early time" (~10^5 yrs.) abundances
are used (Millar 1989). Such conclusions do not imply,
however, that cold clouds are 10^5 years old, since, as the
models point out, initial conditions are quite arbitrary.

Although for the most part theory and observation agree,
there are several problems with ion-molecule models. One of
these difficulties is illustrated in Fig. 1, the spectrum of
H_2S as observed toward TMC-1 and L134N (Mihn, Irvine and
Ziurys 1989). Almost all chemical pathways to H_2S are
thought to contain large activation barriers (e.g. Watson and
Walmsley 1982); thus, its presence in cold gas cannot readily
be explained. Another "problem" molecule is NH_3. Like H_2S,
many simple pathways to its synthesis contain barriers and/or

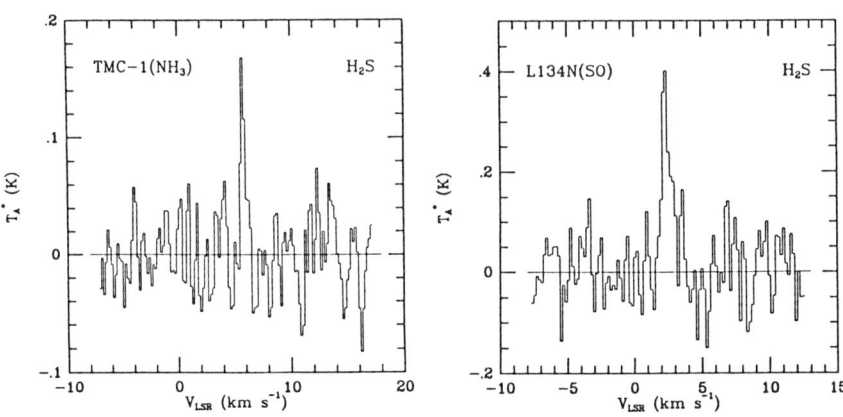

Fig. 1. Spectra of $1_{10}-1_{01}$ transition of H_2S observed
toward TMC-1 and L134N (from Mihn, Irvine, and Ziurys
1989).

are endothermic, and ion-molecule models cannot produce
enough of it to account for its observed abundance (Millar
1989). Nor can they reproduce the relatively high observed
concentrations of large organic molecules such as H_2CCO,
HC_5N, and HC_7N. Discrepancies concerning the complex species
may lie in the lack of understanding of radiative association
and dissociative recombination reactions (Herbst and Bates
1988; Millar et al. 1988).

Yet another difficulty with the theory is its failure to
account for chemical differences between TMC-1 and L134N, and
chemical variations within the clouds themselves (Irvine et
al 1987). Part of this problem may be in differences in
elemental abundances between the clouds, i.e. TMC-1 is
thought to be more carbon-rich than L134N. On the other
hand, a certain amount of cloud-to-cloud chemical variation
is probably to be expected.

CHEMISTRY IN ENERGETIC REGIONS

Theories of Shock Chemistry

The concept of "shock", or simply "high temperature"
chemistry began as a result of the detection of broad line
wings on molecular spectra observed toward dense clouds.
These broad wings suggested the presence of outflowing
material and interstellar shock waves. The first models of
shock chemistry (Iglesias and Silk 1978; Hartquist,
Oppenheimer, and Dalgarno 1980) investigated the effect of a
shock wave propagating into a dense cloud, and the subsequent
rise in temperature and pressure on gas phase chemistry.
Their basic result was that the elevated temperatures caused
by the shocks allowed endothermic reactions, and ones with
activation barriers, to occur that otherwise would not take
place at low temperatures. Opening the new chemical pathways
changed expected "ion-molecule" abundances. Although details
of the models, including the more recent ones (e.g., Mitchell
1984a, Leen and Graff 1988, Hollenbach and McKee 1989,
Neufeld and Dalgarno 1989) differ, they in general agree on
major points: CO is unaffected by the shock, while the
abundances of certain oxygen compounds (OH, H_2O, SiO) and
sulfur containing-species (H_2S, SO, and SO_2) are enhanced
relative to those predicted by ion-molecule models.

Observational Studies of Shock Chemistry

There are many dense clouds that show evidence of energetic
activity whose chemical evolution is of interest. However,
only two such regions have been studied in detail for
evidence of shock chemistry. These are the Orion-KL region,
where high mass star formation has radically changed a once

quiescent cloud, and the clouds associated with SNR IC443, where a recent supernova shock has highly perturbed surrounding molecular material.

Effect of High Mass Star Formation: Orion-KL
The basic scenario for the Orion-KL region is shown in Fig. 2. A young ~25 $M_\odot$ star (IRc2) has formed in a quiescent

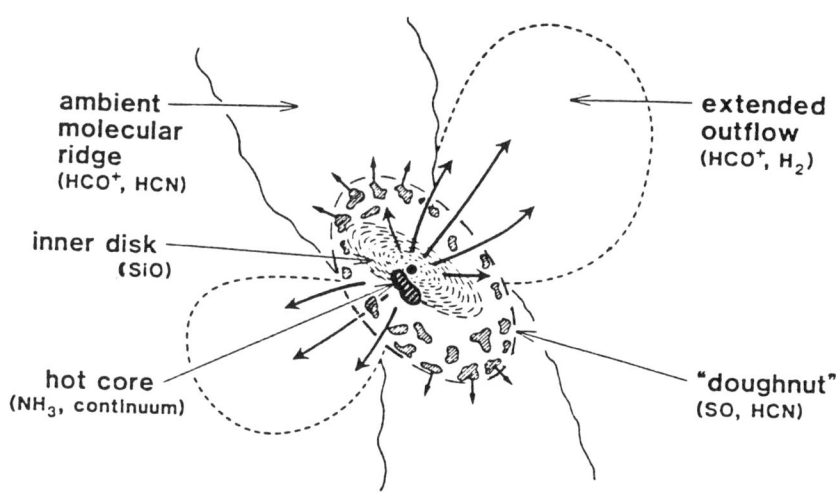

Fig. 2. Model for Orion-KL. IRc2 is the filled circle at the center (from Plambeck et al. 1985).

cloud OMC-1. In the course of its evolution, it has undergone mass loss, thus initiating several outflows into the ambient cloud, which shock and heat the surrounding material (e.g., Plambeck et al. 1985). The Orion-KL region can be separated into several distinct kinematic and chemical components (e.g., Wilson, Downes, and Bieging 1979; Johansson et al. 1984; Plambeck et al. 1985; Blake et al. 1987). First of all, there is the so-called "spike" or "ridge" component, which is the as yet unperturbed material in OMC-1, with $T_K \lesssim 30$ K and $n(H_2) \sim 10^5 cm^{-3}$. Close into the KL/IRc2 region, however, this material is heated and somewhat compressed to form the "spatially-confined ridge", "compact ridge", or "northern condensation" (T ~ 70 K, $n(H_2) \sim 10^6$ cm^{-3}). Centered on IRc2 there is the "low velocity outflow" region, with $T_K \sim 100$ K and $n(H_2) \gtrsim 10^6$ cm^{-3}, sometimes called the "plateau" or "doughnut"; this material is thought to be material swept up in an initial outflow from IRc2. A few arc seconds distant from IRc2 lies the "hot core" - a hot, dense clump of gas with T ~ 200 K and $n(H_2) \gtrsim 10^7$ cm^{-3}. Finally, a high velocity, bipolar flow, the "extended" outflow, exists perpendicular to the plateau, exhibiting T ~ 100 K and $n(H_2)$ ~ 10^5 cm^{-3}. Thus, several hot, presumably shocked regions exist with the KL/IRc2 region, on a scale of ~ 50,000 AU.

There have been many observational studies of the chemistry of Orion, several in the form of broad-band, mm-wave spectral line surveys (e.g., Johansson et al. 1984, Jewell et al. 1989). A thorough chemical analysis of Orion has been carried out by Blake et al. (1987), based on a 1.2 mm spectral survey. The basic results of the Blake et al. work can be summarized as follows. First, abundances in the spike or ridge are found to be well-predicted by models of ion-molecule chemistry. This is perhaps expected, since the physical conditions in this region are similar to the cold, dark clouds. However, the compact ridge, which is heated by KL/IRc2, is found to have large abundances of complex organic O-rich species such as CH_3OCH_3 and $HCOOCH_3$ - molecules <u>not</u> observed in cold clouds. Blake et al. explain that these abundances result from temperature-dependent radiative association reactions, an O-rich environment and a chemical network shown in Fig. 3. In contrast, the hot core appears to be dominated by fully-hydrogenated nitrogen-bearing compounds such as HC_3N, CH_3CN and CH_3CH_2CN which

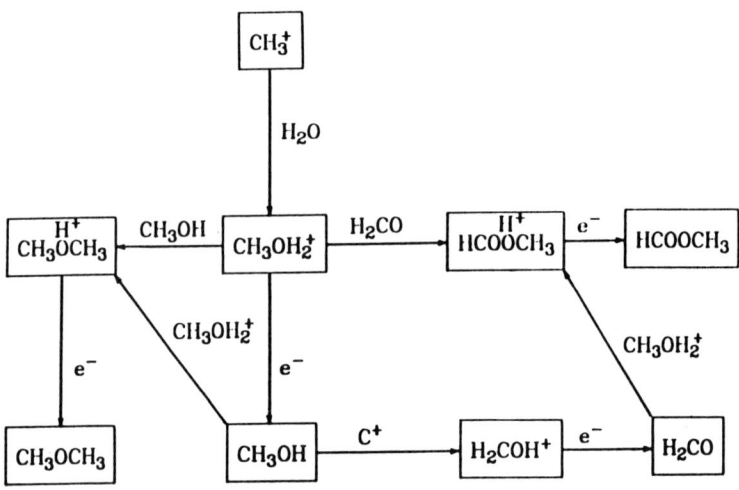

Fig. 3. Reaction network used to model the oxygen-based Orion compact ridge abundances (from Blake et al. 1987).

Blake et al. attribute to grain processing. The plateau, on the other hand, shows enhanced abundances of sulfur species such as SO, SO_2 and H_2S, as well as SiO - in good agreement with shock chemistry models.

From the Blake et al. work, as well as other observations, it is evident that varied physical conditions exist in the Orion regions, resulting in the production of different sets of molecules. Such chemical inhomogeneities have become particularly obvious in studies of Orion-KL done with the Hat Creek interferometer, made with synthesized beams of ~ 3"-4", or 2000 AU at the distance of Orion. The variation in the spatial and chemical distributions of the molecules is shown in Fig. 4. As the figure illustrates, HC_3N, OCS, and CH_3CN are concentrated in the hot core near IRc2, although HC_3N and CH_3CN appear to be present in the northern condensation as well. In contrast, HCN and SO_2 trace the plateau or doughnut. SiO, on the other hand, is only present close into IRc2, in the "hole" of the doughnut, where SO_2 and HCN are not very abundant (Plambeck and Wright 1988a).

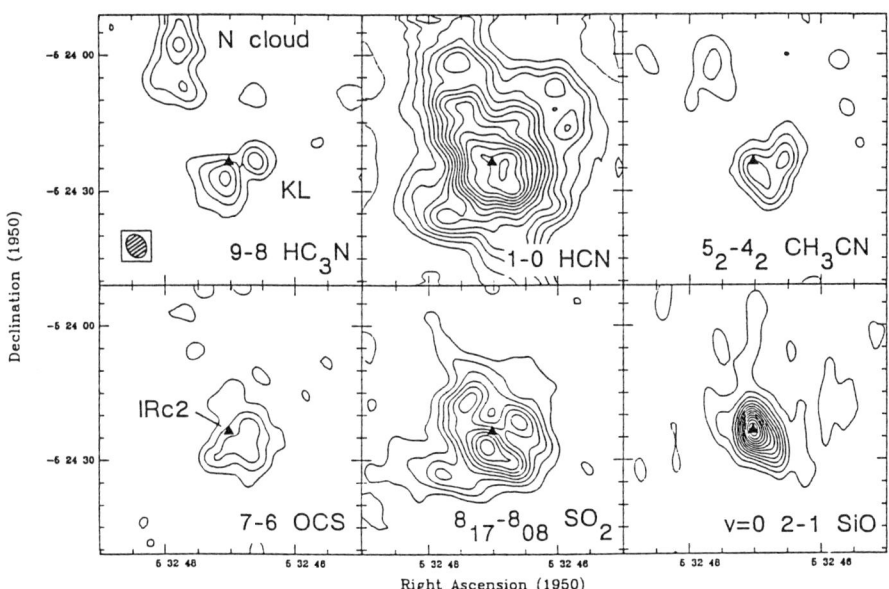

Fig. 4. Hat Creek interferometer maps of 6 molecular transitions observed toward Orion-KL. For the SiO and SO_2 maps, the contour level is 10 K; for the other maps, it is 5 K (from Plambeck and Wright (1988a)).

Another aspect of Orion chemistry has been the usually large abundance measured for HDO in the hot core (Plambeck and Wright 1987); in this region, HDO/H_2O was found to be $\sim 10^{-3}$, vs. the cosmic H/D ratio of $\sim 10^{-5}$. Similar deuterium enhancements have been found for NH_2D and CH_3OD in the hot core as well (Walmsley et al. 1987; Mauersberger et al. 1988). Such deuterium enhancements are predicted by ion-molecule chemistry to occur through H_2D^+, which becomes heavily deuterated in <u>cold</u> gas due to the reaction:

$$H_3^+ + HD \rightleftarrows H_2D^+ + H_2 + \Delta E$$

$$\Delta E \sim 100\text{-}200 \text{ K.}$$

Thus, one would not expect such deuterium enrichment in the hot core, where T $\sim$ 200 K. These results have lead to the conclusion that HDO, NH_2D, and CH_3OD are products of earlier low temperature, ion-molecule chemistry that were subsequently frozen out onto grains. The elevated temperatures in the hot core have now evaporated the grain mantles, therefore releasing these "fossil" molecules back into the gas phase. Because these species show such a high degree of deuterium enhancement it has been claimed they have not yet been chemically processed in the hot core, which indicates that the age of the region must be $\sim 10^3\text{-}10^4$ yrs. old. This value agrees well with the dynamical age of Orion. On the other hand, the spatial distributions of HDO and CH_3OH appear to be poorly correlated (Plambeck and Wright 1988b), suggesting that other chemical processes may actually be involved in the deuterium enrichment.

<u>Effect of a Supernova Shock: The Clouds Near IC443.</u>
The supernova remnant IC443 is thought to be relatively young ($\sim 10^3\text{-}10^4$ yrs.); it is rather unique in that it is associated with small, dense clouds of only a few $M_\odot$'s that appear to be highly perturbed by the SN shock (e.g., Dickman et al. 1989). The large, asymmetric velocity dispersions observed in spectra of CO, OH, HCN, and HCO^+ (DeNoyer 1979, Dickinson et al. 1980; DeNoyer and Frerking 1981) are evidence for this shock perturbation.

A chemical analysis of the shocked clouds associated with IC443 has recently been carried out by Ziurys, Snell and Dickman (1989). They concentrated on studying the most perturbed regions, the so-called B and G clouds. Ziurys, Snell and Dickman detected several new species towards clouds B and G, including SiO, SO, N_2H^+, HNC, CN, NH_3, and $H^{13}CO^+$ (see Fig. 5); they observed the J=3 $\rightarrow$ 2 line of HCO^+ as well. Analysis of NH_3 and HCO^+ transitions showed that $T_K \geq$ 40K and $n(H_2) > 3\times10^5$ cm^{-2} towards both clouds, strong evidence for shock heating and compression.

A summary of the chemical abundances measured towards IC443 B and G is shown in Table II. For comparison, abundances for TMC-1 are shown. Because of the large dipole moment differences between CO and the molecules listed in the table (0.1 D vs. 1-4 D), abundances are referenced to HCN, whose dipole moment is 2.98 D (see Ziurys, Snell and Dickman 1989). As the table shows, the only molecule whose abundance

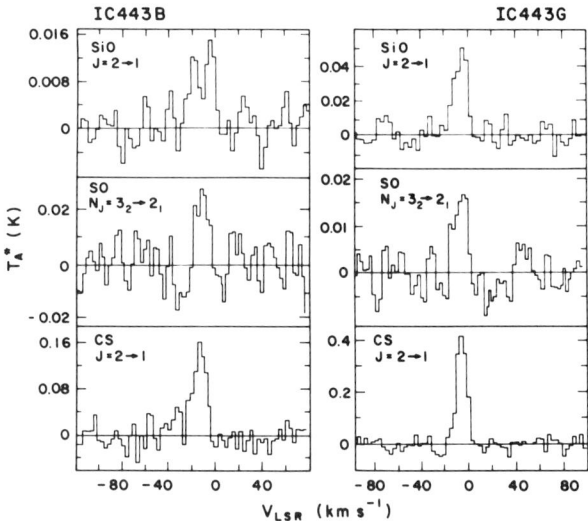

Fig. 5. Spectra of SiO, SO, and CS observed towards the B and G clouds in IC443. SiO and SO are newly detected species towards this region (from Ziurys, Snell, and Dickman 1989).

is significantly enhanced in the shocked gas towards IC443 is SiO, whose concentration is 100 times higher than the upper limit in TMC-1. The SO abundance in IC443, on the other hand, is only enhanced by factors of a few, if at all. The other species observed towards IC443 have nearly identical concentrations to those in TMC-1, with the exception of HNC, whose abundance is less in the shocked clouds by an order of magnitude. This is not unexpected. A value of HNC/HCN ~ 1 is predicted by ion-molecule chemistry (Watson and Walmsley 1982), but appears to fall significantly below that value in hot gas (e.g. Goldsmith et al. 1986).

It is useful to compare abundances found in IC443 vs. those measured in Orion-KL. IC443, however, does not appear as chemically complex as the Orion region. As far as can be deduced from observations, it consists of essentially "one" outflow. Also, the IC443 clouds are not nearly as massive as OMC-1 so overall column densities are lower, making complex species difficult to detect. As an attempt at comparison, some Orion-plateau abundances are also given in Table II, relative to HCN. Although not many species are listed in this table, some interesting points do emerge. As is in the case of IC443, the one molecule in the plateau whose abundance is at least several orders of magnitude higher than

TABLE II Abundances in IC443 Relative to TMC-1 and Orion Plateau

Molecule	Cloud	X/HCN[a)] IC443	X/HCN[b)] TMC-1	X/HCN[c)] Orion Plateau
HCO^+	B	1.2-1.5	0.4	-
	G	0.80-1.3	-	-
N_2H^+	B	0.035	0.025	-
	G	0.013	-	-
CS	B	0.8	0.5	0.08
	G	0.7	-	-
SO	B	0.82	0.25	1.87
	G	0.55	-	-
SiO	B	0.038	<0.0005	0.1
	G	0.063	-	-
HCN	B	1	1	1
	G	1	-	-
HNC	B	0.14	1	0.005[d)]
	G	0.11	-	-
CN	B	1.7	1.5	-
	G	0.89	-	-
NH_3	B	1.3	1	0.5[e)]
	G	0.4	-	-
CO[d]	B	20,000	4,000	430
	G	7,500	-	-

a) From Ziurys et al. (1989)
b) From Irvine et al. (1987)
c) From Blake et al. (1987), unless otherwise indicated
d) From Goldsmith et al. (1986)
e) From Watson and Walmsley (1982)

that found in TMC-1 is SiO. Also, the HNC/HCN ratio found in
the plateau is ~ 0.005-significantly less than one and
typical for hot gas. In contrast to the SNR clouds, however,
the abundance of SO in the plateau is about an order of
magnitude higher than in TMC-1, while that of CS is about an
order of magnitude lower. For IC443, these sulfur-bearing
molecules have abundances very close to those in TMC-1.
While the abundance differences concerning SO and CS do not
involve several orders of magnitude, they may be significant.

Observation vs. Theory: Is SiO Unique for Shock Chemistry?
Observations toward the Orion-KL and IC443 regions do
confirm, to some extent, the predictions of shock
chemistry. The best agreement is for SiO, which appears to
undergo a dramatic abundance enhancement in both shocked
sources, as predicted by theory. In fact, this enhancement
is found in other energetic regions as well. In a recent
survey of SiO (Ziurys, Friberg and Irvine 1989), it was found
that the species was readily observed in all dense clouds
where T_K > 30 K, and was most abundant in the warmest gas.
Also, SiO was not detected towards several cold, dark clouds
down to stringent lower limits (T_A^* ~ 10 mK). These
observations suggest a temperature dependence for the SiO

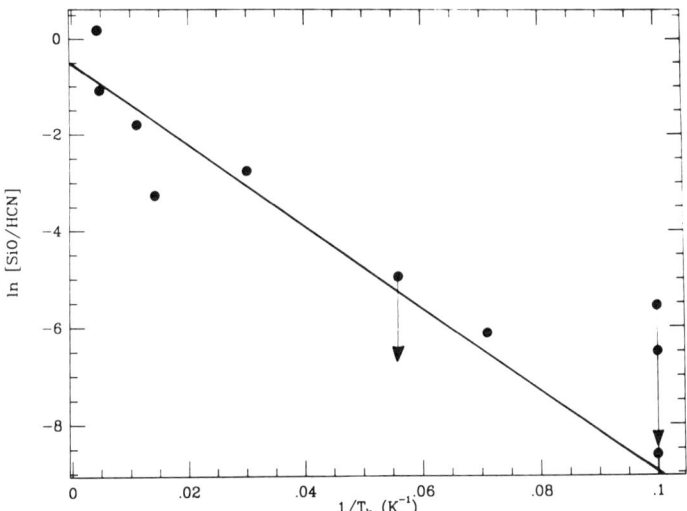

Fig. 6. Plot of the natural log of the SiO abundance
vs. the inverse of the kinetic temperature. The
apparent linear relationship suggests SiO is formed by a
process involving an activation energy near 90 K (from
Ziurys, Friberg, and Irvine 1989).

abundance, which is shown in Fig. 6. The apparent linear relationship between the natural log of the SiO concentration vs. $1/T_K$ suggests that the process leading to the formation of this species has an exponential temperature dependence with an activation energy of ~ 90 K, the slope of the line.

Langer and Glassgold (1989) have suggested an explanation for this temperature dependence, based on the work of Graff (1989). Graff postulated that neutral-neutral reactions with an atom as a reactant may occur with the atom in an excited fine structure state. Langer and Glassgold applied this idea to silicon, whose first excited fine structure state lies ~ 111 K above ground state. Working out a chemical network for SiO, they found that the species' abundance is proportional to $\exp(-\Delta E/T)$, where ΔE, the activation energy, is 111 K. In addition, they determined that the production of SiO requires gas densities $n(H_2) \geq 10^5$ cm^{-3}, in order to excite the fine structure line. These findings fit extremely well with observations.

It would be interesting to see if other silicon-containing species have abundances that are shock-enhanced. Only one other silicon compound, SiS, has been detected in dense clouds. The species has been observed towards Orion and appears to exist only in the outflow region where SiO is present (Ziurys 1988). It also has been detected towards several other outflow regions where SiO is observed as well (Ziurys 1989), and has yet to be seen in quiescent regions. These limited observations do suggest that SiS has a similar chemical behavior to SiO.

For sulfur-bearing species, the shock chemistry situation is not so clear. SO, SO_2 and H_2S appear to undergo significant abundance enhancements in the low velocity plateau outflow in Orion. Yet, SO seems to have a concentration in IC443 comparable to dark clouds. Moreover, the detection of H_2S in cold clouds TMC-1 and L134N adds additional complications to the picture. H_2S is predicted by theory to be particularly sensitive to high temperature production (e.g. Prasad and Huntress 1982; Mitchell 1984b). Almost all chemical pathways to the species are thought to contain large activation energies. The presence of H_2S in cold gas may point to errors in rate estimates or indicate some unusual rate behavior as $T \to 0$.

Another possible explanation for the apparent lack of consistency concerning the shock chemistry of sulfur-bearing species may lie in the H/H_2 ratio. Production of sulfur-bearing species behind shock waves may be quite sensitive to this ratio (e.g. Leen and Graff 1988). The clouds associated with IC443 do not appear as dense as the Orion plateau. Consequently, the H/H_2 ratios in these sources may be different, causing their chemistries to vary. Such an explanation does not account for the appearance of H_2S of dark clouds. As proposed by Charnley et al (1988), however, H_2S may be a residual of previous shocks and a chemistry that has not yet relaxed to equilibrium.

For molecules other than silicon and sulfur-bearing species, it is difficult to speculate on their shock behavior. Additional clouds need to be chemically analyzed, especially sources where the morphology is well characterized so that their chemistry is tractable.

THE ROLE OF GRAINS IN DENSE CLOUD CHEMISTRY

Ion-molecule and shock chemistry in general concern only gas-phase processes. However, surface chemistry on grains is also important. Surface reactions are thought to be the primary pathway to H_2. Gas phase processes leading to the formation of H_2 are far too slow, while the high mobility of H atoms on surfaces, and the high volatility of H_2, make surface formation efficient (e.g. Herbst and Klemperer 1976).

The role of grains and surface reactions in the synthesis of species other than H_2, however, is unclear. Observations at infrared wavelengths towards dense clouds indicate that grains have molecular mantles containing species such as CO, H_2O, and NH_3, among others (e.g. Knacke 1988). From observations of highly-deuterated compounds in the Orion hot core region, there is also evidence of molecules leaving grain mantles. Accretion timescale for molecules condensing out onto grains is through to be $\tau_{accr} \sim 3 \times 10^9/n$ yrs., where n is the particle density (e.g. Millar 1989). Typically for a dense molecular cloud, $n \sim 10^5$ cm^{-3}, which implies that all molecules should be frozen out in $\sim 3 \times 10^4$ years-well within a cloud lifetime. The problem then becomes, why aren't <u>all</u> gas phase molecules condensed out onto grain surfaces?

It is quite obvious from observations towards dense clouds that <u>many</u> molecules have remained in the gas phase. In fact, outside of the presence of molecular mantles, there is little observational evidence for molecule depletion. Wootten et al. (1978) reported to have observed such depletions, but the results were based on a radiative trapping analysis which is suspect. Frerking et al. (1982) did a study which compared dust extinction to CO column densities, and found no evidence for depletion. The Frerking et al. work may have not probed high density gas, but it can certainly be said that high density regions ($n > 10^6$ cm^{-3}) such as the Orion hot core have considerable molecular abundances.

Various mechanisms have been proposed for circumventing this "accretion catastrophe". It has been postulated that UV radiation removes molecules from grain surfaces, either through direct desorption, sublimation of mantles after grain heating, or grain "explosions" or "sputtering" (e.g. Barlow 1978; Draine and Salpeter 1979; d'Hendecourt et al 1982). X-ray and cosmic ray heating of grains may also cause grain mantle evaporation (Leger et al 1985). In addition, shock processing of grains could help return solid material back into the gas phase (e.g. Seab and Shull 1983). It is uncertain, however, if any of these processes work efficiently enough to solve the accretion problem, especially in dense clouds.

The exact role of surface chemistry on grains in the production of molecules other than H_2 remains an unanswered question. The problem is further compounded by uncertainties in grain composition, as well as a lack of general chemical understanding of surface processes. Observations thus far have only conclusively shown that molecular mantles exist on grains, and in energetic regions of dense clouds, these mantles are removed.

CONCLUSIONS

As a dense cloud evolves from a cold quiescent object to an energetic region perturbed by star formation, mass outflow, and shock waves, its chemistry also changes. Molecule formation in a cloud's nonenergetic phases can be fairly well explained by gas-phase ion-molecule reactions. Although there are exceptions, most observed abundances in these objects can be reproduced by ion-molecule theory. When events associated with the birth and death of stars occur, however, the production of molecules becomes a new problem. Theories of shock chemistry offer an explanation as to how the chemistry changes as a result of these events. Measurements of molecules in hot, shocked gas show that, at least as far as some silicon and sulfur-bearing species are concerned, there is agreement between theory and observation. However, the chemistry of other dense clouds showing energetic activity needs to be examined before generalizations can be made.

REFERENCES

Allamandola, L. J., Tielens, A. G. G. H., and Barker, J. R. 1985, Ap. J., **290**, L25.
Barlow, M. J. 1978, M.N.R.A.S., **183**, 367.
Blake, G. A., Sutton, E. C., Masson, C. R., and Phillips, T. G. 1987, Ap. J., **315**, 621.

Charnley, S. B., Dyson, J. E., Hartquist, T. W., and
 Williams, D. A. 1988, M.N.R.A.S., **235**, 1257.
DeNoyer, L. K. 1979, Ap. J. (Letters), **232**, L165.
DeNoyer, L. K., and Frerking, M. A. 1981, Ap. J. (Letters),
 246, L37.
Dickinson, D. F., Rodriquez-Kuiper, E. N., Dinger, A. S., and
 Kuiper, T. B. H. 1980, Ap. J. (Letters), **237**, L43.
Dickman, R. L., Snell, R. L., Ziurys, L. M., and Huang, Y.-L.
 1989, Ap. J., submitted.
Draine, B. T., and Salpeter, E. E. 1979, Ap. J., **231**, 438.
Frerking, M. A., Langer, W. D., and Wilson, R. W. 1982,
 Ap. J., **262**, 590.
Geballe, T. R., Baas, F., Greenberg, J. M., and Shuttle, W.
 1985, Astr. Ap., **146**, L6.
Goldsmith, P. F., Irvine, W. M., Hjalmarson, A, and Ellder,
 J. 1986, Ap. J., **310**, 383.
Graff, M. M. 1989, Ap. J., **339**, 239.
Hartquist, T. W., Oppenheimer, M., and Dalgarno, A. 1980,
 Ap. J., **236**, 182.
d'Hendecourt, L. B., Allamandola, L. J., Baas, F., and
 Greenberg, J. M. 1982, Astr. Ap. 109, L12.
Herbst, E., and Klemperer, W. 1973, Ap. J., **185**, 505.
Herbst, E., and Klemperer, W. 1976, Phys. Today, **29**, 32.
Herbst, E., and Leung, C. M. 1986, M.N.R.A.S., **222**, 689.
Herbst, E., and Bates, D. R. 1988, Ap. J., **329**, 410.
Hollenbach, D. J., and McKee, C. F. 1989, Ap. J., **342**, 306.
Iglesias, E. R. and Silk, J. 1978, Ap. J., **226**, 851.
Irvine, W. M., Goldsmith, P. F., and Hjalmarson, A. 1987, in
 Interstellar Processes, ed. D. J. Hollenbach and H. A.
 Thronson, Jr., Dordrecht: Reidel, p. 561.
Irvine, W. M. et al. 1989, Ap. J., **342**, 871.
Jewell, P. R., Hollis, J. M, Lovas, F. J., and Synder, L. E.
 1989, Ap. J. (Suppl.) 70, 833.
Johannson, L. E. B., et al. 1984, Astr. Ap., **130**, 227.
Knacke, R. F., McCorkle, S., Puetter, R. C., Erickson, E. F.,
 and Krätschmer, W. 1982, Ap. J., **260**, 51.
Knacke, R. F. 1988, in Molecular Clouds in the Milky Way and
 External Galaxies, ed. R. L. Dickman, R. L. Snell, and
 J. S. Young, New York: Springer-Verlag, p. 141.
Kwan, J., and Valdes, F. 1983, Ap. J. **271**, 604.
Langer, W. D., and Glassgold, A. E. 1989, Ap. J., in press.
Leen, T. M., and Graff, M. M. 1988, Ap. J., **325**, 411.
Léger, A., Jura, M., and Omont, A. 1985, Astr. Ap., **144**, 147.
Léger, A., and Puget, J. L. 1984, Astr. Ap., **137**, L5.
Mauersberger, R., Henkel, C., Jacq, T., and Walmsley, C. M.
 1988, Astr. Ap. (Letters), **194**, L1.
McKee, C. F. and Hollenbach, D. J. 1980, Ann. Rev. Astron.
 Astrophys., **18**, 219.

Mihn, Y., Irvine, W. M., and Ziurys, L. M. 1989, Ap. J. (Letters), in press.
Millar, T. J., and Nejad, L. A. M. 1985, M.N.R.A.S., **217**, 507.
Millar, T. J., DeFrees, D. J., McLean, A. D., and Herbst, E. 1988, Astr. Ap., **194**, 250.
Millar, T. J. 1989, in Molecular Astrophysics, ed. T. W. Hartquist, Cambridge University Press, in press.
Mitchell, G. F., Ginsburg, J. L., and Kuntz, P. J. 1978, Ap. J. (Suppl.), **38**, 39.
Mitchell, G. F. 1984a, Ap. J. (Suppl.), **54**, 81.
Mitchell, G. F. 1984b, Ap. J., **287**, 665.
Neufeld, D. A., and Dalgarno, A. 1989, Ap. J., **340**, 869.
Plambeck, R. L., Vogel, S. N., Wright, M. C. H., Bieging, J. H. and Welch, W. J. 1985, Symposium on MM and SubMM-Radio Astronomy, ed. J. Gomez-Gonzales, pub. by URSI, p. 235.
Plambeck, R. L., and Wright, M. C. H. 1987, Ap. J. (Letters), **317**, L101.
Plambeck, R. L., and Wright, M. C. H. 1988a, in Molecular Clouds in the Milky Way and in External Galaxies, ed. R. L. Dickman, R. L. Snell, and J. S. Young, New York: Springer-Verlag, p. 182.
Plambeck, R. L., and Wright, M. H. C. 1988b, Ap. J. (Letters), **336**, L61.
Prasad, S. S., and Huntress, W. T., Jr. 1980, Ap. J. (Suppl.), **43**, 1.
Prasad, S. S., and Huntress, W. T., Jr. 1982, Ap. J., **260**, 590.
Seab, C. G., and Shull, J. M. 1983, Ap. J., **275**, 652.
Swade, D. A. 1987, Ph.D. Thesis, University of Massachusetts.
Walmsley, C. M, Winnewisser, G., and Toelle, F. 1980, Astr. Ap. **81**, 245.
Walmsley, C. M., Hermsen, W., Henkel, C., Mauersberger, R., and Wilson, T. L. 1987, Astr. Ap., **172**, 311.
Watson, W. D. 1973, Ap. J. (Letters), **183**, L17.
Watson, W. D. 1976, Rev. Mod. Physics, **48**, 513.
Watson, W. D., and Walmsley, C. M. 1982, in Regions of Recent Star Formation, ed. R. S. Roger and P. E. Dewdney, Dordrecht: Reidel, p. 357.
Wilson, T. L., Downes, D., and Bieging, J. H. 1979, Astr. Ap., **71**, 275.
Wootten, A., Evans, N. J., Snell, R. L., and Vanden Bout, P. A. 1978, Ap. J., **225**, L143.
Ziurys, L. M. 1988, Ap. J., **324**, 544.
Ziurys, L. M., Snell, R. L., and Dickman, R. L. 1989, Ap. J., **341**, 857.
Ziurys, L. M., Friberg, P., and Irvine, W. M. 1989, Ap. J., **343**, 201.
Ziurys, L. M. 1989, in preparation.

THEORIES OF MOLECULAR CLOUD FORMATION

BRUCE G. ELMEGREEN
IBM Research Division, T.J. Watson Research Center, P.O. Box 218, Yorktown Heights, NY 10598 USA

ABSTRACT   Cloud formation mechanisms including random collisional agglomeration of small clouds into large complexes, the Parker instability, and the gravitational instability are reviewed. The most likely mechanism for the formation of cloud complexes with intermediate and large masses appears to be a combination of the self-gravitational, magnetic Rayleigh-Taylor, and thermal instabilities, applied to the *cloudy* interstellar medium rather than to thermal media as originally proposed for each separate case. The growth rate for this instability is fast for typical densities, magnetic field strengths, and rotation rates, and in relatively quiescent regions the instability can form clouds on all scales down to the cloud collision mean free path. This process may also give the observed cloud mass spectrum. The random collisional build-up model, on the other hand, has problems with collisional cloud destruction, too long a collision time for molecular clouds, and magnetic constraints to cloud motions; the mass spectrum produced by this model is also steeper than the observed spectrum if the clouds are self-gravitating. Cloud formation in swept-up shells and by the disruption of larger clouds is also discussed. Emphasis is placed on the likely difference between the molecular fraction of the interstellar medium, including both diffuse and self-gravitating molecular clouds, and the fraction of the gas that is in the form of self-gravitating clouds, whether or not they are molecular.

INTRODUCTION

There are essentially three distinct mechanisms that have been proposed for the formation of giant cloud complexes: 1. random collisional agglomeration of small clouds, with a return to small clouds following star formation; 2. instabilities that form large cloud complexes from an initially uniform (but cloudy) fluid, and 3. forced accumulation of shells, sheets, and filaments in high-pressure environments such as OB associations, isolated supernova, and spiral density wave shocks.

This review summarizes the strong and weak points of the first two of these models, and discusses some of their history. An overview of cloud formation is in the final section.

THE RANDOM COALESCENCE MODEL

History

The first of the three cloud formation mechanisms discussed above considers that random motions and mutual gravity occasionally bring two independent clouds together, whereupon the clouds stick and form a larger cloud.

The current viewpoint began with a suggestion by Oort (1954) and a theory for the cloud mass spectrum by Field and Saslaw (1965). Further general theoretical developments were made by Field and Hutchins (1968), Penston et al. (1969), Taff and Savedoff (1973), Goldstein and Mazzella (1974), Handbury, Simon and Williams (1977), Kwan (1979), Scoville and Hersh (1979), Cowie (1980), Hausman (1982), Yuan and Wang (1982), Scalo and Struck-Marcell (1984, 1986), Struck-Marcell and Scalo (1984, 1987), and others. Recent reviews are in Scalo (1985) and Kwan (1988).

Numerical simulations of cloud coagulation in N-body fluids illustrate various principles of the proposed model; examples are in Casoli and Combes (1982), Kwan and Valdes (1983, 1987), Tomisaka (1984, 1986), Combes and Gerin (1985), Johns and Nelson (1986), and Roberts and Steward (1987, and references therein). These simulations show an enhanced collision rate in spiral density wave arms (cf., Norman and Silk 1980).

Theoretical studies of the cloud collision process itself may be found in Kahn (1955), Stone (1970), Chieze and Lazareff (1980), Hausman (1981), Gilden (1984), Lattanzio et al. (1985), and Lattanzio and Henriksen (1988). Magnetic collisions were considered by Clifford and Elmegreen (1983) and Clifford (1985).

The random collisional build-up model for molecular cloud formation followed soon after the observation of a high molecular fraction in the 5 kpc region of the Galaxy (Scoville and Solomon 1975; Burton et al. 1975). This observation implied that a large fraction of the gas is in the form of molecular clouds, which were assumed to be strongly self-gravitating like the local molecular clouds. Such clouds are generally thought to be highly mobile (i.e., ballistic) because they are much denser than the ambient interstellar medium (hereafter ISM). Moreover, the initial failure to detect spiral structure in CO, combined with the high molecular fraction in the inner Galaxy, was taken as evidence that molecular clouds are long lived, with ages exceeding 10^8 or possibly 10^9 years (Scoville and Hersh 1979; Xiang and Lou 1985). Then the presumed billiard-ball clouds collide with each other many times before they get destroyed by internal star formation.

Random Coalescence Model: Strong Points

The traditional strong point of the model is that it explains the observed cloud mass spectrum, which is approximately $n(M) \sim M^{-1.5}$ over at least five orders of magnitude, as observed on the smallest scales by Scheffler (1966, 1967a,b), Bhatt, Rowse, and Williams (1984), and Casoli, Combes, and Gerin (1984), and on the largest scales by Solomon, Sanders, and Scoville (1979), Liszt, Xiang, and Burton (1981), Sanders, Solomon, and Scoville (1984), Bhatt and Williams (1986), Terebey et al. (1986), Dame et al. (1986), and Solomon et al. (1987). A comparison of the diffuse and molecular cloud mass spectra is in Dickey and Garwood (1989).

Another strong point is that the cloud formation rate can be enhanced in regions of converging gas flow, such as spiral density wave arms. This enhancement has been used to explain the relatively larger sizes of molecular clouds in spiral arms compared to clouds in the interarm regions (Stark 1979; Stark 1986; Dame et al. 1986). The model has also been used to explain enhanced star formation rates in spiral arms, assuming that cloud collisions trigger star formation (Scoville, Sanders, and Clemens 1986).

Random Coalescence Model: Weak Points
This section reviews the assumptions and indicates possible weak points for the model. The observation of a highly molecular ISM in the inner part of the Galaxy and in other galaxies depends on the assumed conversion factor between CO luminosity and cloud mass, and has been challenged by different groups (e.g., Blitz and Shu 1980; MacLaren, Richardson and Wolfendale 1988). A lower molecular fraction implies that molecular clouds are a less important constituent of interstellar matter than originally thought, and that molecular cloud lifetimes are shorter than originally assumed.

The additional assumption that a highly molecular ISM is one composed largely of Orion-type, self-gravitating clouds should also be questioned. All highly molecular ISMs that have been observed so far are also high *pressure* ISMs because their densities are large. Diffuse clouds that are atomic in the low pressure environment of the solar neighborhood are expected to turn molecular at slightly higher pressures. This is illustrated in figure 1, which shows the molecular fraction of a population of diffuse clouds with a mass spectrum $n(M) \sim M^{-4/3}$ from Knude (1979), an equation of state inside the clouds given by $P \sim \rho^{2/3}$ (from Jura 1976) and a criterion for H_2 molecule formation given by $S = (n/60\text{cm}^{-3})^{5/3}(R/1.36 \text{ pc })^{2/3}(\phi/\phi_\odot)^{-1}(Z/Z_\odot) > 1$ from Federman, Glassgold and Kwan (1979), as written by Elmegreen (1989a). Here $\phi/\phi_\odot$ is the radiation field normalized to the value in the solar neighborhood and $Z/Z_\odot$ is the normalized metallicity. With these assumptions, the molecular fraction of the diffuse cloud ensemble varies with ϕ, Z and P as (Elmegreen 1989d)

$$f = 1 - (1 - f_\odot)(\phi/\phi_\odot)^3(Z/Z_\odot)^{-3}(P/P_\odot)^{-6.5}, \qquad (1)$$

which is plotted in Figure 1 assuming $f_\odot = 0.01$. This expression is a simplification of an integral result under the assumption of certain appropriate limits and, when written this way, is valid only for $f > 0$. Evidently a small increase in pressure or metallicity or a small decrease in radiation field will shift the molecular fraction of diffuse clouds from nearly zero to nearly unity. Thus, highly molecular ISMs need not be composed entirely of self-gravitating clouds; they could have a substantial mass in the form of molecular diffuse clouds (e.g., Magnani 1987; Polk et al. 1988; Miyawaki et al. 1988; Lada and Blitz 1988). The role of the galactic metallicity gradient in determining the radial variation of the molecular fraction was also discussed by Franco and Cox (1986), and the role of the pressure gradient was discussed by Bohigas (1988). The galactic radial variation of the atomic-molecular shielding layer on the surfaces of clouds was also discussed by Shaya and Federman (1987).

Other regions of high pressure, such as the environment of an OB association, could convert atomic diffuse clouds into molecular diffuse clouds also. This may account for the local high latitude molecular diffuse clouds, which are near the Sco-Cen association (Magnani, Blitz, and Mundy 1985; Elmegreen 1988).

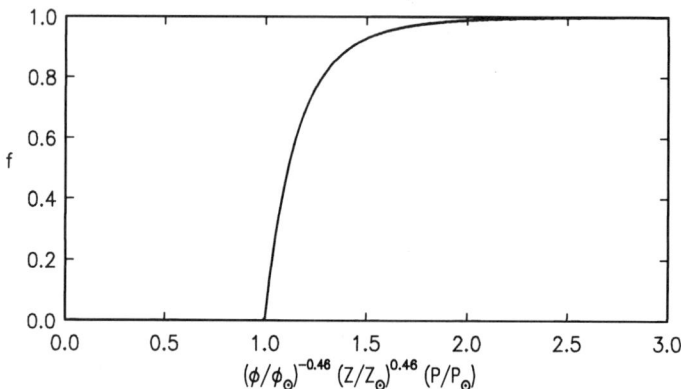

Fig. 1. The molecular fraction of the diffuse cloud population is plotted as a function of the interstellar pressure, normalized to the values in the solar neighborhood. A small change in pressure (or radiation field or metallicity) can change the molecular fraction in the non-self-gravitating diffuse cloud population by a large amount.

An assumption that enters into the calculation of the cloud mass spectrum is that the product of the cloud cross section and the velocity dispersion is nearly independent of cloud mass. This assumption is a remnant of the original theory due to Field and Saslaw (1965), which was applied to diffuse (non-self-gravitating) clouds. For self-gravitating clouds, the assumption need not be true. Self-gravitating molecular clouds have a nearly constant column density (e.g., Solomon et al. 1987, and references therein) and a velocity dispersion that is nearly independent of mass (Stark and Brand 1989, and references therein). Thus the product of the cross section and the dispersion scales directly with mass. In this case, one can show that the mass spectrum expected for a randomly coalescing ensemble of clouds should be $n(M) \sim M^{-2}$ (Elmegreen 1989e), not $M^{-1.5}$, as formerly derived.

The spectrum could be shallower than M^{-2}, possibly $M^{-1.75}$, if the coagulation process is two-dimensional, as in a galactic disk with a small cloud scale height. Then the effective cross section is proportional to only the cloud radius and not the radius squared. The observed spectrum could also be steeper than the usually quoted $M^{-1.5}$. This mass spectrum often comes from a directly observed size spectrum, $n(R)dR \sim R^{-2.5}dR$, after assuming $M \sim R^3$; if $M \sim R^2$, however, then $n(M) \sim M^{-1.75}$ for the same size spectrum.

Another concern of the random coalescence model is that computer simulations of cloud collisions show destruction rather than coalescence at typical velocities (e.g., Lattanzio et al. 1985; see other references above). This property of collisions has been used to suggest a regulatory mechanism for star formation in galaxy disks by Scalo and Struck-Marcell (1987) and Vasquez and Scalo (1989). If the threshold velocity for mutual cloud destruction is low compared to the observed velocity dispersion, then random cloud agglomeration would be uncommon.

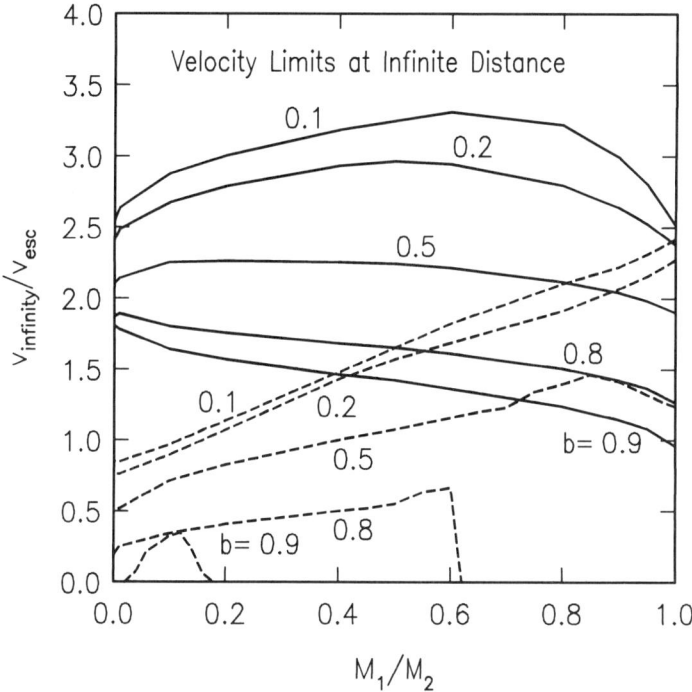

Fig. 2. The threshold velocities for completely fragmenting collisions (solid lines) and completely coalescing collisions (dashed lines) are plotted as a function of cloud mass ratio for four different values of the impact parameter, normalized to the sum of the cloud radii. Cloud velocities larger than the limits given by the solid lines lead to an eventual separation of all three post-collision pieces, and cloud velocities less than the limits given by the dashed lines lead to complete coalescence. Between these limits, the collision produces partial coalesce or orbiting pieces.

In an effort to determine a threshold velocity for coalescence, I calculated the orbits of the non-overlapping and overlapping pieces of two clouds following a collision, with the assumption that the initial post-collision velocities of each piece are determined from momentum conservation. The cloud pieces are treated as point particles moving under their mutual gravitational forces. Both clouds are assumed to have the same column density before the collision, as observed for molecular clouds. The results are two limits to the velocity at infinite separation (with gravitational focussing and acceleration before the collision included). There is an upper limit at which two clouds touch each other and eventually have at least two of the three post-collision pieces stick together, and a lower limit at which the two original clouds fragment (i.e., one or more of the post-collision pieces leaves the system on a hyperbolic orbit). Between these velocity thresholds, partial coalescence of the two initial clouds results; above the upper limit no two pieces stick together; below the lower limit all three pieces stick together. Magnetic fields have been included in some of the calculations but found to have relatively little significance to the overall results. The results discussed here are for the calculations without magnetic fields.

The results are shown in Figure 2. The solid lines are the maximum velocities for coalescence of any two of the three post-collision pieces, drawn as functions of the initial cloud mass ratio for four different impact parameters, b, normalized to the summed radii of the two clouds ($b = 0$ is for head-on collisions, $b = 1$ for grazing collisions). The velocities are normalized to the escape velocity of the largest cloud before the collision. The dashed lines are the minimum velocities for disruption. The results are relatively independent of mass ratio because of the assumption of a constant initial cloud column density. Evidently two clouds with a relative velocity exceeding 2 to 3 times the escape velocity of the larger cloud will not coalesce at all, and two clouds with relative velocities less than approximately 1 times this escape velocity will completely coalesce (see also Vazquez and Scalo 1989).

We can compare these theoretical limits to the observed ratios of velocity dispersion to escape velocity for molecular clouds. The observed *total* dispersion (Stark and Brand 1989), including streaming motions (which is relevant for collisions if cloud streams intersect) is approximately 8 km s^{-1} for clouds smaller than $3 \times 10^5 M_\odot$ and 3 km s^{-1} for clouds larger than $3 \times 10^5 M_\odot$. We multiply these velocities by $3^{0.5}$ to obtain three-dimensional dispersions, and by $2^{0.5}$ to obtain rms relative velocities between two colliding clouds. The escape velocities for molecular clouds come from the size-linewidth relation (Solomon *et al.* 1987), which gives

$$v_{esc} = (2GM/R)^{0.5} = 1.3R^{0.5} = 0.34M^{0.25} km\ s^{-1} \qquad (2)$$

for R in pc and M in $M_\odot$. These numbers and Figure 2 suggest that two clouds moving at the random dispersion will partially stick together if the largest cloud mass exceeds several times $10^5 M_\odot$ for a nearly head-on collision ($v_{rel} \lesssim 3\ v_{esc}$; $b = 0.1$), and $10^6 M_\odot$ for a nearly grazing collision ($v_{rel} \lesssim 1.5\ v_{esc}$; $b = 0.9$). Clouds less massive than $10^5 M_\odot$ should completely fragment ($v_{rel} > 3v_{esc}$). For the one-dimensional dispersion of 3 km s^{-1} suggested by Clemens (1985), the mass thresholds decrease by a factor $(3/8)^4 = 0.02$. Evidently the outcome of a collision depends strongly on the assumed velocity dispersion, but clouds smaller than $10^4 M_\odot$ should not coalesce significantly for either dispersion after a partially overlapping collision, unless magnetic fields significantly affect the collision. Note that the usual derivation of the mass spectrum assumes complete coalescence of the two clouds; partial coalescence gives a steeper spectrum (Hausman 1982).

Another consideration of the random collisional build-up model of molecular cloud formation is that the time scale for the collision of the observed molecular clouds is very long, on the order of 2 to 5 times 10^8 years. Massive star formation and the resulting cloud dispersal apparently limit the lifetime of a molecular cloud *following* star formation to less than 10^8 years, possibly 4×10^7 years (Bash, Green and Peters 1977; Blitz and Shu 1980; Elmegreen 1985a,b; Leisawitz *et al.* 1989). Moreover, the low fraction (25 per cent; Mooney and Solomon 1988) of massive molecular clouds which contain no massive star formation limits the lifetime of the clouds *before* massive star formation to approximately 1/3 of this post-star-formation value, or some 10^7 years. Thus the total lifetime of the clouds is approximately 5×10^7 years. Note that a high molecular fraction in the inner galaxy does not imply long cloud lifetimes (as previously assumed) if a substantial fraction of the *diffuse* gas is molecular or if the cloud formation time is short (cf. Eq. 3). The collision time is therefore $\sim$ 10 times longer than the likely cloud lifetime.

The usual solution to the problem of time scales is to assume that cloud formation occurs primarily in spiral arms. There the cloud density can be sufficiently large that the collision rate exceeds the destruction rate due to internal star formation. This does not explain why galaxies without spiral arms have the same molecular abundances and star formation rates as galaxies with spiral arms, however (Elmegreen and Elmegreen 1986; Stark, Elmegreen and Chance 1987). There is also another problem in that the velocity dispersion of clouds in the arms might be high as a result of the adiabatic compression of the cloud fluid; this increases the likelihood of collisional cloud destruction. For example, the clouds that are streaming into an arm can have very large velocities relative to the clouds that are already in the arm (e.g. the 30 km s^{-1} "shock" velocity in a spiral density wave). The collisional build-up model assumes either that these high velocity clouds stick upon impact with the arm clouds, which seems unlikely at these velocities, or that non-destructive scattering randomizes their incident motions and decreases the velocity dispersion, with cloud agglomeration following. The non-destructive nature of such randomizing collisions may be a problem for the model. If these collisions result from magnetic entanglements (Elmegreen 1988), which may be reasonably non-destructive, then most of the kinetic energy from the spiral arm streaming motion will be conserved during randomization, and the relative collision velocities inside the arm will still be comparable to the incident streaming velocities, as in a conventional adiabatic shock front. Then even the random collisions will be too fast to permit coalescence. The randomizing cloud collisions have to be dissipative to reduce the relative cloud velocity in an arm, permitting coalescence without disruption, but it is hard to imagine dissipative collisions in which the individual clouds remain in tact until their dispersion decreases. More likely, a large fraction of the clouds entering a spiral density wave arm, especially a strong arm as in M51, will disperse as a result of collisions, and new clouds will form from the remnants. Such dispersal may contribute to the observed uniformity of galactic dust lanes. The dispersed cloud pieces need not be atomic.

For some applications, a distinction should be made between the random and systematic components of the clouds' motions in the collision model. The systematic part of the motion leads to a net convergence of the cloud fluid (as in a density wave crest or swept-up shell); this convergence might be viewed as a cloud formation process itself if the whole dust lanes and shells that result are viewed as distinct clouds. Some of the physical processes involved with such cloud collision fronts have been discussed by Elmegreen (1988). On the other hand, the convergence may be viewed as only a precursor to cloud formation, which is more directly the result of a secondary coalescence of the individual clouds that enter the front, or the result of gravitational instabilities that bring these clouds together inside the front (see below).

Another primary assumption of the random collisional build-up model is that the component clouds come from different directions and from distances of up to several mean free paths before they coalesce. If the model applies to the observed molecular clouds, then this distance can exceed one kiloparsec. Because each cloud should carry its magnetic field lines as it moves in space, the magnetic structure in the composite cloud should be very irregular, with field lines emerging in all of the directions from which the incident clouds arrived. Such irregular structure appears to be inconsistent with observations of an approximately uniform field inside and outside of giant cloud complexes, as revealed by star light polarization maps, for example (Cleary, Heiles, and Haslam 1979). Perhaps magnetic diffusion and reconnection inside the cloud complexes is fast enough to smooth out these irregularities, but isolated clouds may have a sufficiently high ionization fraction on their borders to limit the diffusion.

Magnetic fields are rarely considered in discussions of the random collision model. If the magnetic field pressure is comparable to the kinematic pressure in the ISM, which appears to be the case locally, then magnetic forces increase the collision cross section significantly and they permit a higher fraction of the cloud collisions to be purely magnetic (Clifford and Elmegreen 1983; Elmegreen 1987a). But magnetic fields also tie the molecular clouds to lower density surrounding gas, which makes even the densest clouds non-ballistic (Elmegreen 1981a). The result is an ISM composed of strongly interacting magnetic clouds, generally confined to move along the mean field direction. This differs from the basic state of the ISM that is usually assumed for the random collisional build-up model, and it differs from the ISM of N-body fluid codes. The implications of such a magnetic ISM for the random collisional build-up model of molecular cloud formation are not obvious.

The frequency of cloud collisions may be checked observationally by searching among a sample of completely mapped clouds for examples where the CO linewidth is greatly in excess of the value usually obtained from the size-linewidth relation. If the linewidth is comparable to the one-dimensional rms relative velocity dispersion of such clouds, or if the spectral line has three components with two at likely velocities for clouds that have not yet touched and one at the average velocity between the other two, then the object could be two clouds in collision separated by a molecular shocked interface. At any one time, the fraction of clouds in a sample that are undergoing a collision should equal approximately 3 times the volume filling factor for all clouds of that type, provided the clouds move randomly in space. This is because the grazing collision rate is $n_c 4\pi R^2 v$ for cloud number density n_c, radius R, and mean relative velocity v, and the average duration of a collision is approximately R/v, so the product, which is the fraction of clouds undergoing a collision, is $4\pi n_c R^3 \sim 3 f_c$ for filling factor f_c. For example, if, in a sample of 100 clouds, the duration of a collision, t_d, is 0.1 times the collision time, t_c, then at any typical time, 10 clouds will be colliding with 10 other clouds, giving 10 composite clouds and 80 other, non-interacting clouds. The fraction of these 90 observed clouds undergoing a collision is therefore $10/90 \sim t_d/t_c$. The fraction of the primary 100 clouds undergoing a collision is twice as large, 20/100.

Random Coalescence Model: Summary
The modern version of the random collisional build-up model of cloud formation, especially as applied to highly molecular interstellar media, consists of a random sticking and coalescence of dense, self-gravitating molecular clouds that are able to move in a ballistic fashion over a distance of approximately one kiloparsec. The above discussion suggests, however, that highly molecular ISM are not necessarily dominated by self-gravitating molecular clouds; that the mass spectrum expected for the model is steeper than the observations suggest; that most colliding clouds, especially at the low mass end, will not stick efficiently; that internal star formation destroys a typical molecular cloud before it gets a chance to collide with another molecular cloud; that the magnetic field line structure in a randomly assembled cloud should be more irregular than is typically observed, and that cloud motions are not ballistic when magnetic forces are considered.

Nevertheless, the basic point of the model, that cloud formation necessarily involves the collection of small clouds into large complexes, must still be true. The problems outlined above result mostly from an incomplete approach to the collision process itself (which usually ignores magnetic forces and cloud disruption for example), and from the likelihood that collective interactions between many clouds, and not just binary collisions, contribute to the growth of a complex (see discussion of the Jeans instability below).

Several obvious revisions to the standard model might be considered. We suggest that the collision partners for the most *common* types of random aggregates are diffuse clouds rather than self-gravitating clouds, because of the larger collision rate for diffuse clouds and the relatively short lifetime of molecular clouds. This is possible even in highly molecular regions because many of the molecular clouds there could be diffuse molecular clouds, as discussed above. Diffuse clouds can have large masses; the essential point in stressing diffuse clouds is that they generally have a large cross section per unit mass. We also suggest that random binary collisions in the *ambient* ISM may be important for forming low mass clouds, possibly less than 10^4 $M_\odot$, and that larger complexes may form by this mechanism primarily in the converging flow of a density wave. The small size constraint for the *ambient* medium is the result of a dominance of collective interactions among clouds on larger scales (e.g., gravitational instabilities; see Elmegreen 1989c and below). Also on large scales, the magnetic field should limit the peculiar motions of individual clouds. Moreover, the time scale to build a large cloud complex from random collisions in the ambient medium is much longer than the cloud lifetime, which is probably limited by internal star formation. Forced accumulations, however, (galactic dust lanes, swept up shells, etc.) can combine clouds into large dense regions arbitrarily fast, and the magnetic field follows this general compression. The formation of individual cloud complexes inside the compressed regions may or may not involve random collisions, however. We note also that magnetic fields are an essential part of the model; they may permit the efficient dissipation of energy and momentum during a collision via Alfven wave radiation, which would allow clouds to collide without severe disruption; unfortunately the outcome of a

THE PARKER INSTABILITY MODEL

History

The second general type of cloud formation mechanism discussed in the introduction follows from Jeans (1902) in the case of the gravitational instability, and Parker (1966) for the magnetic Rayleigh-Taylor instability. We consider first the Parker instability and defer the Jeans instability to the next section. The basic point of these two sections is that the instabilities probably operate together, along with the thermal instability, with relative contributions depending sensitively on the parameter values.

The Parker instability assumes an equilibrium state (e.g., Parker 1966; Badhwar and Stephens 1977) in which the magnetic field and cosmic ray pressures add a substantial amount of support to the gas perpendicular to the galactic plane. A perturbation with relatively low mass-to-flux ratio is then buoyant and rises above the plane, causing the mass to drain off the field lines from the high points to the low points. This decreases the mass to flux ratio in the high points and causes the highest field lines to rise even faster. Eventually a cloud forms in the magnetic valley and the field lines on either side extend high above the galactic plane. The component of the perturbed magnetic pressure that drives the motion of gas parallel to the plane depends on the square of the z-component of the field, so most of the cloud driving force occurs after the field lines are buckled.

Detailed analyses of the Parker instability, and applications to cloud formation, were made by Parker (1966, 1967a,b, 1968a,b,c), Lerche (1967a), Lerche and Parker (1968), Mouschovias, Shu and Woodward (1974), Shu (1974), Zweibel and Kulsrud (1975), Asseo *et al.* (1978), Lachieze-Ray *et al.* (1980), Kuznetsov and Ptuskin (1983), Schlickeiser and Lerche (1985), Kuznetsov (1986), Chagelishvili *et al.* (1988), Matsumoto *et al.* (1988), Horiuchi *et al.* (1988), and others mentioned below. Equilibrium states of the clouds that form were investigated by Lerche (1967b,c) and Mouschovias (1974). The thermal and Parker instabilities were combined by Ames (1973), and the Jeans and Parker instabilities were combined by Elmegreen (1982a,b). Reviews of the instability are in Cesarsky (1980) and Mouschovias (1981).

The Parker instability was one of the most popular cloud formation mechanisms in the early 1970s. It forecast the discovery of giant cloud complexes and explained interstellar turbulence. This was a time when self-gravity was thought to be unimportant, both for the ambient medium and for individual clouds, and when the interstellar medium was thought to be in approximate pressure equilibrium. At this time also, radio telescopes were mapping giant (10^4 - 10^5 $M_\odot$) HI clouds near many of the visible OB associations. A remark made by Bridle and Kesteven (1970) in a paper entitled "A Massive HI Cloud Surrounding Some Compact HII Regions" reflects the sympathy of the era: "The gravitational potential energy of the cloud is 20 per cent of the total internal energy, assuming the mean molecular weight of the gas to be 1.5 and a uniform spherical distribution of material in the cloud. This suggests that the cloud is not gravitationally bound, and is consistent with theories of interstellar cloud formation based on thermal or magnetic instabilities (Goldsmith et al. 1969; Parker and Lerche 1969), rather than on a gravitational instability." This statement was made before the molecular cloud era. After CO was discovered, the clouds near HII regions were found to be strongly self-gravitating and at a high internal pressure.

Parker Instability: Strong Points
A strong point of the Parker instability model is that two clouds on the same magnetic flux tube attract each other more because of the induced field line curvature than because of mutual gravity (Parker 1966). Thus the Parker instability model was initially preferred over the gravitational instability model. The Parker instability also operates well even in the presence of galactic rotation (Lerche 1967a; Shu 1974; Zweibel and Kulsrud 1975), unlike early models of the gravitational instability (without a magnetic field) which required a threshold column density (see below).

The model also predicted and explained the origin of clouds with masses on the order of $10^5 M_\odot$, as observed near OB associations (Mouschovias, Shu and Woodward 1974), and it accounted for the observed field line structure around some cloud complexes, as in the "magnetic sac" discovered by Appenzeller (1971).

Parker Instability: Weak Points
A weak point of the Parker instability model is that adjacent flux tubes are expected to move and buckle separately, suggesting that the primary result of the instability is a chaotic or turbulent structure in the interstellar medium and not the formation of *giant*, well-defined cloud complexes (Parker 1967b; Asseo et al. 1978; Lachieze-Ray et al. 1980; Cesarsky 1980).

There is also a question as to whether the cosmic rays can stream along the field lines as rapidly as originally assumed (Kuznetsov and Ptuskin 1983; Nelson 1985). The cosmic rays might generate waves as they stream, and then scatter off these waves, or they may have other problems diffusing through the gas. Pre-existing clouds on the field lines and magnetic irregularities slow the growth too (Zweibel and Kulsrud 1975).

Moreover, the growth time for the instability, even without these impediments, may be too long to make this a competitive mechanism for cloud formation. Early discussions tended to ignore the increase in growth time that results from galactic rotation. As an example, consider the parameters $\alpha = B^2/(8\pi\rho c^2) = 0.2$ and $\beta = P_{CR}/(\rho c^2) = 0.2$, using standard values for the local ISM of $B=3\times10^{-6}$ Gauss, $\rho = 1.5 m_H$, $c=7$ km s^{-1}, and $P_{CR}=4\times10^{-13}$ erg cm^{-3} (from Spitzer, 1978; Mayer 1969). Consider also a polytropic index of $\gamma=0.8$, indicating that higher density gas is cooler. Then the growth time with no rotation and an infinite perturbation height is 4×10^7 years, and the growth time with the local rate of rotation is 1.0×10^8 years. The evaluation here is from Elmegreen (1982c) and is based on the analysis in Zweibel and Kulsrud (1975). On such long time scales, the interstellar gas is not likely to slide continuously down smoothly curve field lines as in the idealize model of the instability, but to become stirred and distorted by frequent supernovae and random hot stars. The result may be numerous instabilities on small scales but relatively few coherent instabilities forming cloud complexes on large scales.

The original argument favoring the Parker instability as a mechanism for giant cloud formation, especially when triggered by a spiral density wave (e.g., Mouschovias, Shu and Woodward 1974), assumed $\alpha=\beta=1$ or larger, which gives faster growth rates and less of a dependence on rotation. Such large parameter values may in fact be reasonable, but we do not know B, ρ, c and P_{CR} well enough to evaluate α and β to within a factor of ~ 5. For example, Parker (1966) assumed $B = 5x10^{-6}$ Gauss. Moreover, the *effective* values of α and β, considering existing and inevitable non-uniformities in the gas, are not known either.

The primary strong point of the Parker instability as a cloud formation model was that the magnetic force that attracts two clouds on the same field line exceeds the gravitational force by a factor $g^2/GB^2 \sim 5$, for acceleration g toward the galactic plane and field strength B (Parker 1966). This factor may be irrelevant for the linear phase of the instability, however. The initial growth of the instability involves an attraction not between pairs of clouds but between a large region of unperturbed interstellar gas (containing many clouds) at the ambient density and an equally large region of perturbed magnetic and gravitational force at the *same* location. For example, in the case of the gravitational part of the instability, the gravitational force density exerted by a perturbation on the unperturbed mass in that region is $4\pi G\rho_0\rho_1/k$ for ambient density ρ_0, perturbed density ρ_1, and wavenumber k. Gravitational forces exceed magnetic curvature forces in this realistic case when the dimensionless parameter $s = 8\pi G\rho H^2/c^2$ exceeds the dimensionless parameters β and α (Elmegreen 1982a,b). Here ρ is the mean density, H is the scale height and c is the one-dimensional velocity dispersion. This inequality, $s > \alpha + \beta$, could be satisfied in regions compressed by density waves, in fact s/α should increase with the compression because, although α increases as $\alpha \sim \rho$ for $B \sim \rho$, s increases as $s \sim \rho^2$ for $H \sim v_A \sim \rho^{0.5}$ with Alfven velocity v_A. If the scale height does not increase in a density wave shock, then s/α could be constant. Parker's dimensionless ratio of forces, g^2/GB^2, also decreases with the compression, because in a narrow density wave shock B increases with ρ faster than g, which is the perpendicular acceleration mostly from underlying stars. Thus a density wave compression may trigger what is essentially a gravitational instability, even though α and β increase in the wave. Moreover, the inequality $s > \alpha + \beta$ could also be true in the local interstellar medium if $\alpha \sim \beta \sim 0.2$ because $s \sim 1.2$ with the above parameters and a scale height of $H = 100$ pc. Larger values of $\alpha \sim \beta \sim 1$ could make the Parker instability dominate in the initial state unless the diffusion of cosmic rays is slow or there are other problems, as discussed above.

The Parker instability has an additional difficulty in explaining molecular clouds because it operates entirely in pressure equilibrium. This problem vanishes when self-gravity is included, but then the amount of self-gravity required to give the observed compression in molecular clouds is so large that the role of field line curvature as a force for driving the gas motions at intermediate and late times becomes minor compared to the role of self-gravity, even if field line curvature dominates the initial growth.

Parker Instability: Summary

Although the Parker instability is likely to have an important role in driving interstellar turbulence and in relieving magnetic flux and cosmic ray pressure from a galaxy disk, it may not to be able to form strongly self-gravitating clouds fast enough and with enough cohesion by itself to compete with gravitational instabilities, unless α and β are large. In any case, the force from field line curvature in the Parker instability should contribute at a level of $(\alpha + \beta)/(s + \alpha + \beta) \sim 0.2$ to 0.6 to the early stages in the development of more general instabilities involving also the self-gravity of the interstellar medium.

THE JEANS INSTABILITY MODEL

History

The conventional Jeans (1902) analysis for the gravitational instability suggests that in an infinite uniform polytropic medium, perturbations larger than $c/(G\rho)^{1/2} \sim 500$ pc for $c \sim 7$ km s^{-1} and $\rho \sim 1$ cm^{-3}, will be unable to establish an internal pressure gradient large enough to support themselves against gravitational collapse. Modifications to the analysis including magnetic fields (Chandrasekhar and Fermi 1953), rotation (Chandrasekhar 1954a; Fricke 1954; Bel and Schatzman 1958), rotation with a magnetic field (Chandrasekhar 1954b; Sharma and Singh 1988), and a 2-dimensional geometry (Ledoux 1951; Simon 1965) all give essentially the same result, a minimum size for the instability which, when evaluated for parameters typical of the Galaxy, generally exceeds the mass of a typical giant molecular cloud. Because of the large scales and masses involved with this instability, applications were initially limited to the stability of the whole disk against forming concentric rings (Safronov 1960; Toomre 1964) and spiral arms (Goldreich and Lynden Bell 1965). The disk, especially the stellar disk, is generally thought to be stable against ring formation, which implies that $Q_{STAR} = \kappa c/(\pi G\sigma) > 1$ throughout. Here κ is the epicyclic frequency, c is the one-dimensional velocity dispersion, and σ is the mass column density in the disk. The parameters of the local gas disk, $\kappa = 30$ km s^{-1} kpc^{-1}, $c = 7$ km s^{-1}, $\sigma = 10$ $M_\odot$ pc^{-2} for HI (Kulkarni and Heiles 1987) and 2 $M_\odot$ for H$_2$ (Dame et al. 1987) suggest stability as well: $Q_{GAS} = 1.3$. Without such general stability, parts of the disk would presumably collapse catastrophically.

Aside from early discussions of gravitational instabilities as a general feature of star formation (e.g., Chandrasekhar 1961; Spitzer 1968), specific applications to cloud formation were slow to begin, partly because galactic disks were thought to be stable, but also because the Parker instability was thought to dominate the gravitational instability in the gas under usual conditions (see above). Nevertheless, there were suggestions that gravitational instabilities were involved with cloud or star formation (Kuz'min 1965; Marochnik 1966; Quirk 1972) if the disk was only marginally stable against ring formation. The observational implications of this suggestion were apparently overlooked until recently (Kennicutt 1989, and references therein).

Cloud formation by the Jeans instability became more tenable when aperture synthesis maps of HI in other galaxies showed clouds with masses comparable to the Jeans mass in galaxy disks. In reference to the large clouds found in M33 by Wright et al. (1972), Boulesteix et al. (1974) remarked: "The HI distribution in the arm is far from homogeneous. It is patchy with quite a regular typical size of about 300 pc, a value closely similar to Jean's gravitational length for the gas." Similar HI clouds were found in M101 by Allen, Goss, and van Woerden (1973), and several other clouds of this mass were observed in extinction in these two galaxies and in two others by D. Elmegreen (1980). This latter observation, combined with previous work on shock induced gravitational collapse and star formation, led to an investigation of the gravitational stability of spiral density wave shock fronts (Elmegreen 1979). Such fronts, represented by magnetic cylinders, were found to be grossly unstable, and it was suggested that the observed dust lanes in galaxies should continuously collapse and reform, making giant cloud complexes similar to those observed by Allen *et al.*, Wright *et al.*, and others. This application of the Jeans instability did not encroach on the generally accepted ban against galactic instabilities because it applied only to the gas, which alone created the shock front, and because it occurred in regions of particularly low shear (shear is less in regions compressed by spiral density waves than it is in non-compressed regions – see Elmegreen 1987b). Thus, the instability condition $Q_{GAS} < 1$ could be satisfied in the shock even though $Q_{STAR} > 1$.

A problem with this instability model was that the characteristic mass of an unstable region in both a spiral density wave shock and in the ambient medium is $10^7 M_\odot$, which is much larger than a "typical" giant molecular cloud ($10^5 M_\odot$). Thus the lower mass clouds were required to form by secondary processes such as hierarchical fragmentation of the larger objects (Elmegreen 1979). This places constraints on where the smaller clouds are located relative to the large clouds.

Initial resistance to this instability model by advocates of the Parker instability (e.g., Blitz and Shu 1980) led to an investigation of the combined instabilities at this time (Elmegreen 1982a,b). The result, discussed above, was that the gravitational contribution to the combined instabilities depends on the relative magnitude of the three dimensionless parameters, s, α, and β, and because s was thought to exceed $\alpha + \beta$ in the local interstellar medium and $s/(\alpha + \beta)$ tends to increase in spiral density wave shocks, the importance of self-gravity seemed assured.

The large primary masses implied by the instability was still considered to be a problem, however. A plausible solution came from Cowie (1981), who suggested that cloud-cloud collisions in spiral density wave arms cool the general fluid so much that the velocity dispersion drops to only a few km s^{-1}, thereby decreasing the Jeans mass to values comparable to the masses observed for giant molecular clouds near OB associations. Although such low dispersions were inconsistent with available observations of the molecular cloud scale height (Cohen and Thaddeus 1977) and velocity dispersion (Stark 1979), the supercooled regions could have been unrecognized in the surveys.

This problem with forming $10^7 M_\odot$ clouds soon found an observational solution. In a compilation of properties of star-forming regions, both in our Galaxy and in other galaxies, and following a survey of the long-discussed "beads on a string" phenomenon regarding star formation in galactic spiral arms, Elmegreen and Elmegreen (1983) proposed that most star formation in galaxies really does occur on the scale of $10^7 M_\odot$ and 1-2 kpc, and that Orion-type clouds are in fact usually only the core regions of the larger objects. The Orion cloud itself was noted to be part of Gould's Belt, which is the more fundamental star-forming feature in the solar neighborhood (see also Efremov and Sitnik 1988, and references therein). The reason why $10^5 M_\odot$ appeared to be so common as a scale for molecular clouds was recognized several years later. By then it was known that the mass spectrum for molecular clouds shows no maximum or other feature at around $10^5 M_\odot$, so this impression of a characteristic mass (e.g., Stark and Blitz 1978) was thought to be an observational selection effect. The reason for this selection effect (Elmegreen 1987b) is apparently that $10^5 M_\odot$ is the smallest cloud likely to form an O star, based on statistical sampling of the initial mass function and on typical star formation efficiencies. Because the mass distribution is a sharply decreasing function of mass, $10^5 M_\odot$ is therefore the most common mass for a cloud near an OB association. Thus most of the molecular cloud maps, which were guided by previously known locations of OB associations and HII regions, contained several times $10^5 M_\odot$. This is not the most common mass for all molecular clouds however, nor is it likely to be a fundamental mass in terms of cloud formation.

Growing evidence in the mid-1980s in favor of enormously large scales for regions of star formation, combined with the CO survey work and discovery of several $10^6 M_\odot$ molecular clouds and clusters of these clouds by Solomon and collaborators, led Jog and Solomon (1984a,b) to reinvestigate the general stability of the combined star+gas disk with the purpose of applying the results to cloud formation (see also Min 1988). Because the gas has a lower velocity dispersion than the stars and is highly compressible, the gas makes the stars considerably less stable than they would be alone (see also Lubow, Balbus and Cowie 1986), but because of the incompressibility of the stellar fluid, the stars do not initially make the gas much less stable. Nevertheless, Jog and Solomon pointed out that an initially mild, spiral-like instability in the gas+star system can increase the gas density by a factor of ∼3.5 locally, and that after this increase, the gas can be sufficiently unstable all by itself to collapse into cloud complexes. This two-step process allows a gas disk to fragment into clouds even if the *average* value of $Q_{GAS} > 1$.

The instability of the compressed gas in a spiral density wave was then reinvestigated by Balbus and Cowie (1985) who included the time dependence of the down-stream flow in the stability condition. They found that the dimensionless parameter Q, while still required to be small for instability, should have its mass column density replaced by the geometric mean of the spiral arm peak value and the mean azimuthal value. This implied a weaker instability than found by Elmegreen (1979), but instabilities could still be present if the gas was sufficiently compressible (i.e., low γ in the expression $P \sim \rho^\gamma$), as in the model by Cowie (1980).

A basic limitation to the results of Jog and Solomon (1984a,b) and Balbus and Cowie (1985), as well as to Toomre's (1964) calculation, was that each could be applied only to one-dimensional instabilities in a galactic disk, i.e., to rings or spiral strips. Cloud formation requires at least two-dimensions in the galactic disk (three dimensions if the Parker instability is included), and two-dimensions introduce shear (Goldreich and Lynden-Bell 1965). The instability parameter for the radial dimension, Q, is still useful for two dimensions, but not absolute because transient shear instabilities can occur in two-dimensions even when the fluid is stable in the radial direction.

Two dimensional instabilities also introduce magnetic effects not present in one dimension. The additional magnetic forces arise because a growing density perturbation begins to twist with the Coriolis force, and this twist leads to a restoring tension from an azimuthal magnetic field. The result is that the gas follows the field lines and loses angular momentum as the density increases. For one-dimensional instabilities in the radial direction only, an azimuthal magnetic field merely adds a term to the effective sound speed, i.e., c is replaced by

$(c^2 + v_A^2)^{1/2}$ for Alfven speed v_A; this addition makes the gas more stable. In two dimensions, however, the azimuthal field makes the gas less stable because of its resistance to the Coriolis force. This tendency toward instability was first demonstrated by Chandrasekhar (1954b) for an infinite, uniformly rotating cylinder with a non-aligned magnetic field. The application to a disk with solid body rotation was made by Lynden-Bell (1966). Specific applications to cloud formation, considering two dimensional instabilities in shearing, magnetic galaxy disks, were made by Elmegreen (1987b). The latter result suggested that the gaseous component of a disk should always be unstable, regardless of Q, as long as an azimuthal magnetic field of modest strength is present.

The instability model with a magnetic field finally overcame the earliest objections to the Jeans instability as a cloud formation mechanism, i.e., that rotation might stabilize the disk; rotation apparently stabilizes only the stellar component, leaving the magnetic gas unstable to form clouds. A density wave shock is not even required to trigger the collapse, although any such compressions would speed up the process because the instability growth rate increases with density.

In another discussion of two-dimensional instabilities in spiral arms, using a polytropic gas and no magnetic field, Balbus (1988) found that mild cloud-forming instabilities could lead to the formation of spiral arm spurs. This is consistent with the presence of multiple arms and star-forming spurs in late type galaxies, which tend to be gas rich.

Although by this time the gravitational instability model could explain the largest clouds and star-forming regions, and it was beginning to be reasonably robust in the presence of rotation and shear, there were still several problems with the model that were not faced by other models of cloud formation. These included the direct formation of low mass clouds from the ambient medium and the origin of the mass spectrum.

These problems were apparently related to an oversimplified equation of state. As long as there is a time-independent relation between pressure and density, such as a polytropic equation of state, there will always be a minimum scale for the instability at which a pressure gradient overcomes the gravitational force. A time-dependent energy equation, with heating and cooling terms considered explicitly, may solve this problem, giving a result that is essentially a combination of the gravitational and thermal instabilities (Tomisaka 1987; Graziani and Black 1987; Elmegreen 1989d). This combined instability has a small, or possibly no, minimum length scale and it has no Q-type instability criterion when rotation and shear are included (Elmegreen 1989c).

In considering the pressure and equation of state for the macroscopic cloud fluid in a galaxy, it is important to know the distribution of heat sources for random cloud motions, such as supernovae and OB associations. These sources are likely to be highly non-uniform in space and time, and there should be regions where the stirring of clouds has temporarily dropped below the average rate. The shocked regions in spiral density waves are an example because there is a large amount of gas in the shock but not much star formation (yet). This differs from the situation for the thermal energy in a gas. Thermal heat sources, such as starlight, cosmic rays, and magnetic waves, are likely to be much more uniformly distributed than macroscopic energy sources for bulk cloud motions. Cloud formation by thermal instabilities in a *uniform* gas was discussed by Field (1965) and many others (e.g., Kritsuk 1985; Parravano 1987; Balbus and Soker

1989). Cloud formation in a non-uniform, or cloudy, gas is likely to be very different. In cloudy regions temporarily without cloud stirring, the gravitational (plus thermal) instability operates with no minimum length scale (down to at least the cloud collisional mean free path) on a time scale that is midway between the conventional instability time and the cloud-cloud collisional cooling time (Elmegreen 1989d).

A curious feature of this combined gravitational+thermal instability is that the gravitational instability time and the cloud-cloud collisional cooling time are approximately the same for diffuse cloud collisions. These are the type of cloud collisions that may dominate the energy dissipation in the interstellar medium. If they do, then the growth rate of the instability is nearly independent of length scale. This suggests that all scales should collapse simultaneously, possibly giving a hierarchical structure with no preferred mass scale. It also suggests that the resulting cloud mass spectrum might be a power law. For example, if the probability of forming a cloud of mass M is proportional to the growth rate and phase space density on that scale (Di Fazio 1986), $P(M)dM \sim \omega(k)k^i dk$, where $i=2$ for three dimensions and $i=1$ for two dimensions, then $P(M) \sim k^i dk/dM$ which is $\sim M^{-2}$ for both two ($M \sim k^{-2}$) and three dimensional ($M \sim k^{-3}$) instabilities. Such a mass distribution is steeper than the observed cloud mass spectrum, but a decrease in $\omega(k)$ with k, as obtained for a cloudy medium composed of self-gravitating clouds rather than diffuse clouds, or for a medium with a small amount of cloud stirring (cf. Fig 1 in Elmegreen 1989d), could give agreement.

This explanation for the molecular cloud mass spectrum differs from the explanation given in the random collisional build-up model, as reviewed above, because in the instability model the clouds that collide do not have to stick together; they can even destroy each other and the instability still collects mass into cloud complexes at the calculated rate. Moreover, the spectrum arises from the geometry of the fragmentation process simultaneously for all masses, rather than from a cascade of collisions from the small to the high mass clouds. This latter process gives a mass spectrum that depends critically on the break-up spectrum of the large clouds, which is generally assumed to be peaked at the lowest mass.

Evidently, the gravitational instability envisioned by Jeans has progressed to the point where it can account for most of the requirements of a cloud formation model. Early objections to the model, such as those based on rotational stabilization in the galaxy or an unrealistically large minimum cloud size, have disappeared with advances in the theory, particularly those including magnetic effects and a time-dependent energy equation. The next two sections review the strong and weak points of the model as it now stands.

Gravitational Instability: Strong Points
A strong point of the gravitational instability model, now including magnetic forces and a time-dependent energy equation, is that the interstellar gas is expected to be unstable for all rotation and shear rates, and on all length scales in regions temporarily without cloud stirring. The instability operates at about the same rate regardless of the magnetic field strength and shear rate (Elmegreen (1989c), making it easy to apply in global models of star formation. The rate for a medium composed largely of diffuse clouds, and in a region without much cloud stirring is $\sim 2\pi G\sigma/c$ for mass column density σ and one-dimensional velocity dispersion c. Note that this is twice the rate usually found in a disk using

a polytropic equation of state. The corresponding growth time equals 1.3×10^7 years for $\sigma=20$ $M_\odot$ pc^{-2} and $c=7$ km s^{-1}. This is a much shorter time than either the growth time in the Parker instability or the random collisional build-up time, although the total cloud formation and collapse time may be twice this initial exponential growth time. The instability is faster in spiral density wave arms because σ is larger there. An approximate density-squared dependence for the star formation rate results (Elmegreen 1989d).

Another strong point is that the gravitational instability model explains the origin of giant cloud complexes with masses comparable to the classical Jeans mass in the galactic disk (e.g., $10^7 M_\odot$). Such clouds are now commonly observed in our Galaxy and in other galaxies (see tabulation and review in Elmegreen 1987c; see also the review by Lo at this conference). A remarkable property of these giant clouds is that, even in the cases where they are seen primarily in HI, they are gravitationally bound, virialized objects, similar to giant molecular clouds but larger (Elmegreen 1987c; Elmegreen and Elmegreen 1987).

The reason why the largest clouds in our Galaxy are atomic rather than molecular may be related to H_2 self-shielding requirements for virialized objects in a common pressure environment (Elmegreen 1989a): self-shielding requires that the recombination rate of hydrogen atoms on the surfaces of grains balance the incident photo-dissociation rate. This gives a critical column density that varies approximately inversely with density. But because virialized clouds with a common pressure boundary have densities that vary inversely with radius, the larger clouds should have lower densities. It turns out that for typical pressures and metallicities, the densities in clouds more massive than several times $10^6 M_\odot$ become so low that the overall cloud cannot shield itself from ambient radiation, except in the densest cores. Thus the most massive clouds in galaxies with pressures and metallicities similar to those in the Milky Way should be atomic even though they are dynamically similar to giant molecular clouds. Galaxies with higher pressures, such as M51, should have their most massive clouds in molecular form (see Lo, this conference), and galaxies with low metallicities should have even lower mass clouds in atomic form, as observed for Magellanic-type irregular galaxies (Thronson 1988; Dettmar and Heithausen 1989). It follows that *the molecular or atomic nature of clouds depends more on the interstellar pressure, metallicity and radiation field than on the cloud formation mechanism* . This conclusion is consistent with the study of the diffuse and molecular cloud mass spectra by Dickey and Garwood (1989).

A third strong point is that cloud formation by gravitational instabilities leads naturally to star formation by continued gravitational collapse. This may explain the observed hierarchy of star formation sizes, from the largest clouds and star formation regions (Efremov 1978; Shevchenko 1979; Elmegreen and Elmegreen 1983; Elmegreen 1987c; Avedisova 1988; Efremov and Sitnik 1988) to OB associations and individual stellar clusters. All of the clouds that form in the model are also virialized so they satisfy the size-linewidth relation, or constant-column-density relation, as soon as they become dense.

Because the material that forms a cloud complex accretes along the prevailing magnetic field lines in this model, the magnetic field structure in the final object should be somewhat regular. The weight of the cloud in the galactic background gravitational field can cause the gas to bend down the attached magnetic field lines, as observed in Perseus by Appenzeller (1971) and in M31 by Beck et al. (1989); such field line structure does not imply that the cloud formed exclusively by the Parker instability, as previously suggested.

Gravitational Instability: Weak Points
The applicability of the gravitational instability model to the formation of low and intermediate mass clouds is not yet known even though the model allows clouds of this mass to condense directly from the ambient medium in regions with lower than average cloud stirring. The low mass clouds might form primarily as broken pieces of larger clouds, however, or as gravitationally condensed fragments of larger clouds. Low mass clouds could also form in shells (see below) or in a small-scale version of the Parker instability, for which self-gravity may be relatively unimportant.

The applicability of the instability model to the mass spectrum is also uncertain, although simple arguments given above suggest a possible connection. Non-linear calculations or N-body simulations would be useful for checking this idea.

The most obvious weakness of the instability model is the general uncertainty about macroscopic heating and cooling processes in the ISM. We do not know much about the macroscopic energy sources and how this energy is distributed among the cloud population (e.g., Tarrab 1983; McKee 1988). Cloud stirring may or may not be intermittent, for example, as assumed to get the low mass end of the instability spectrum. Alfven waves and other processes may redistribute the available energy quickly, making the actual heat input more uniform than the heat sources. In some respects, these uncertainties are more important for the instability model than for other models of cloud formation.

CLOUD FORMATION IN SHELLS

Although the primary emphasis in this paper has been in a discussion of *spontaneous* cloud formation in the interstellar medium, the stimulated processes, such as pressurized shell formation, are certainly at work too, and may even dominate cloud formation in some situations. As discussed above, the formation of distinct clouds *inside* the shells still requires a primary mechanism, such as a gravitational instability in the shell, or a direct coalescence of pre-existing clouds that enter the accumulation front. The Parker instability has also been proposed as a cloud or fragment formation mechanism inside swept-up shells (Baierlein 1983).

Reviews of cloud formation in swept-up shells may be found in Elmegreen (1981b, 1985c, 1987a, 1987c), Elmegreen and Wang (1988), and Tenorio-Tagle and Bodenheimer (1989). Recent progress in understanding the physical processes involved with this model may be found in Chiang and Prendergast (1985), Katz and Green (1986), McCray and Kafatos (1987), Chiang and Bregman (1988), Elmegreen (1989b), and Palous, Franco, and Tenorio-Tagle (1989).

OVERVIEW

Cloud formation apparently involves a wide range of physical processes, including collisional agglomeration of pre-existing clouds, both in the ambient medium and in swept-up shells, and a combination of the magnetic Rayleigh-Taylor, gravitational and thermal instabilities in both the ambient and shocked cloud fluids. These processes presumably also combine to give the observed cloud mass spectrum and hierarchical structure, and because of the self-gravitational forces, they are intimately involved with star formation too. Many or most of the smallest clouds, such as Bok globules and small dark clouds, are probably condensed fragments or disrupted pieces of larger clouds. Cloud formation is therefore also a part of cloud collapse and dispersal.

The observation of diffuse molecular clouds and giant, self-gravitating atomic clouds implies that cloud formation processes are distinct from molecule formation processes. The presence of molecules in a cloud apparently depends more on the ambient pressure, metallicity and radiation field than on the mechanism of cloud formation.

The time scale for gravitationally driven cloud formation is on the order of $(4\pi G\rho)^{-1/2}$ or $c/(\pi G\sigma)$ for three and two dimensional densities ρ and σ and one-dimensional rms velocity dispersion c. These two expressions are approximately the same if the midplane density is used in the second case, after changing σ to $2\rho H$ for scale height H. The time scale corresponds to approximately $2 \times 10^7 (n/1 \text{ cm}^{-3})^{-1/2}$ years for an average hydrogen density n; the time scale can be very short in spiral density wave crests or in regions of high ambient density. The time scale for cloud formation by random cloud collisions is approximately $\sigma_c/(\rho c)$ for average cloud mass column density σ_c. For diffuse clouds, σ_c is small and the collision time is small, comparable to the gravitational time scale. For self-gravitating clouds, σ_c is large and the collision time is large, possibly much larger than the time for gravitational processes to operate. Such large times apparently limit the applicability of the collision model, as discussed above.

The lifetime of a self-gravitating cloud is presumably determined by the destruction from internal star formation. This lifetime is on the order of 10 to 50 million years for clouds that contain OB stars. If t_D is the destruction time and t_F is the formation time of dense self-gravitating clouds, then the average fraction of the ISM in the form of these clouds is approximately the ratio of time scales,

$$p_{SG} \sim t_D/(t_F + t_D), \qquad (3)$$

which is approximately 0.5 to 0.8 for $t_D \sim 4 \times 10^7$ years and $t_F \sim (4\pi G\rho)^{-1/2} \sim 2 \times 10^7$ years, depending on density. This is a different quantity than the molecular fraction of the ISM, although it may be similar in actual value in some regions. The molecular fraction is

$$f \sim p_{DIF} f_{DIF} + p_{SG} f_{SG} \qquad (4)$$

where p_{DIF} and p_{SG} are the mass fractions for diffuse and self-gravitating clouds respectively ($p_{DIF} = 1 - p_{SG}$), and f is the molecular fraction in each population (cf. Eq. 1 for f_{DIF}). If $f_{SG} \sim 1$, as expected for normal metallicities, pressures and radiation fields, then $f > p_{SG}$, i.e., the molecular fraction of the total ISM exceeds the mass fraction in the form of self-gravitating clouds.

The ultimate goal of cloud formation theories is to explain the mass range, mass spectrum, and geometrical configuration of the observed cloudy structures in the ISM. The relative importance of the various theories is likely to depend on cloud mass, in addition to the parameters of the ISM, such as density, magnetic field strength, heat input, rotation and shear rate, etc.. While some progress has been made in explaining the mass range and, to a lesser extent, the mass spectrum, very little is understood yet about the geometric configuration of interstellar clouds.

REFERENCES

Allen, R. J., Goss, W. M., and van Woerden, H. 1973, *Astr. Ap.*, **29**, 447.
Ames, S. 1973, *Ap. J.*, **182**, 387.
Appenzeller, I. 1971, *Astr. Ap.*, **12**, 313.
Asseo, E., Cesarsky, C. J., Lachieze-Ray, M., and Pellat, R. 1978, *Ap. J. (Letters)*, **225**, L21.
Avedisova, V. S., 1988, in *Structure of Galaxies and Star Formation*, ed. J. Palous, Prague: Czechoslovakian Academy of Sciences, p.171.
Badhwar, G. D., and Stephens, S. A. 1977, *Ap. J.*, **212**, 494.
Baierlein, R. 1983, *M.N.R.A.S.*, **205**, 669.
Balbus, S. A. 1988, *Ap. J.*, **324**, 60.
Balbus, S. A., and Cowie, L. L. 1985, *Ap. J.*, **297**, 61.
Balbus, S. A., and Soker, N. 1989, *Ap. J.*, **341**, 611.
Bash, F. N., Green, E., and Peters, W.L. 1977, *Ap. J.*, **217**, 464.
Beck, R., Loiseau, N., Hummel, E., Berkhuijsen, E.M., Grave, R., and Wielebinski, R. 1989, *Astr. Ap.*, , submitted.
Bel, N., and Schatzman, E. 1958, *Rev.Mod.Phys.* **30**, 1015.
Bhatt, H. C., Rowse, D. P., and Williams, I. P. 1984, *M.N.R.A.S.*, **209**, 69.
Bhatt, H. C., and Williams, I. P. 1986, *M.N.R.A.S.*, **219**, 217.
Blitz, L., and Shu, F. H. 1980, *Ap. J.*, **238**, 148.
Bohigas, J. 1988, *Astr. Ap.*, **205**, 257.
Boulesteix, J., Courtes, G., Laval, A., Monnet, G., and Petit, H. 1974, *Astr. Ap.*, **37**, 33.
Bridle, A. H., and Kesteven, M. J. I. 1970, *Astr. J.*, **75**, 902.
Burton, W. B., Gordon, M. A., Bania, T. M., and Lockman, F. J. 1975, *Ap. J.*, **202**, 30.
Casoli, F., and Combes, F. 1982, *Astr. Ap.*, **110**, 287.
Casoli, F., Combes, F., and Gerin, M. 1984, *Astr. Ap.*, **133**, 99.
Cesarski, C. J. 1980, *Ann.Rev.Astr.Ap.*, **18**, 289.
Chagelishvili, G. D., Lominadze, J. G., and Sokhadze, Z. A. 1988, *Ap.Sp.Sci.*, **141**, 361.
Chandrasekhar, S. 1954a, *Vistas in Astronomy*, ed. Beer, London: Pergamon.
Chandrasekhar, S. 1954b, *Ap. J.*, **119**, 17.

Chandrasekhar, S. 1961, *Hydrodynamic and Hydromagnetic Stability*, Oxford: Clarendon Press.
Chandrasekhar, S., and Fermi, E. 1953, *Ap. J.*, **118**, 116.
Chiang, W. H., and Prendergast, K. H. 1985, *Ap. J.*, **297**, 507.
Chiang, W. H., and Bregman, J. N. 1988, *Ap. J.*, **328**, 427.
Chieze, J. P., and Lazareff, B. 1980, *Astr. Ap.*, **91**, 290.
Cleary, M. N., Heiles, C., and Haslam, C. G. T. 1979, *Ap. J. Suppl.*, **36**, 95.
Clemens, D. P. 1985, *Ap. J.*, **295**, 422.
Clifford, P. 1985, *M.N.R.A.S.*, **216**, 93.
Clifford, P., and Elmegreen, B. G. 1983, *M.N.R.A.S.*, **202**, 629.
Cohen, R. S., and Thaddeus, P. 1977, *Ap. J. (Letters)*, **217**, L155.
Combes, F., and Gerin, M. 1985, *Astr. Ap.*, **150**, 85.
Cowie, L.L. 1980, *Ap. J.*, **236**, 868.
Cowie, L.L. 1981, *Ap. J.*, **245**, 66.
Dame, T. M., Elmegreen, B. G., Cohen, R. S., and Thaddeus, P. 1986, *Ap. J.*, **305**, 892.
Dame, *et al.* 1987, *Ap. J.*, **322**, 706.
Dettmar, R. J., and Heithausen, A. 1989, *Ap. J. (Letters)*, **344**, L61.
Dickel, H. 1976, *Ap. J.*, **141**, 1306.
Dickey, J. M., and Garwood, R. W. 1988, *Ap. J.*, **341**, 89.
Di Fazio, A. 1986, *Astr. Ap.*, **159**, 49.
Efremov, Yu. N., and Sitnik, T. G. 1988, *Sov.Astr.Letts.*, **14**, 347.
Efremov, Yu. N. 1978, *Sov.Astr.Letts.*, **4**, 66.
Elmegreen, B. G. 1979, *Ap. J.*, **231**, 372.
Elmegreen, B. G. 1981a, *Ap. J.*, **243**, 512.
Elmegreen, B. G. 1981b, in *Submillimeter Wave Astronomy*, ed. J. Beckman and J. Phillips, Cambridge: Cambridge Univ. Press, p. 1.
Elmegreen, B. G. 1982a, *Ap. J.*, **253**, 634.
Elmegreen, B. G. 1982b, *Ap. J.*, **253**, 655.
Elmegreen, B. G. 1982c, in *Formation of Planetary Systems*, ed. A. Brahic, Toulouse: Cepadues, p. 61.
Elmegreen, B. G. 1985a, in *Protostars and Planets, II*, ed. D.C. Black and M.S. Matthews, Tucson: Univ. of Arizona Press, p. 33.
Elmegreen, B. G. 1985b, *Ap. J.*, **299**, 196.
Elmegreen, B. G. 1985c, in *Birth and Infancy of Stars*, ed. R. Lucas, A. Omont, and R. Stora, Amsterdam: North Holland, p. 215.
Elmegreen, B. G. 1987a, in *Interstellar Processes*, ed. D. Hollenbach and H. Thronson, Dordrecht: Reidel, p. 259.
Elmegreen, B. G. 1987b, *Ap. J.*, **312**, 626.
Elmegreen, B. G. 1987c, in *Star Forming Regions*, ed. M. Peimbert and J. Jugaku, Dordrecht: Reidel, p. 457.
Elmegreen, B. G. 1988, *Ap. J.*, **326**, 616.
Elmegreen, B. G. 1989a, *Ap. J.*, **338**, 178.
Elmegreen, B. G. 1989b, *Ap. J.*, **340**, 786.
Elmegreen, B. G. 1989c, *Ap. J. (Letters)*, **342**, L67.
Elmegreen, B. G. 1989d, *Ap. J.*, **344**, 306.
Elmegreen, B. G. 1989e, *Ap. J.*, **347**, in press.
Elmegreen, B. G. 1982b, *Ap. J.*, **253**, 655.
Elmegreen, B. G., and Elmegreen B. G. 1983, *M.N.R.A.S.*, **203**, 31.
Elmegreen, B. G., and Elmegreen B. G. 1986, *Ap. J.*, **311**, 554.

Elmegreen, B. G., and Elmegreen B. G. 1987, *Ap. J.*, **320**, 182.
Elmegreen, B. G., and Wang, M. 1988, in *Molecular Clouds in the Milky Way and External Galaxies*, ed. R.L. Dickman, R.L. Snell, and J.S. Young, Berlin: Springer-Verlag, p. 240.
Elmegreen, D. M. 1980, *Ap. J. Suppl.*, **43**, 37.
Federman, S. R., Glassgold, A. E., and Kwan, J. 1979, *Ap. J.*, **227**, 466.
Field, G. H. 1965, *Ap. J.*, **142**, 531.
Field, G. H., and Hutchins, J. 1968, *Ap. J.*, **153**, 737.
Field, G. H., and Saslaw, W. J. 1965, *Ap. J.*, **142**, 568.
Franco, J., and Cox, D.P. 1986, *Pub. A.S.P.*, **98**, 1076.
Fricke, W. 1954, *Ap. J.*, **120**, 356.
Gilden, D. L. 1984, *Ap. J.*, **279**, 335.
Goldreich, P., and Lynden-Bell, D. 1965, *M.N.R.A.S.*, **130**, 97.
Goldsmith, D. W., Habing, H. J., and Field, G. B. 1969, *Ap. J.*, **158**, 173.
Goldstein, J. S., and Mazzella, A. J. 1974, *Nouvo Cimento* **21B**, 142.
Graziani, F. R., and Black, D. C. 1987, *Ap.Letters Comm.*, **25**, 235.
Handbury, M. J., Simon, S., and Williams, I. P. 1977, *Astr. Ap.*, **61**, 443.
Hausman, M. A. 1981, *Ap. J.*, **245**, 72.
Hausman, M. A. 1982, *Ap. J.*, **261**, 532.
Horiuchi, T., Matsumoto, R., Hanawa, T., and Shibata, K. 1988, *Pub. A.S.Japan*, **40**, 147.
Jeans, J. H. 1902, *Phil. Trans. Roy. Soc. London*, **A199**, 1.
Johns, T. C., and Nelson, A. H. 1986, *M.N.R.A.S.*, **220**, 165.
Jog, C. J., and Solomon, P. M. 1984a, *Ap. J.*, **276**, 114.
Jog, C. J., and Solomon, P. M. 1984b, *Ap. J.*, **276**, 127.
Jura, M. 1976, *Astr. J.*, **81**, 178.
Kahn, F. 1955, in *Gas Dynamics in Cosmic Gas Clouds*, ed. H.C. van de Hulst and J.M. Burgers, Amsterdam: North Holland, Ch.12.
Katz, J. I., and Green, M. L. 1986, *Astr. Ap.*, **161**, 139.
Kennicutt, R. C. 1989, *Ap. J.*, **344**, 685.
Knude, J. 1979, *Astr. Ap. Suppl.*, **38**, 407.
Kritsuk, A. G. 1985, *Sov.Astr.*, **29**, 85.
Kulkarni, S. R., and Heiles, C. 1987, in *Interstellar Processes*, ed. D.J. Hollenbach, Dordrecht: Reidel, p. 87.
Kuz'min, G. G. 1965, *Kinematics and Dynamics of Stellar Systems*, Alma-Ata, Nauka Press, p. 11.
Kuznetsov, V. D., and Ptuskin, V. S. 1983, *Ap.Sp.Sci.*, **94**, 5.
Kuznetsov, V. D. 1986, *Sov.Astr.AJ*, **30**, 5.
Kwan, J. 1979, *Ap. J.*, **229**, 567.
Kwan, J. 1988, in *Molecular Clouds in the Milky Way and External Galaxies*, ed. R.L. Dickman, R.L. Snell, and J.S. Young, Dordrecht: Reidel, p. 281.
Kwan, J., and Valdes, F. 1983, *Ap. J.*, **271**, 604.
Kwan, J., and Valdes, F. 1987, *Ap. J.*, **315**, 92.
Lachieze-Ray, M., Asseo, E., Cesarsky, C. J., and Pellat, R. 1980, *Ap. J.*, **238**, 175.
Lada, E. A., and Blitz, L. 1988, *Ap. J. (Letters)*, **326**, L69.
Lattanzio, J. C., Monagahan, J. J., Pongracic, H., and Schwarz, M. P. 1985, *M.N.R.A.S.*, **215**, 125.
Lattanzio, J. C., and Henriksen, R. N. 1988, *M.N.R.A.S.*, **232**, 565.
Ledoux, P. 1951, *Ann.d'Ap.*, **14**, 438.

Leisawitz, D., Bash, F.N, and Thaddeus, P. 1989, *Ap. J. Suppl.*, **70**, 731.
Lerche, I. 1967a, *Ap. J.*, **148**, 415.
Lerche, I. 1967b, *Ap. J.*, **149**, 395.
Lerche, I. 1967c, *Ap. J.*, **149**, 553.
Lerche, I., and Parker, E. N. 1968, *Ap. J.*, **154**, 515.
Liszt, H. S., Xiang, D., and Burton, W. B. 1981, *Ap. J.*, **249**, 532.
Lubow, S. H., Balbus, S. A., and Cowie, L. L. 1986, *Ap. J.*, **309**, 496.
Lynden-Bell, D. 1966, *Observatory*, **86**, 57.
MacLaren, I., Richardson, K. M., and Wolfendale, A. W. 1988, *Ap. J.*, **333**, 821.
Magnani, L. A. 1987, PhD. Thesis, Univ. of Maryland.
Magnani, L. A., Blitz, L., and Mundy, L. 1985, *Ap. J.*, **295**, 402.
Marochnik, L. S. 1966, *Sov.Astr.AJ*, **10**, 738.
Matsumoto, R., Horiuchi, T., Shibata, K., and Hanawa, T., 1988, *Pub. A.S.Japan*, **40**, 171.
Mayer, P. 1969, *Ann.Rev.Astr.Ap.*, **7**, 1.
McCray, R. A., and Kafatos, M. 1987, *Ap. J.*, **317**, 190.
McKee, C. 1988, *Ap.Sp.Sci.*, **118**, 383.
Min, K. W. 1988, *Ap.Sp.Sci.*, **145**, 167.
Miyawaki, R., Hasegawa, T., and Hayashi, M. 1988, *Pub. A.S.Japan*, **40**, 69.
Mooney, T. J., and Solomon, P. M. 1988, *Ap. J. (Letters)*, **334**, L51.
Mouschovias, T. C. 1974, *Ap. J.*, **192**, 37.
Mouschovias, T. C. 1981, in *Fundamental Problems in the Theory of Stellar Evolution*, ed. D. Sugimoto, D.Q. Lamb, and D.N. Schramm, Dordrecht: Reidel, p. 27.
Mouschovias, T. C., Shu, F. H., and Woodward, P. 1974, *Astr. Ap.*, **33**, 73.
Nelson, A. H. 1985, *M.N.R.A.S.*, **215**, 161.
Norman, C., and Silk, J. 1980, in IAU Symposium No. 87, *Interstellar Molecules*, ed B. H. Andrew, Dordrecht: Reidel, p. 137.
Oort, J. H. 1954, *B.A.N.*, **12**, 177.
Palous, J., Franco, J., and Tenorio-Tagle, G. 1989, *Astr. Ap.*, submitted.
Parker, E. N. 1966, *Ap. J.*, **145**, 811.
Parker, E. N. 1967a, *Ap. J.*, **149**, 517.
Parker, E. N. 1967b, *Ap. J.*, **149**, 535.
Parker, E. N. 1968a, *Ap. J.*, **154**, 49.
Parker, E. N. 1968b, *Ap. J.*, **154**, 57.
Parker, E. N. 1968c, *Ap. J.*, **154**, 875.
Parker, E. N., and Lerche, I. 1969, *Comments Ap.Sp.Sci.*, **1**, 215.
Parravano, A. 1987, *Astr. Ap.*, **172**, 280.
Penston, M. V., Munday, A., Strickland, D. J., and Penston, M. J. 1969, *M.N.R.A.S.*, **142**, 355.
Polk, K. S., Knapp, J. G., Stark, A. A., and Wilson, R. W. 1988, *Ap. J.*, **332**, 432.
Quirk, W. J. 1972, *Ap. J. (Letters)*, **176**, L9.
Roberts, W. W., and Steward, G. R. 1987, *Ap. J.*, **314**, 10.
Safronov, V. S. 1960, *Ann. d'Ap.*, **23**, 979.
Sanders, D. B., Solomon, P. M. and Scoville, N.Z. 1984, *Ap. J.*, **276**, 182.
Scalo, J. M. 1985, in *Protostars and Planets II*, ed D. C. Black and M. S. Matthews, Tucson: University of Arizona Press, p. 200.
Scalo, J. M., and Struck-Marcell, C. 1984, *Ap. J.*, **276**, 60.
Scalo, J. M., and Struck-Marcell, C. 1986, *Ap. J.*, **301**, 77.

Scheffler, H. 1966, *Zf.f.Ap.*, **63**, 267.
Scheffler, H. 1967a, *Zf.f.Ap.*, **65**, 60.
Scheffler, H. 1967b, *Zf.f.Ap.*, **66**, 33.
Schlickeiser, R., and Lerche, I. 1985, *Astr. Ap.*, **151**, 151.
Scoville, N.Z., and Solomon, P. M. 1975, *Ap. J. (Letters)*, **199**, L105.
Scoville, N., and Hersh, K. 1979, *Ap. J.*, **229**, 578.
Scoville, N. Z., Sanders, D. B., and Clemens, D. P. 1986, *Ap. J. (Letters)*, **310**, L77.
Sharma, R. C., and Singhe, B. 1988, *Ap.Sp.Sci*, **143**, 233.
Shaya, E. J., and Federman, S. R. 1987, *Ap. J.*, **319**, 76.
Shevchenko, V. S. 1979, *Sov.Astr.*, **23**, 163.
Shu, F. H. 1974, *Astr. Ap.*, **33**, 55.
Simon, R.. 1965, *Ann.d'Ap.*, **28**, 40.
Solomon, P. M., Sanders, D. B., and Scoville, N.Z. 1979, in IAU Symposium No. 84, *Large Scale Characteristics of the Galaxy*, ed W. B. Burton, Dordrecht: Reidel , p. 35.
Solomon, P. M., Rivolo, A.R., Barrett, J., and Yahil, A. 1987, *Ap. J.*, **319**, 730.
Spitzer, L. Jr. 1968, in *Diffuse Matter in Space*, New York: Interscience.
Spitzer, L. Jr. 1978, in *Physical Processes in Interstellar Space*, New York: Interscience.
Stark, A. A. 1979, PhD Dissertation, Princeton University.
Stark, A. A. 1986, in *Highlights of Astronomy*, **7**, 507.
Stark, A. A., and Blitz, L. 1978, *Ap. J. (Letters)*, **225**, L15.
Stark, A. A., Elmegreen, B. G., and Chance, D. 1987, *Ap. J.*, **322**, 64.
Stark, A. A., and Brand, J. 1989, *Ap. J.*, **339**, 763.
Stone, M. E. 1970, *Ap. J.*, **159**, 293.
Struck-Marcell, C., and Scalo, J. M. 1984, *Ap. J.*, **277**, 132.
Struck-Marcell, C., and Scalo, J. M. 1987, *Ap. J. Suppl.*, **64**, 396.
Taff, L.G., and Savedof, M.P. 1973, *M.N.R.A.S.*, **164**, 357.
Tarrab, I 1983, *Astr. Ap.*, **125**, 308.
Tenorio-Tagle, G., and Bodenheimer, P. H. 1988, *Ann.Rev.Astr.Ap.*, **27**, 145.
Terebey, S., Fich, M., Blitz, L., and Henkel, C. 1986, *Ap. J.*, **308**, 357.
Tomisaka, K. 1984, *Pub. A.S.Japan*, **36**, 457.
Tomisaka, K. 1986, *Pub. A.S.Japan*, **38**, 95.
Tomisaka, K. 1987, *Pub. A.S.Japan*, **39**, 109.
Toomre, A. 1964, *Ap. J.*, **139**, 1217.
Thronson, H. A. 1988, in *Molecular Clouds in the Milky Way and External Galaxies*, ed. R.L. Dickman, R.L. Snell, and J.S. Young, Berlin: Springer-Verlag, p. 413.
Vazquez, E.C., and Scalo, J. M. 1989, *Ap. J.*, **343**, 644.
Wright, M. C. H., Warner, P. J., and Baldwin, J. E. 1972, *M.N.R.A.S.*, **155**, 337.
Xiang, D.-l., and Lou, G.-f. 1985, *Acta Astron.Sin.*, **26**, 321.
Yuan, C., and Wang, C.Y. 1982, *Ap. J.*, **252**, 508.
Zweibel, E. G., and Kulsrud, R. M. 1975, *Ap. J.*, **201**, 63.

THE EVOLUTION OF GALACTIC GIANT MOLECULAR CLOUDS

LEO BLITZ
Astronomy Program
University of Maryland, College Park, MD 21742

ABSTRACT   After a brief review of the global properties of GMCs in the Galaxy, the angular momentum of local GMCs is discussed. It is shown quantitatively that the specific angular momentum of GMCs is typically a factor of more than four smaller than that of the diffuse interstellar medium; the problem of angular momentum shedding in the process of star formation evidently starts when GMCs form. A notable exception is the Rosette Molecular Cloud. Two clouds in different evolutionary states are described which may be probed to investigate their structural differences. It is also shown that GMCs have atomic envelopes with masses comparable to their molecular masses. These envelopes are likely to pervade the interclump medium and be responsible for most of its mass.

The structure of GMCs is then discussed in some detail with emphasis on the structure of the Rosette Molecular Cloud. It is shown that the clumps have a small volume filling fraction, and that the clump-interclump density contrast is high. The clump ensemble is shown to have undergone dynamical evolution, with the most massive clumps concentrated toward the midplane of the complex and also posessing a lower clump-to-clump velocity dispersion. Like the situation in some other clouds, the majority of clumps do not appear to be gravitationally bound, these clumps can be confined by the pressure of an atomic intercloud medium. The observations are used to suggest an outline for the evolution of GMCs in which sequential star formation is unnecessary.

INTRODUCTION

One of the major astronomical discoveries of the latter half of the twentieth century is that stars form in molecular clouds. While the stellar life cycle is well understood (with some notable gaps), and there has been considerable progress related to the formation of single stars from dense molecular cores (*e.g.* Shu, Adams, and Lizano 1987), there is currently a rather poor understanding of how the interstellar medium collects itself from a diffuse atomic state into dense molecular clouds that form stars. Phrasing the question in this way already presupposes that there is considerable mass transfer between the atomic and molecular phases; a supposition for which there is some contrary evidence from observations of extragalactic molecular clouds (Allen, Atherton and Tilanus 1986; Vogel, Kulkarni, and Scoville 1988).

This paper briefly reviews some of the global properties of giant molecular clouds (GMCs), and then discusses the relationship of the atomic environment of the clouds to the GMCs in the solar vicinity. It is then shown that it is possible to identify GMCs in different evolutionary states, and we discuss two such objects. The structure of one of these, the Rosette Molecular Cloud, is examined in detail, in order to find clues related to the evolution of the molecular material into dense star forming material. Finally, a sketch of the evolution of the GMCs is outlined that suggests a picture different from some of the main ideas that have been in the literature until now.

GLOBAL PROPERTIES OF GIANT MOLECULAR CLOUDS

Giant molecular clouds have been studied as a whole largely through their CO emission in the radio portion of the spectrum. The studies have been mainly of two kinds. 1) Observations of individual objects, where a cloud is identified, often by its association with a visible HII region, and then mapped to its outer boundaries. 2) Surveys of the Galactic plane in CO, where GMCs are identified through some objective criterion. In the second case, GMCs are generally defined down to some contour level because of the possibility of confusion with other gas along the line of sight. Surprisingly, the general properties of GMCs defined in this way are not significantly different form those identified by the first method. Table 1 gives the properties derived for clouds in the solar vicinity (e.g. Blitz 1987b). Inner Galaxy molecular clouds may be somewhat denser and more opaque (see McKee 1989; Solomon, *et al.* 1987), but they do not seem to form a separate population from the local clouds. This important conclusion suggests that detailed studies of the molecular clouds near the Sun can tell us about the ensemble properties of GMCs everywhere in the disk. An exception is likely to be in the innermost regions of the Galaxy such as the molecular disk within 400 pc of the center and the 3 kpc arm of the Galaxy.

TABLE 1. Global Properties of Solar Neighborhood GMCs

Mass	$1 - 2 \times 10^5$ M$_\odot$
Mean diameter	45 pc
Projected surface area	2.1×10^3 pc^2
Volume	9.6×10^4 pc^3
Volume averaged $n(H_2)$	~ 50 cm^{-3}
Mean $N(H_2)$	$3 - 6 \times 10^{21}$ cm^{-3}
Local surface density	~ 4 kpc^{-2}
Mean separation	~ 500 pc

From the study of Local GMCs, the following general conclusions may be drawn:

1) GMCs are discrete objects with well defined boundaries (see e.g. Blitz and Thaddeus 1980 for a quantitative discussion of this point - see especially Appendix B). The well defined boundaries suggest that there is a phase transition at the edges of a molecular cloud. This point is discussed in greater detail below.

2) All OB associations form from GMCs; thus GMCs are the nucleation sites for nearly all star formation in the Milky Way (Zuckerman and Palmer 1974; Blitz 1978, 1980 and references therein).

3) Within the uncertainties, the cloud-to-cloud velocity dispersion of local GMCs (with typical masses of $\sim 2 \times 10^5$ $M_\odot$) is the same as that of the small molecular clouds (with typical masses of ~ 50 $M_\odot$) found at high galactic latitude (Blitz 1978, Magnani, Blitz and Mundy 1985). The velocity dispersion of local molecular clouds therefore appears to be independent of mass over three or four orders of mass. Furthermore, the velocity dispersion of molecular clouds seems to vary only weekly with galactocentric distance (Liszt and Burton 1983, Stark 1984, Clemens 1985) implying that the constancy of velocity dispersion with mass is likely to hold everywhere in the disk of the Galaxy. For the most massive clouds, Stark (1983) has shown that the scale height, and by implication therefore the vertical velocity dispersion of GMCs, exhibits a decrease in the inner Galaxy. All of these results taken together imply that collisional processes are not important in the formation of GMCs except possibly at the highest masses ($\sim 10^6$ $M_\odot$).

4) The survey of local molecular material within 1 kpc of the Sun (Dame *et al.* 1986) finds no GMCs without star formation. In fact, within 3 kpc of the Sun, only one GMC is found without evidence of star formation (Maddalena and Thaddeus 1986). This cloud is discussed in more detail below. In the solar vicinity at least, molecular clouds without star formation are quite rare. It therefore appears that star formation is quite rapid after the formation of a GMC.

5) The above result, when combined with calculations of the destructive processes associated with the formation of massive stars implies that the molecular clouds are quite young, $\sim 3 \times 10^7 y$ (Blitz and Shu 1980). That is, since all GMCs appear to be sites of star formation, especially massive star formation, the clouds do not appear to be able to survive the birth of more than a few generations of massive stars, an argument that is consistent with various other lines of reasoning (Blitz and Shu 1980). Direct observational evidence for lifetimes of this order were first presented by Bash, Green and Peters (1977), and subsequently by Leisewitz, Bash, and Thaddeus (1986) from the association (or lack thereof) of molecular material with young clusters of vaying ages. Cohen (*et al.* 1980), have come to the same conclusion from the confinement of the molecular arms in the outer Galaxy to narrow kinematic ranges.

6) A large fraction of the stars formed in GMCs do not return their material to the interstellar medium in a Hubble time. Therefore, either they were more numerous in the past, or the interstellar medium is replenished by infall or inflow from the outer reaches of the Galaxy (e.g. Lacey and Fall 1985).

Studies of GMCs based on inner Galaxy CO surveys run into the problem of finding an objective way to identify the molecular clouds. The degree of blending of the spectral lines, the large amount of foreground and background emission, and the necessity of using kinematic criteria for identifying the clouds add a significant amount of uncertainty to the identification of GMCs (see *e.g.* Adler 1988, Blitz 1987a). Nevertheless, a number of properties of GMCs are inferred from the Galactic CO surveys, and the searches for molecular clouds in the outer Galaxy. Some of the more important of these are the following:

1) There appears to be a linewidth-size relation for GMCs in the inner Galaxy. From various studies, the results may be given as follows:

Dame et al. (1986)	$\Delta v = 1.20 R^{0.50}$
Solomon et al. (1987)	$\sigma_v = 1.0 S^{0.50}$
Scoville et al. (1987)	$\sigma_v = 0.31 D^{0.55}$
Sanders et al. (1985)	$\Delta v = 0.88 D^{0.62}$

where the quantity on the left is either the full width at half maximum or the one dimensional velocity dispersion in km s^{-1} and the quantity on the right is some measure of the mean linear size of a GMC in pc. The scatter in all of the studies is not terribly large. The close agreement between the various groups in the value of the power law exponent in the linewidth-size relation is so striking, that unless there is a selection effect common to all of the studies (see *e.g.* Blitz 1987a), it suggests that the relation underlies some fundamental property of GMCs.

2) If the clouds are in virial equilibrium, then

$$(\Delta V)^2 = \alpha GM/R.$$

If $(\Delta V) \sim R^{0.5}$, then $M \sim R^2$ which in turn implies that the mean H$_2$ column density of GMCs is constant. This simple conclusion is presumably telling us something important about either how GMCs form or how they regulate themselves. McKee (1989) has assumed the latter in his recent theory of photo-regulated star formation in GMCs.

3) GMCs in the outer Galaxy appear to have lower kinetic temperatures than GMCs in the inner Galaxy (Mead and Kutner 1988). It is unclear at present whether this is due to a lower external heating rate, decreased star formation rate or both.

4) The distribution of masses for GMCs in the inner Galaxy is found to be (for masses in excess of 10^5 M$_\odot$), $dN(M)/dM \propto M^{-1.5}$ (Solomon et al. (1987). The power law exponent is the same as that found for the distribution of clump masses in a GMC. If one tries to determine at what cloud mass half of the H$_2$ mass in the Galaxy resides, the various analyses of the inner Galaxy CO give results that fall within a factor of two of the range $1 - 2 \times 10^5 M_\odot$ when account is made of the different assumptions used. This is very close to the mean value in the solar vicinity (Stark and Blitz 1978).

One of the glaring deficiencies in Galactic studies of GMCs is a quantitative study of how the properties of GMCs vary with galactic radius. For example, we might expect that to be stable against the larger tidal forces and the larger energy density of dissociating radiation, that inner Galaxy clouds would be denser than the clouds at larger galactocentric distance. Such a conclusion was reached from an indirect analysis of Liszt, Burton, and Xiang (1984). However, no direct confirmation of this currently exists. Such a study would be particularly useful in trying to understand how different Galactic environments affect the formation and evolution of GMCs.

ANGULAR MOMENTUM

That GMCs rotate has long been known from the velocity gradients observed across them (e.g. Kutner *et al.* 1977). In order to understand the formation of the GMCs, it is therefore important to know how the specific angular momentum, that is the angular momentum per unit mass, compares to the specific angular momentum of the material from which the clouds formed.

Consider a typical GMC in the solar vicinity with a mass of 2×10^5 $M_\odot$. Such clouds have a typical maximum dimension of 100 pc and are elongated along the Galactic plane (Stark and Blitz 1978; Blitz 1980). If the true shape of the clouds is that of a cigar, the specific angular momentum is $1/3 R^2\Omega$; if it is a disk, then the specific angular momentum is $1/2 R^2\Omega$. The largest known velocity gradient for an entire GMC is 0.18 km s^{-1} pc^{-1}, for the Rosette Molecular Cloud (Blitz and Thaddeus 1980), the next largest is 0.10 km s^{-1} pc^{-1} for the L1641 cloud in Orion (the GMC containing the Orion Nebula), and all of the other gradients are significantly smaller. Typically, the velocity gradient is less than half that of the L1641 cloud over an entire GMC. Thus, if the velocity gradient is entirely due to rotation, Ω has an extreme value for GMCs of 0.18 km s^{-1} pc^{-1}, in the solar vicinity, but has a typical value somewhat less than 0.05 km s^{-1} pc^{-1}. The specific angular momentum of a typical GMC is therefore somewhat less than about 60 pc km s^{-1} or 1.22×10^{26} cm^2 s^{-1}. The specific angular momentum of the Rosette GMC is however 225 pc km s^{-1}. The uncertainty in these numbers is relatively small (see below).

The specific angular momentum of the general ISM is given by $1/2 R^2\Omega$, where Ω can be derived either from the Oort A constant or a flat rotation curve, giving values that differ from one another by 20%. The value of Ω is therefore 0.025 km s^{-1} pc^{-1}. We now ask from what radius must a cloud contract to obtain a mass of 2×10^5 $M_\odot$? The surface density of atomic gas in the vicinity of the Sun is 5 $M_\odot$ pc^{-2} (Henderson, Jackson and Kerr (1982). We may neglect the contribution from the molecular gas because most of that is already due to GMCs (see e.g. Solomon, *et al.* 1987), and the contribution to the midplane density of small molecular clouds can be ignored. Assuming that the local effctive scale height of the atomic gas is 200 pc (Falgarone and Lequeux 1973), then the midplane density of atomic gas is 0.5 cm^{-3} (see McKee, this volume for a similar estimate). Assuming that the initial configuration is cylindrical (this makes little difference) with a diameter equal to the height, the radius of a cylinder that would contract to form a typical GMC is 140 pc. Thus the specific angular momentum of the general ISM in the vicinity of the Sun is 250 km s^{-1} pc, a value 4 or more times higher than that of a typical GMC in the solar vicinity. Note that for many GMCs, there is no overall gradient measured, so the angular momentum shedding may be as much as an order of magnitude or more.

Thus, for a typical GMC, angular momentum is shed even in the process of formation from the diffuse ISM; the shedding of angular momentum in the process of star formation begins even when a molecular cloud forms. On the other hand, there are examples of molecular clouds that have shed almost none of their angular momentum, *e.g.* the Rosette Molecular Cloud. It is perhaps not a coincidence that the two GMCs with among the largest specific angular momenta are also among the richest OB associations in the solar vicinity.

Studies of the angular momentum distribution among the GMCs simply do not exist at present in spite of the importance that they have for understanding how the molecular clouds form from the interstellar medium. Even the results quoted above come partly from unpublished analyses of molecular clouds and from personal communications. There is a wealth of information, however, in the large scale surveys of the CO distribution in the Galaxy. For example, there have been hundreds of GMCs that have been catalogued (Solomon, et al. 1987; Scoville et al. 1987) in the inner Galaxy. One certainly expects that all GMCs that exhibit measurable velocity gradients will have the sense of the angular momentum parallel to the differential angular momentum of the Galaxy. This easily verifiable fact has never been demonstrated. Nor has a detailed study of the angular momentum of GMCs as a function of Galactic radius been done. Since the specific angular momentum of the ISM increases with decreasing Galactic radius, is it also true that the mean or the largest specific angular momenta for GMCs also increase? Many other questions about the distribution of angular momentum are answerable with existing data.

It is also important to recognize that the values of J/M quoted above are reasonably well determined. For a molecular cloud, the velocity differences across a cloud can be measured to an accuracy of about 1 km s^{-1} (measurement errors are considerably smaller). Large velocity gradients are therefore measurable to an accuracy of 10-20%. The major source of uncertainty is probably the distance to a cloud, which, for clouds like Orion and the Rosette are probably as low as about 20%. For the ISM, the largest uncertainties are the value of the Oort A constant, which is probably known to an accuracy of 20%, and the midplane density of the atomic hydrogen gas, which translates into the distance R out to which the ISM must be collected to form a GMC. The uncertainty in the density is probably less than 50%, but in any event, the distance R depends only on the 1/3 power of the midplane density, and thus J/M on the the 2/3 power. The major uncertainty for the specific angular momentum of the ISM is the actual process by which GMCs are formed. For example, if the efficiency of transformation of the ISM into molecular material is low, then R may be bigger, and the resulting J/M may be larger. However, because the efficiency also enters into the problem to the 1/3 power, it is unlikely to alter R by very much.

CLOUDS IN DIFFERENT EVOLUTIONARY STATES

Within one or two kpc of the Sun, all GMCs have been found to have at least some traces of star formation. However, in 1986 Maddalena and Thaddeus found a cloud in the outer Galaxy with a mass in excess of 10^6 $M_\odot$, a longest diameter of about 150 pc (both of these numbers assume that the kinematic distance of 3 kpc is valid), without any obvious traces of star formation activity. Moreover, the cloud has relatively weak broad CO lines which appear to be different from those seen in most of the clouds observed in the local solar neighborhood. Maddalena and Thaddeus speculated that the cloud is so young that it has not yet had time to form stars.

If this hypothesis is correct, there should not be any buried or embedded population of stars within the cloud. It is possible, for example, that there is a large HII region on the far side of the cloud, which is obscured by the intervening dust (Even this hypothesis is unlikely since HII regions are almost always accompanied by strong CO peaks). We thus made maps of the cloud using the IRAS 100 μm data base, and the result is shown in Figure 1a (Puchalsky and

Blitz in preparation). For comparison, a similar map of the Rosette Molecular Cloud is shown in Figure 1b. Both maps have the zodiacal emission removed and are background subtracted. Remarkably, the highest contour in Figure 1a is lower than the *lowest* contour in Figure 1b. The average 100 μm emissivity for the Maddalena-Thaddeus cloud is more than *two orders of magnitude* lower than that of the Rosette Molecular Cloud. Since the far infrared emission is generally thought to come from reradiated starlight, by comparison with the Rosette, which has an infrared emissivity typical of GMCs in the solar vicinity (Boulanger and Perault 1988), the Maddalena-Thaddeus cloud is extremely deficient in embedded stars.

This evidence supports the hypothesis that the Maddalena-Thaddeus cloud is so young that it has not yet had time to form stars. It is therefore an excellent candidate to examine the differences in structure between it and a more evolved cloud like the Rosette Cloud, or the Orion Molecular Cloud. A large effort is now underway to make such a comparison. The current difficulty is that the linear resolution of the observations of the Maddalena-Thaddeus cloud is ten times worse than that of the Rosette, and 30 times worse than that of Orion, a situation we are currently trying to remedy. Furthermore, although good, extensive ^{13}CO observations exist of both Orion and the Rosette Molecular clouds, no such observations currently exist of the Maddalena-Thaddeus cloud.

RELATIONSHIP TO ATOMIC HYDROGEN

The relatively sharp boundaries of GMCs are inferred from the quantitative ananlysis of Blitz and Thaddeus (1980), and the simple appearance of many GMCs on the Palomar Observatory Sky Survey prints. In the latter case, we observe a well defined region of dust obscuration for many GMCs, and these follow the contours of the CO emission quite closely. This is especially true for GMCs within 1 kpc of the Sun where the contrast between the foreground dust obscuration and the background stars is the highest. Good examples are found for the Orion Molecular Cloud (especially the boundary at the lowest galactic latitudes - see the CO maps of Kutner *et al.* 1977 and the ^{13}CO maps of Bally *et al.* 1987), the Mon OB1 molecular clouds (see Blitz 1980), and the Ophiuchus molecular clouds (see the CO maps made by Loren 1989, and deGeus 1988). There are many other examples that illustrate this point.

These sharp boundaries suggest that there is a phase transition that takes place at the boundaries of the clouds, but what then is the state of the gas at the other side of the phase transition of the GMC? In an unpublished analysis of the atomic clouds associated with local GMCs, Blitz and Terndrup have analyzed the HI emission from the velocity range observed in CO for a number of GMCs in the solar vicinity. One such map was shown in Blitz (1987a). We show in Figure 2, another such map. This map is of the HI column density associated with the molecular accompanying the Per OB2 association (this is the cloud that contains the NGC 1333 star forming region. The map contours are in units of 10^{20} cm^{-2}; the contour marked 10 therefore is associated with an extinction (A_v) of about 0.5 mag. The map itself shows the HI emission integrated over the velocity range 4.2 - 12.7 km s^{-1} , the velocity range associated with the Per OB2 cloud, which is shown as the shaded area in the figure. The association of the atomic gas with the molecular cloud is quite obvious.

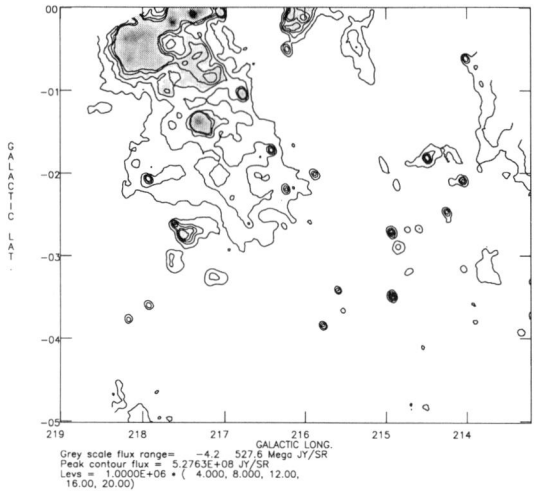

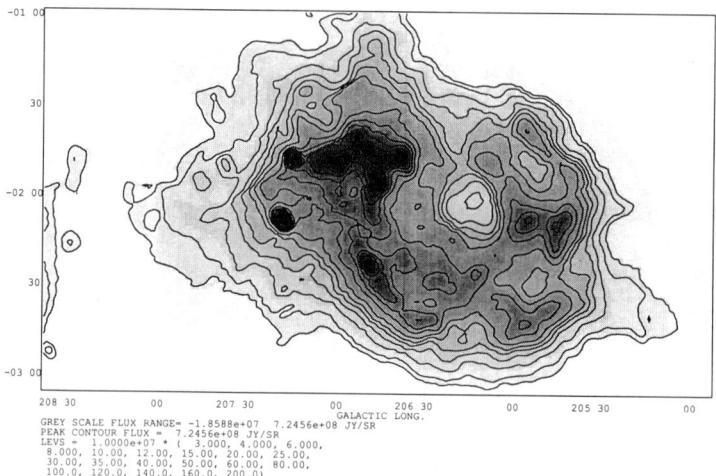

Fig. 1. IRAS 100 μm emission from two GMCs in apparently different evolutionary states. **1a.** Emission associated with the Maddalena-Thaddeus GMC. The molecular cloud is located in the range $-1°45' < b < -3°30'$ and $214° > l > 219°$. There is no IRAS flux detectable in the figure that can be definitely associated with the molecular cloud. The largest flux detected in the direction of the cloud is about 20 MJy/sr. **1b.** IRAS emission associated with the Rosette Molecular Cloud. Note that the lowest contour in the emission is 50% higher than the highest contour in Figure 1a. The difference in the mean emission of the two figures is at least two orders of magnitude.

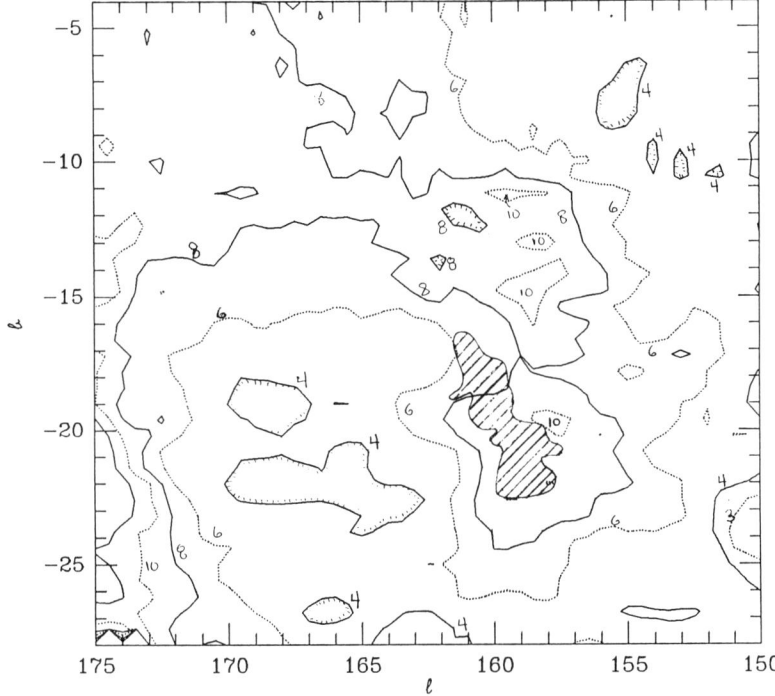

Fig. 2. Map of the atomic hydrogen emission associated with the Per OB2 molecular cloud. The emission is integrated over the velocity range 4.2-12.7 km s^{-1}, and the contours are in units of 10^{20} cm^{-2}. The molecular cloud is shown as the shaded region. There is clearly enhanced emission in the vicinity of the molecular cloud. This map is typical of all of the GMCs in the solar vicinity.

In order to quantify the relationship between the atomic and molecular gas, Blitz and Terndrup estimated the mass of atomic gas associated with the molecular clouds they studied by estimating out to what distance the atomic gas shows an excess over the background in the relevant velocity range. These masses are then plotted as a function of $M(H_2)$ derived from the CO maps using the CO/H_2 conversion ratio of Bloemen *et al.* (1986). The results are plotted in Figure 3. The error bars are from estimates of the uncertainty in defining the background level for the atomic clouds, and in determining the CO/H_2 conversion ratio for an individual molecular cloud. What the figure clearly shows is that the molecular clouds have atomic envelopes that are as massive as the molecular clouds in most cases. The individual maps show a very small range in the peak column density of the atomic gas associated with the molecular clouds. That is, only $1 - 2 \times 10^{20}$ cm^{-2}, or an A_v of 0.5 - 1.0 mag. A similar result has been found for diffuse molecular clouds in the Milky Way (Savage, *et al.* 1977), and for a GMC in M31 (Lada *et al.* 1988).

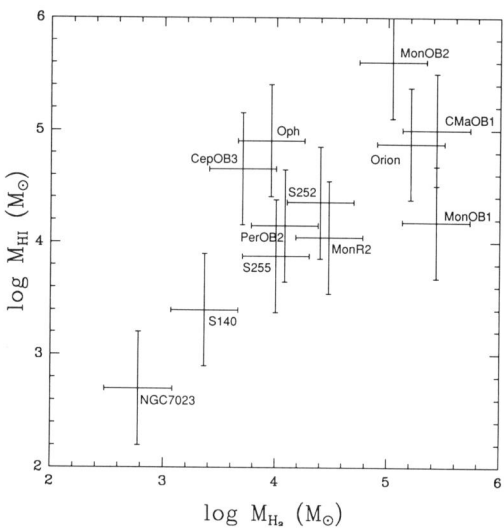

Fig. 3. Plot of the mass of atomic hydrogen emission associated with a number of GMCs in the solar vicinity. The atomic hydrogen mass associated with the GMC is taken over a narrow velocity range centered on the molecular cloud velocity and the molecular mass is taken form the CO emission integated over the area of the molecular cloud.

THE CLUMPY STRUCTURE OF GMCS

As early as 1980, Blitz and Shu noted that all GMCs that have been observed up until that time have exhibited clumpy structure. The evidence at the time suggested that the density contrast between the clumps and the interclump medium is large because the difference between the density required to detect CO at the observed antenna temperatures are an order of magnitude greater than the mean densities inferred if the clouds have dimensions along the line of sight comparable to their dimensions in the plane of the sky. Thus, if the H_2 is clumped so that the volume filling fraction is ~ 0.1 or less, implying a clump/interclump density contrast ≥ 10, the difficulty disappears. This conclusion was qualitatively confirmed from partial ^{13}CO mapping of the dense ridge of the Rosette Molecular Cloud (Blitz and Thaddeus 1980).

In principle, an understanding of the clumpy structure of molecular clouds holds a great deal of information about their evolution. For example, is the clumpy structure a result of the process of star formation, or is the clumpiness primordial? If the clumpiness is primordial, one should be able to see the process of clump formation taking place in the diffuse interstellar medium. In either case, detailed observations of the kinematics of the clumps should be able to resolve the issue. Does the interclump medium play a role in the structure and evolution of the GMCs? In what form is the interclump medium? How do the clumps eventually form stars? OB associations and massive star clusters must eventually form from large massive clumps. Can we find these clumps? Because collisions between clumps should be highly inelastic, the kinematics of the clump ensemble should provide a great deal of information on the history of a GMC.

Observations of the density structure of GMCs is best seen in a pervasive optically thin molecule. The best current candidate is ^{13}CO, but receivers sensitive enough to observe ^{13}CO over the large angular extents needed to fully map a GMC have been available only since the early 1980s. Thus, fully quantitative information on the density structure of GMCs has not become available until fairly recently. Large scale ^{13}CO maps are now available of the Rosette Molecular Cloud (Blitz and Stark 1986), the Orion Molecular Cloud (Bally, et al. 1987), parts of the Cep OB3 molecular cloud (Carr 1987), parts of the Ophiuchus molecular cloud (Loren 1989), and some anonymous CO clouds toward $l = 90°$ (Perault, Falgarone and Puget 1985).

We review here some of the results from the ^{13}CO observations of the Rosette Molecular Cloud to be submitted for publication shortly (Blitz, Stark and Long in preparation). Blitz and Stark (1986) showed that the clumpy structure of the Rosette Molecular Cloud becomes especially clear when analyzed using position-velocity diagrams. Such plots are superior to channel maps (maps made in two spatial dimensions at a particular velocity interval) because small differences in clump velocities are not as apparent in the channel maps. Because the velocity differences between clumps are frequently smaller than the line width of the CO emission from an individual clump, channel maps tend to blend the emission from several adjacent clumps into one large entity. On the other hand, because the velocity resolution of the channel maps is generally high enough that the individual line profiles and the velocity differences between clumps is easily seen, the identification of separate kinematic units is easier to accomplish.

For the Rosette Molecular Cloud, analysis by means of channel maps has made it possible to produce a catalogue of individual clumps because in that cloud the position and separation between the clumps is large enough to make the identifications of individual objects possible by eye. A similar procedure has been carried out by Loren (1989) for the streamers in Ophiuchus, and by Carr (1987) for a piece of the Cep OB3 molecular cloud. What makes the Rosette study different is that enough of the molecular cloud has been observed to make conclusions about the structure of the cloud as a whole. It is unclear, however, to what degree the procedure of identifying clumps by eye will work for other GMCs. This is simply another way of saying that it is unclear at present to what degree the structure of the Rosette Molecular Cloud is typical of GMCs as a whole. It should be pointed out, however, that in terms of its overall properties, the Rosette Cloud is quite typical of ther GMCs in the solar vicinity.

Blitz, Stark and Long reach the following conclusions from an analysis of their data:

1) Between 60% and 90% of the H_2 mass resides in the clumps they have identified in their catalogue. Therefore, most of the molecular mass is in the clumps and is not in a distributed component.

2) The mean H_2 density for all of the clumps is $\sim 1 \times 10^3\ cm^{-3}$. The volume averaged density for the entire complex is $\sim 25\ cm^{-3}$ (Blitz and Thaddeus 1980); therefore the volume filling fraction of the clumps is $\sim 2.5\%$. If we assume that 10 - 50% of the H_2 mass is in a diffuse component not related to the clumps, then the interclump density is $\sim 2.5 - 12.5\ cm^{-3}$. The existence of the HI envelopes suggests that the interclump medium in GMCs also contains an atomic component, but because the envelopes are much more distended than the molecular clouds, the average HI density within the volume of the molecular

cloud is unlikely to be more than about $\sim 15\ cm^{-3}$. Thus, the clumps are not small density enhancements in a relatively smooth substrate, but are dense blobs moving through a tenuous interclump medium held together by their mutual gravitational attraction.

3) The distribution of mass follows a power law such that

$$dN(m) = N_0 M^{1.54} dM$$

where $dN(m)/dM$ is the number of clouds per solar mass interval, and N_0 is 460 $M_\odot^{-1}$. Although most of the mass is in clumps that are at or near the point of being gravitationally bound, most of the clumps by number are not gravitationally bound. That is, half the mass is in the 10 most massive clumps of the 86 catalogued, however, the 40 or so lowest mass clumps do not appear to be gravitationally bound.

Carr (1987) and Loren (1989) show that none of the clumps in the clouds they observed are gravitationally bound. However, because the clouds they observed are an order of magnitude closer than the Rosette Cloud, they are sensitive to a much smaller mass range of clumps than are the Rosette observations. In fact, with one exception, their most massive clumps are all within the mass range for which only unbound clumps are found in the Rosette. Their observations are therefore compatible with the Rosette observations. They have concluded, however, that the lack of gravitational boundedness implies that the clumps in these clouds are expanding. Blitz (1987b) has shown that the clumps in a GMC can, however, be reasonably confined by the pressure of the interclump medium. What is required is that the gas either be warm (since P/k needs to have a value near 1×10^5 K cm^{-3}), or that the pressure comes from bulk motion of the gas (such as a "turbulent" pressure). In this case, the one dimensional velocity dispersion of the gas must be about 7-8 km s^{-1} assuming that the interclump density is 15 cm^{-3}. For the Rosette Molecular Cloud, a value close to this was observed by Raimond (1966) in his HI observations of the HI associated with the Rosette Nebula.

4) The most massive clumps in the Rosette Molecular Cloud lie close to the midplane. This can be seen from Figure 4 which shows the eight most massive clumps shaded on the map of the velocity integrated ^{13}CO emission published by Blitz and Stark. These clumps as a group are the most gravitationally bound of all of the clumps as measured by the ratio of their gravitational to internal kinetic energy. Thus, the next cluster of stars to form in the cloud complex is most likely to form from one of these clumps.

5) The clumps show a strong velocity segregation with mass. Figure 5 is a plot of the clump-to-clump velocity dispersion as a function of mass for the 86 catalogued clumps. The most massive clumps, the ones that lie closest to the midplane of the complex also show the smallest velocity dispersion. *These results srongly suggest that the clumps have undergone dynamical evolution in the time since the complex has formed.* That is, the clump maps and the clump kinematics show a clear signature of inelastic collisions that indicate that the cloud has evolved considerably since it first formed. *Because the observations were made in a part of the complex that has not been affected by the Rosette Nebula itself, and because there is little evidence for star formation in the mapped area, it appears that the structure and kinematics that are observed is the result of the dynamics of the clumps themselves without the intervention of energetic*

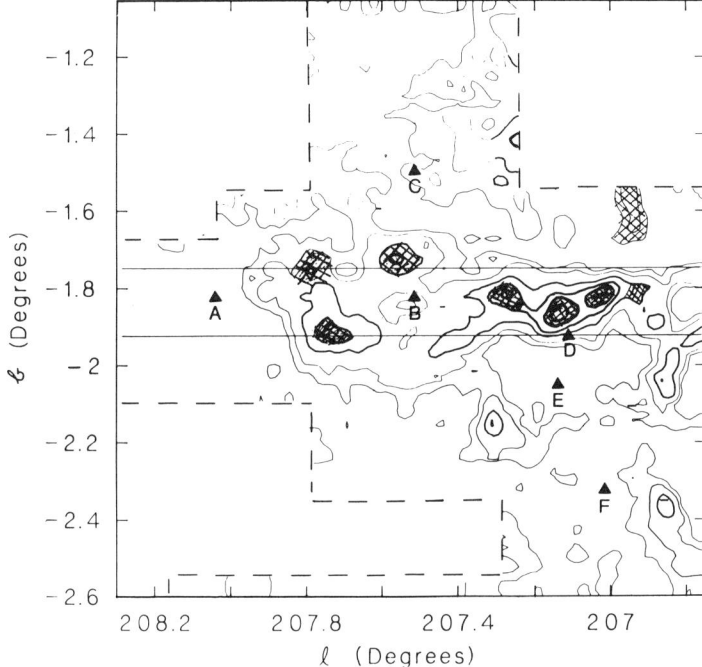

Fig. 4. Map of the ^{13}CO emission from the Rosette Molecular Cloud taken from Blitz and Stark (1986). The shaded areas are the locations of the eight most massive clumps associated with the molecular cloud complex.

phenomena associated with star formation (e.g. HII regions, stellar winds, protostellar outflows, supernova explosions.)

THE EVOLUTION OF THE ROSETTE MOLECULAR CLOUD

The observations, taken together, suggest a partial picture of the evolution of a GMC. We imagine first, that by some process, the interstellar medium collects enough mass together to form a GMC. Two methods by which we know that this can be done is through spiral arm shocks (Roberts 1969; Cowie 1981; Elmegreen 1982), and through "supershells" driven by multiple supernovae (Heiles 1979, McCray and Kafatos 1987). Evidence that some other method is also at work comes from observations of other galaxies which suggest that the star formation rate is not strongly dependent on spiral arm morphology; even galaxies without coherent spiral arms form stars quite efficiently (Stark, Elmegreen and Chance 1987). In any event, the dominant process by which the diffuse interstellar medium is collected into clouds in the solar vicinity has not been identified.

As enough material is collected together to form a cloud, when the column density of atomic hydrogen exceeds a threshold of about 10^{21} cm^{-2}, the associated visual extinction of about 0.5 mag is sufficient to shield the gas from the

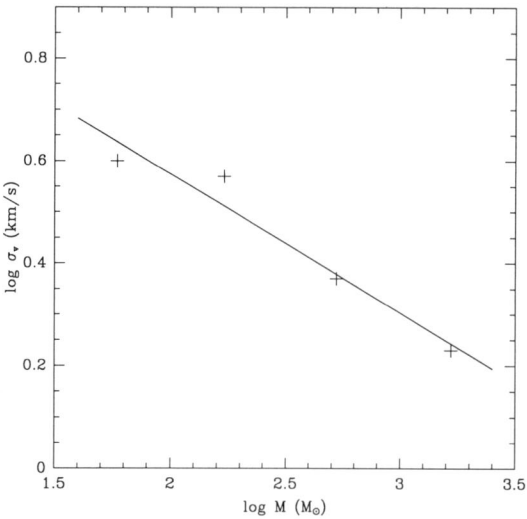

Fig. 5. Plot of the clump-to-clump velocity dispersion as a function of mass in the Rosette Molecular Cloud. The straight line is a least squares fit to the date and has a power law index of -0.27.

UV radiation that dissociates the molecules that form, and the cloud begins to turn molecular. It is therefore early in the process of turning molecular that the clumps have begun to form. Evidence for this comes not only from the Rosette Molecular Cloud, where the kinematics suggests that clumpiness precedes the star formation process and then evolves, but from observations of high latitude molecular clouds very close to the Sun. Many of these clouds appear to be very young ($\leq 2 \times 10^6 y$), and all show evidence of clumpiness that must surely be primordial. Furthermore, the properties of these clouds are quite similar to the gravitationally unbound clumps in the Rosette Cloud.

It is not necessary that the interstellar medium produces proto-GMCs that are gravitationally bound. A gravitationally neutral cloud will become gravitationally bound, because the clumps that have formed will collide, and the kinetic energy of the ensemble of clumps can be radiated away because the clump collisions are so inelastic. The details of such a process are amenable to numerical modeling. If a GMC forms in this way, then it is natural that the atomic gas from which it formed remains as the interclump gas. All of the gas will remain in pressure equilibrium, and as the cloud becomes gravitationally bound, the interclump medium will respond to the gravitational potential well of the GMC. The pressure of the interclump medium will therefore increase above the nominal interstellar value, putting the clumps within the cloud under higher pressure than they had initially.

Angular momentum, must, of course be preserved, but if the cloud is initially tied to the diffuse ISM through the magnetic field, as appears to be the case (see *e.g.* Heiles 1988), then at least some angular momentum can be transfered to the ISM through Alfven waves (e.g. Mouschovias 1985). This seems to be the only way to produce the decrease in specific angular momentum that has been observed, but the details have yet to be worked out.

The magnetic field can support a clump against gravity up to a certain critical mass (*e.g.* Mouschovias 1987; McKee 1989). Once a clump has become magnetically supercritical, then even the magnetic field cannot prevent collapse (see *e.g.* McKee 1989). Thus star formation will take place in the clumps that have grown to be the largest and densest though collisions. Some support of the clumps themselves can also come from the energy input from protostellar and neostellar winds (see *e.g.* Margulis, Lada and Snell (1988). It may very well be that a process of self-regulating star formation similar to that described by McKee (1989) may then take place, until the cloud is ultimately consumed from the dissociating effects of the HII regions, stellar winds, and supernova remnants of the stars that formed within it.

Note, however, that the process of molecular cloud evolution described above has no need for sequential star formation. Aside from the requirement in the initial suggestion of Elmegeen and Lada (1977) that an ionization front propagate into an initially homogeneous molecular cloud, and that no such cloud has yet been identified, a troubling aspect of this picture has been the identification of young OB subgroups that lie *between* older subgroups such as that found in Cep OB3 (Sargent 1977), and the Sco-Cen OB association (Blaauw 1964; deGeus 1988). Furthermore, sequntial star formation requires a first generation of star formation to initiate the process. In the outline of GMC evolution described here, star formation proceeds in a quasi-random way within a molecular cloud. That is, the collisions between clumps in a molecular cloud proceed until one of the clumps becomes unstable to the process of formation of an OB subgroup. The stellar activity will ionize the moleuclar gas in its vicinity, and in a time probably less than a million years (depending on a number of variables), this subgroup and its attendent HII region will appear to be at the edge of the molecular cloud. The next large clump to form an OB association will then be determined by the collisional processes between it and the other clumps in the complex, and have no bearing on the previous star formation history of the cloud. In this way, there is no need to postulate a special event to produce the first generation of stars, and the apparently random orientaion of the subgroups associated with most GMCs appears quite naturally. Cases like the orientation of the subgroups of the Orion complex therefore are the result of chance (and are not very unlikely).

The degree to which this evolutionary picture is true will depend on the results obtained from the ^{13}CO mapping of a number of GMC complexes, and the identification of more complexes in different evolutionary states. Once the process of the formation and evolution of GMCs is better understood in the Milky Way, we can then apply our understanding to other galaxies where the environmental conditions are markedly different.

ACKNOWLEDGEMENTS This work was partially supported by NSF grant AST-8918912, and funding from the state of Maryland to the Laboratory for Millimeter-wave Astronomy.

REFERENCES

Adler, D., 1988, Ph.D. Dissertation, University of Virginia.
Allen, R.J., Atherton, P.D., and Tilanus, R.P.J., 1986, *Nature*, **319**, 296.
Bally, J., Langer, W.D., Stark, A.A., and Wilson, R.W., 1987, *Ap. J. (Letters)*, **312**, L45.
Bash, F.N., Green, E., and Peters, W.L., 1977, *Ap. J.*, **217**, 464.
Blaauw, A., 1964, *Ann. Rev. Astron. Ap.*, **2**, 213.
Blitz, L., 1978, Ph.D. Dissertation, Columbia University.
Blitz, L., 1980, in *Giant Molecular Clouds in the Galaxy*, Solomon and Edmunds, eds., Pergammon:Oxford, p.1.
Blitz, L., 1987a, in *Millimetre and Submillimetre Astronomy*, Wolstencroft and Burton, eds., (Kluwer:Dordrecht), p.269.
Blitz, L., 1987b, in *Physical Processes in Interstellar Clouds*, Morfill and Scholer, eds., (Reidel:Dordrecht), p.35.
Blitz, L. and Shu, F.H., 1980, *Ap. J.*, **238**, 148.
Blitz, L. and Thaddeus, P., 1980, *Ap. J.*, **241**, 676.
Blitz, L. and Stark, A.A., 1986, *Ap. J. (Letters)*, **300**, L89.
Blitz, L. Stark, A.A., and Long, K., 1991, in preparation.
Bloemen, J.B.G.M., et al. , 1986, *Astron. Ap.*, **154**, 25.
Boulanger, F., and Perault, M., 1988, *Ap. J.*, , **330**, 964.
Carr, 1987, *Ap. J.*, , **323**, 170.
Cohen, R.S., Cong, H-I., Dame, T.M., and Thaddeus, P., 1980, *Ap. J. (Letters)*, **239**, L53.
Clemens, D.P., 1985, *Ap. J.*, **295**, 402.
Cowie, L.L., 1981, *Ap. J.*, **245**, 66.
Dame, T.M., Elmegreen, B.G., Cohen, R.S., and Thaddeus, P., 1986, *Ap. J.*, **305**, 892.
deGeus, E., 1988, Ph.D. Dissertaiton, Leiden University.
Elmegreen, B.G., and Lada, C.J., 1977, *Ap. J.*, **214**, 725.
Elmegreen, B.G., 1982, *Ap. J.*, **253**, 655.
Falgarone, E., and Lequeux, J., 1973 *Astron. Ap.*, **25**, 253.
Heiles, C. 1979, *Ap. J.*, **229**, 533.
Heiles, C., 1988, *Ap. J.*, **324**, 321.
Henderson, A.P., Jackson, P.D., and Kerr, F.J., 1982, *Ap. J.*, **263**, 116.
Kutner, M.L., Tucker, K.D., Chin, G., and Thaddeus, P., 1977, *Ap. J.*, **215**, 521.
Lacey, C.G., and Fall, S.M., 1985, *Ap. J.*, **290**, 154.
Lada, C.J., Margulis, M., Sofue, Y., Nakai, M., and Handa, T., 1988, *Ap. J.*, **328**, 143.
Leisewitz, D., Bash, F.N., and Thaddeus, P., 1989, *Ap. J. Suppl.*, **70**, 731.
Liszt, H.S., Burton, W.B., and Xiang, D.,1981, *Astron. Ap.*, **140**, 303.
Liszt, H.S., and Burton, W.B., 1983, in *Kinematics, Dynamics, and Structure of the Milky Way*, W.L. Shuter, ed., (Reidel:Dordrecht), p.135.
Loren, R.B., 1989, *Ap. J.*, **338**, 902.
Maddalena, R., and Thaddeus, P., 1985, *Ap. J.*, **294**, 231.
Magnani, L., Blitz, L., and Mundy, L., 1985, *Ap. J.*, **295**, 402.
Margulis, M., Lada, C.J., and Snell, R.L., 1988, *Ap. J.*, **333**, 316.
McCray, R., and Kafatos, M., 1987, *Ap. J.*, **317**, 190.
McKee, C.F., 1989, *Ap. J.*, **345**, 782.

Mead, K.M., and Kutner, M.L., 1988, *Ap. J.*, **330**, 399.
Mouschovias, T., 1985, *Astron. Ap.*, **142**, 41.
Mouschovias, T., 1987, in *Physical Processes in Interstellar Clouds*, Morfill and Scholer, eds., (Reidel:Dordrecht), p.453.
Perault, M., Falgarone, E., and Puget, J.L., 1985, *Astron. Ap.*, **152**, 371.
Raimond, E., 1966, *B.A.N.*, **18**, 191.
Sanders, D.B., Scoville, N.Z., and Solomon, P.M., 1985, *Ap. J.*, **289**, 323.
Sargent, A.I., 1977, *Ap. J.*, **218**, 736.
Savage, B.D., Bohlin, R.C., Drake, J.F., and Budich, W., 1977, *Ap. J.*, , **216** 291.
Scoville, N.Z., Yun, M.S., Clemens, D.P., Sanders, D.B., and Waller, W.H., 1987, *Ap. J. Suppl.*, **63**, 821.
Shu, F.H., Adams, F.C., and Lizano, S., 1987, *Ann. Rev. Astron. Ap.*, **25**, 23.
Solomon, P.M., Rivolo, A.R., Barrett, J., and Yahil, A., 1987, *Ap. J.*, **319**, 730.
Stark, A.A., and Blitz, L., 1978, *Ap. J. (Letters)*, **225**, L15.
Stark, A.A., 1983, in *Kinematics, Dynamics, and Structure of the Milky Way*, W.L. Shuter, ed., (Reidel:Dordrecht), p.127
Stark, A.A., 1984, *Ap. J.*, **281**, 624.
Stark, A.A., Elmegreen, B.G., and Chance, D., 1987, *Ap. J.*, **322**, 64.
Vogel, S. Kulkarni, S.R., and Scoville, N.Z., 1988, *Nature*, **334**, 402.
Zuckerman, B. and Palmer P., 1974, *Ann. Rev. Astron. Ap.*, **12**, 279.

FORMATION OF HIGH MASS STARS

Wm. J. Welch
Radio Astronomy Laboratory
University of California
Berkeley, CA 94720

ABSTRACT. We briefly review the evidence regarding the evolution of the dense interstellar medium into massive stars. Studies of star formation on large scales based on low angular resolution observations tend to lead to ambiguous conclusions regarding such questions as the cloud density dependence of the star formation rate. Observations of individual clouds and cloud cores show evidence of massive clumps that may become stars, disks around young massive stars, progressive star formation, and infall onto the cores that may be left over from the early stages of star formation.

1. INTRODUCTION

The most massive of the main sequence stars, the OB stars, are bright and prominent in the Milky Way. Their ultraviolet light ionizes the gas in their surroundings, and they and their HII regions delineate the arms in the grand design spiral galaxies. They are generally found in groups and associations. Since they are short lived, they must have formed recently, and yet until about two decades ago, it was not known where they are born. Following the discovery in the late sixties and early seventies that the interstellar medium is filled with massive clouds of molecular hydrogen (Rank, Townes, and Welch, 1971; Scoville and Solomon, 1975), it became clear that these molecular clouds were the birth places for the massive stars.

Principally from infrared and radio studies, much has been learned in the past twenty years about the properties of these clouds and their connections with the stars (cf Dame et al., 1987), and yet many basic questions about the course of star formation remain. For example, how do the clouds themselves form? Is there a single mechanism for star formation? Related questions are: Why do the OB stars form in groups and associations, and is the Initial Mass Function the same in every cloud? What cloud property determines the mass of a star and the IMF? Is the star formation process essentially spontaneous within a cloud, or is some trigger, such as the collision between clouds, necessary? The formation and evolution of stars is an important component in the large scale evolution of galaxies, and one may therefore study star formation both on galactic scales and within individual molecular clouds. In the following we review briefly evidence on how OB stars form from observations of the process on different scales.

2. LARGE SCALE EVIDENCE

With the assumption that the mechanism of OB star formation is universal, one can hope to understand its general characteristics by comparing its effects in galaxies of the same and different types, by observing its possible enhancement in structures such as spiral arms, by measuring its dependence on galactic radius, and by measuring its dependence on distance normal to the galactic plane. Two possible hypotheses that might be tested, for example, are that (a) OB star formation occurs if enough cloud material is present, or that (b) it is the result of collisions between clouds. In the former case, the star formation rate is expected to be simply proportional to the mass of molecular clouds; in the latter, it would be proportional to the square of cloud density. Three possible measures of star formation rate are typically used: blue luminosity, Hα luminosity, and farIR luminosity. Each has its limitations. The first two may be affected by dust extinction, and the latter may be the result of more than current star formation and is significantly model dependent (Hunter et al., 1989).

2.1 LARGE SCALE OBSERVATIONS OF GALAXIES

Recent improvements in instrumental sensitivities at millimeter wavelengths and the availability of farIR fluxes from the IRAS database have enabled studies of star formation and the distributions of molecular material in nearby galaxies. In the largest compilation of galaxy observations in CO, Hα, and farIR, Young et al (1989) found a linear correlation between Hα, farIR, and CO(1-0) luminosities for 124 galaxies. Assuming that the first and second are good measures of OB star formation rates and the third is a measure of total molecular hydrogen mass, we may conclude that the rate of OB star formation is proportional to cloud mass, supporting the first of the above two hypotheses. Some caution is necessary, however, because there is a large scatter in the data, perhaps due to the effects noted above, and, while there is a good correlation of molecular hydrogen mass with CO(1-0) luminosity in the Milky Way, there is evidence that the correlation factor is different in other galaxies (Israel, 1989).

The distributions of OB star formation with radius in our galaxy and others and the distribution normal to the galactic plane in our galaxy offer some clues to the molecular mass dependence of the OB star formation rate. In the Milky Way the density of CO falls off significantly with radius, particularly outside the solar circle, but the apparent rate of star formation, as shown by OB stars, falls at about the same rate (Terebey and Fich, 1987). Normal to the galactic plane the scale heights of both the CO (molecular gas) and the young stars are approximately equal (Solomon and Sanders, 1985). Both of these facts suggest that the rate of OB star formation is proportional to molecular gas density. On the other hand, Scoville et al. (1986), using a measure of cloud mass density based on a survey of the CO emission of the galaxy, find that the local number of HII regions in the disk of the Milky Way is proportional to the square of the mean molecular hydrogen density; they conclude, therefore, that the high mass star formation rate is proportional to the square of the density of clouds and may be due to cloud-cloud collisions, in contrast with the above speculation. Thus, conclusions based on the large scale distributions are somewhat ambiguous.

2.2 THE EFFECTS OF SPIRAL DENSITY WAVES

The arms of the Grand Design spiral galaxies stand out in the light of OB stars and their HII regions. The density wave theory for spiral galaxies has enjoyed considerable success in explaining the existence of the arms (Lin and Shu, 1964), and the expected shocks associated with the arms have been suggested to enhance the star formation rate and so produce the OB stars (Roberts, 1969). Vogel *et al.* (1988), in a high resolution CO study of M51, found large streaming motions across the spiral arms with the molecular gas upstream of the HII regions, providing evidence for enhancement of OB star formation in the spiral density waves. Allen *et al.* (1986) found HI down stream of the dust lanes in M83 and interpreted this as a result of dissociation of molecular hydrogen due to star formation in the spiral arm shock (Figure 1). Both of the above studies argue for the enhancement of OB star formation by spiral density wave shocks. Whether there is real enhancement depends on the arm-interarm density contrast, for if it is large, the abundance of OB stars may simply be due to the relatively large amount of gas in the arms and have nothing to do with the shock. The CO(2-1) study of M51 at 12″ resolution by Guelin *et al.* (1989) found a variable contrast, up to a factor of 17 in some places, suggesting that the apparent enhancement may be largely due to arm-interarm density contrast. Furthermore, Elmegreen (1987) pointed out that the rates of star formation in messy spiral galaxies where the effects of density waves seem less important are about the same as in the grand design spirals, indicating that the spiral density waves do not play a major role in enhancing star formation.

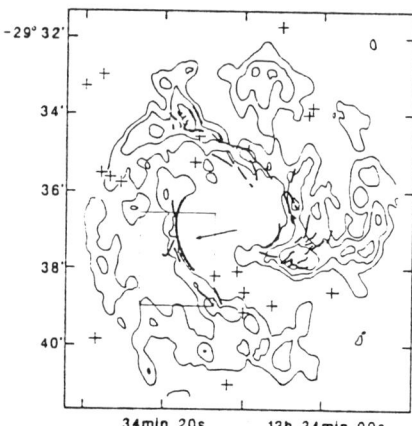

Fig.1 Contours show the distribution of atomic hydrogen, and dark lines show the dust lanes upstream of the hydrogen in the density wave for the spiral galaxy M83 (Allen *et al*, 1986).

Overall, the large scale evidence for the form of the dependence of OB star formation on molecular gas density or spiral density waves is unclear. The difficulty may simply be due to the uncertainties in the interpretation of the indicators of star formation. Alternatively there may be competing processes which cannot be sorted out on the large scale.

3. EVIDENCE FROM CONNECTIONS BETWEEN MOLECULAR CLOUDS AND OB ASSOCIATIONS.

The most convincing evidence that OB stars form from molecular clouds is the close proximity and the similar radial velocities between the stars and the parent clouds (*e.g.* Sargent, 1979; Blitz and Thaddeus, 1980). For example, the Rosette Nebula is an HII region ionized by a rich OB association adjacent to its parent molecular cloud (Blitz and Thaddeus, 1980). The age of the HII region is estimated to be about 3×10^5 years, roughly 10% of the MS lifetimes of the O stars. Thus, the powerful winds and radiation of the O stars have quickly dispersed most of the material of that part of the cloud within which the stars have formed. Evidently, this is typical. The HII regions surrounding the embedded OB stars are much more dense and compact. Wood and Churchwell(1989) found from a combined VLA and IRAS study of compact HII regions that the time an OB star spends imbedded in a molecular cloud is only 10-20% of its main sequence lifetime. Thus, if the OB association contains lower mass stars, they presumably formed before the O stars, since once the latter have formed, they quickly disperse the cloud in their local neighborhood (Herbig, 1962).

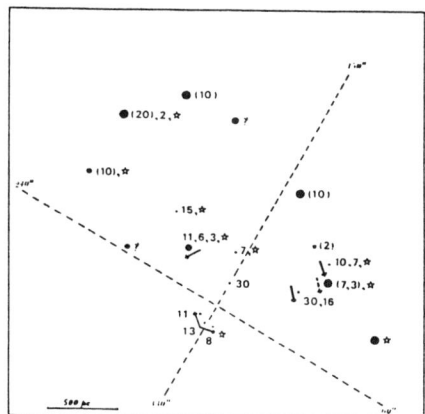

Fig.2 Subgroups of the OB-associations in the Orion arm. Arrows indicate the projected direction of the progression in the formation process within each association. Numbers are photometric ages in millions of years (Blaauw, 1985).

In some cases, for example the I OR Association (Blaauw, 1964), the sub-groups of the OB association show a linear progression in ages and positions. When it was also noticed that in some cases, such as in Orion. the youngest sub-group was adjacent to an elongated parent molecular cloud in which there appeared to be incipient star formation, Elmegreen and Lada (1977) proposed that the sub-groups were formed by sequential or progressive star formation. In this view, the oldest OB subgroup forms at the end of an elongated cloud, dispersing and ionizing the gas locally. Then, the pressure from the HII region drives shocks into the adjacent part of the cloud. compressing the gas and triggering further star formation and producing another subgroup. And so on, with subgroups forming in succession along the length of the could until it is totally used up. A number of examples which fit this picture very well have been found (*e.g.*, Lada *et al.*, 1978). Blaauw (1985) has shown examples of progressive star formation in OB associations in the Orion arm, where the ages of subgroups fit a linear progression (Figure 2).

There are a number of circumstances which fit the scenario of sequential or progressive star formation nicely, and it seems most likely to be the correct explanation for what has happened in these cases. On the other hand, there are many more cases that do not fit (*e.g.*, Blaauw, 1985; Sargent, 1979), and this explanation cannot be regarded as the most general.

One interesting dynamical point should be noted. The massive molecular clouds appear to be gravitationally bound, whereas the OB associations which form from them are not. Evidently, when a few OB stars form. they quickly blow away most of the dust and gas of the cloud, leaving the stars with random velocities similar to those of the cloud material but with little residual total mass. Thus. the remaining star cluster is unbound (Deurr *et al.*, 1982). In this view, the unbound cluster is a natural outcome of the low OB star formation efficiency (Mathieu, 1983).

One might begin to conclude that there may be a number of distinct mechanisms of star formation at work. Garmany *et al.* (1982) and Humphreys (1985) found convincing evidence that the IMF for O stars is definitely flatter for stars inside the solar circle in the galaxy than outside. This suggests that whatever determines the distribution of stellar masses is not the same everywhere. It may, for example. be a function of mean gas density. In any case, the mechanics may be rather different in different environments, and this is a strong argument against global investigations and for detailed studies in particular regions. On the other hand, mass functions for imbedded clusters with sufficient accuracy to permit comparisons are difficult to obtain. In a detailed study of 11 clusters imbedded in the NGC 6334 molecular cloud, Straw *et al.* (1989) found mass functions which were indistinguishable from one another and the original Salpeter function (1955) within the large scatter of the data. These results are similar to those for exposed young clusters. One could conclude either, as did Straw *et al.* (1989), that they are all the same or that more accurate measurements (if possible) are necessary to find possible differences.

4. STUDIES OF INDIVIDUAL CLOUDS CONTAINING IMBEDDED YOUNG OB STARS.

Advances in techniques at both infrared and radio wavelengths have permitted the study of individual active clouds with improved sensitivity and improved angular resolution. The most striking properties of clouds where massive stars are forming are really the effects of the star formation process rather than its cause: massive outflows of material and associated strong chemical enrichment (Snell, 1985; Lada, 1985; Welch, 1988). Essentially all young stellar objects show outflows. They have wind velocities up to 100 km/ sec, wind mechanical luminosities of up to a few percent of the stellar luminosities, and mass loss rates of up to 10^{-3} M$\odot$ per year. The outflows are typically bipolar, and they are the most energetic for the most luminous stars. Especially for the young OB stars, the outflows stir up the clouds enormously, significantly affecting the prestellar distribution of gas and dust. The winds also have a major effect on the chemical composition of the region near the young star. The molecular composition shifts from that of a mainly carbon rich environment in the large cloud to being sulfur and silicon enhanced and apparently oxygen rich close to the outflow.

In addition to the evidence of outflows from young stellar objects and their effects, the high resolution studies of high-mass star-forming clouds also show features that may be associated with the early phases of the star formation. These include the presence of massive clumps, disks around the stars, and infalling gas that may be left over from the first phases of star formation.

4.1 Disks

Indirect evidence for disks around massive embedded young stars is provided by the fact that the outflows are typically bipolar, suggesting that the flows may be collimated by disks (Snell *et al.*, 1980). There is also now direct evidence for the presence of a disk around the luminous young stellar object IRC2 in Orion/KL, the nearest region of massive star formation. Plambeck *et al.* (1990) measured the spatial/velocity structure of SiO maser emission from this source at a linear resolution of 7 AU, finding that it shows a rotating, expanding disk centered on the star, with inner and outer radii of 40 AU and 80 AU respectively. This confirms in detail the interpretation of Barvainis (1984) based on lower angular resolution linear polarization observations of the SiO emission.

Disks are likely to play an important role in the formation of stars of all masses (Shu *et al.*, 1987). For high mass stars, the disk provides a means for material to fall onto the star, by first falling into the disk, and not be blown away by the intense radiation pressure of the already ignited massive core. The disk is also a natural consequence of the collapse of material with substantial angular momentum. Disks, such as that around Orion/KL above, probably formed along with the stellar core and may further evolve into a companion star or planets.

4.2 Clumps

The interstellar medium is clumpy on all the angular scales at which observations have been made (cf. Myers and Benson, 1983; Wilson and Walmsley, 1989). In regions of massive star formation, the luminosity of the OB stars greatly warms their surroundings and permits the study of clumps on scales as small as a few hundredths of a pc, even in distant regions (Pratap et al, 1990; Rudolph et al, 1990). In these regions, most of the mass appears to be in clumps, many of which have masses of the order of 100 $M_\odot$, sizes of the order of 0.1 pc, and narrow linewidths, suggesting that they are bound. These clumps may be the sites of further star formation. On the other hand, it is not clear whether these relatively massive clumps were present when the massive stars began to form or are produced by the action of the winds and radiation pressure of the stars now present.

4.3 Evidence for Cloud Collapse

Although it is generally believed that stars must form as a result of the collapse of cloud material, there is little observational evidence to date which shows this unambiguously. The main reason for the lack of evidence is practical. The free fall velocity onto a one solar mass core exceeds the local sound speed only at radii of a few hundredths of a pc. Thus, only at very high angular resolution will the infalling gas exhibit motions that can be distinguished from ambient gas. The situation is made more confusing by the ubiquitous presence of outflows at tens of km/sec..

The prospects are somewhat better for massive stars, and indeed Keto *et al.* (1987; Ho and Hashick, 1986) have found evidence for residual infall of material on to the HII region of a massive star. They observed the emission and absorption of ammonia toward the compact HII region G10.6-0.4 using the VLA to provide the necessary high angular resolution (3″). They detected red-shifted absorption and blue-shifted emission, indicating that material behind the HII region is moving toward us and foreground material is moving away from us. They also found evidence of rotation in the gas. The simplest plausible interpretation is that material is infalling and spinning up as it falls toward the HII region. This may well be the remnant accreting material from the formation of the O star (or sub-group) in the middle of the HII region. The velocity of infall is consistent with the mass of the OB stars needed to ionize the HII region, 50-100 $M_\odot$. They found also that the highest velocity gas, presumably the closest to the HII region, showed the highest excitation temperature, as expected. Although the interpretation based on the available evidence is convincing, the velocity differences between emitting and absorbing gas are small, 2-4 km.sec, and there is some danger of confusion with the absorption and emission of possibly unrelated clumps of gas with random motions. Further observations of the structure of the whole region, perhaps in other molecular lines, will strengthen the picture.

On larger scales, Welch *et al.* (1987) and Rudolph *et al.* (1990) have reported similar results, but for the much more massive cores in W49 and W51, respectively. In W49, high angular resolution (3″) millimeter wavelength spectra of HCO+ toward several compact HII regions in the dense core of the cloud show broad blue-shifted emission and red-shifted absorption. The absorption is observed in the spectra of all of the HII regions that are sufficiently bright. The blue-shifted emission is clearly in the core but behind the HII regions. The foreground red-shifted absorption is also in the core: the same gas is seen in emission in directions away from the HII regions requiring molecular hydrogen densities greater than 10^4/cc and therefore placing the gas in the dense core. The simplest interpretation is again that there is a spherical collapse, in this case of very large proportions, onto the cloud core. An important fact is that the absorption is very broad, nearly 15 km/sec, which fits very well free fall collapse onto about 50,000 M⊙, the amount of mass that is known to be in the core. In W51 there are (at least) two such "inverse P-Cygni" profiles toward compact HII regions, suggesting collapse onto one core with a mass of about 50,000 M⊙ and onto another of about 10,000 M⊙(Figure 3). In each case the spectra are very broad and not likely to be due to chance positioning of small clouds in front of the HII regions. Nevertheless, further high angular observations in other molecular lines, especially in higher energy transitions, will provide more details and clarify the interpretation.

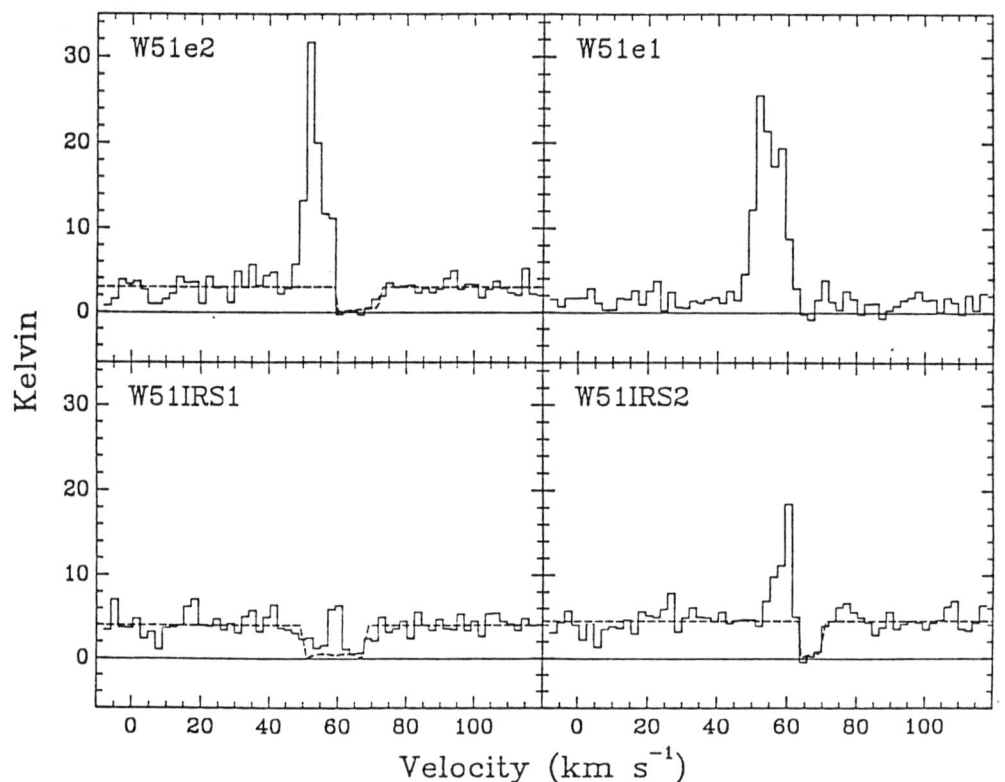

Fig. 3 Spectra of HCO+ toward the molecular cloud W51 showing blue-shifted emission and red-shifted absorption in e2 and IRS2 (Rudolph et al., 1990).

If the interpretation for the above massive cores is correct, there are a number of interesting implications. There are a dozen compact HII regions in the W49 core. Their dimensions are 0.1-0.3 pc, and each is probably an entire OB association like the I Ori association. The coherent large scale collapse provides a natural explanation for the coeval formation of the associations. The ages of the associations are not more than $1\text{-}3 \times 10^5$ years, and yet they are spread out over an extent of about 2 pc. The HII regions are near the center of the molecular cloud, where the density is highest. The collapse can be modeled as an "inside-out" collapse (Shu, 1977), where a massive quasi-stable isothermal sphere develops in the cloud core and then begins to collapse at its center, the collapse envelope propagating outward at the sound speed in the gas. The remnant high density in the cloud core fits this idea, and the observed absorption spectra are produced in the gas that is still falling inward in the outer part of the core. The large spatial scales and small time scales require a large sound speed, several km/sec. This would have to be the Alfven speed, suggesting that magnetic fields play an important dynamical role. The required magnetic fields of a few milligaus are detected in the OH masers in the cores of these two regions. It may be that these magnetic fields support the cloud allowing it to accumulate a large mass before it collapses (Shu et al., 1989). Thus, the magnetic field may be a key ingredient in the formation of massive stars.

5. SUMMARY AND CONCLUSIONS.

From the many observations of OB stars and the molecular clouds in which they form, very briefly summarized here, many intriguing facts relevant to star formation have emerged. Probably the most amazing result is the discovery that apparently all the young stellar objects exhibit powerful winds even as they are forming, especially the most massive young stars. This unexpected circumstance dominates cloud dynamics and, unfortunately, somewhat obscures the study of the early stages of the star formation process itself. Nevertheless, there are some convincing findings about star formation, such as: OB stars form in molecular clouds but spend only 10-20% of their main sequence lives in the clouds, and the IMF for OB stars is not the same everywhere, being flatter in the inner part of our galaxy. Studies of individual regions have produced some results of value to the understanding of OB star formation, although it is not yet clear how general these findings are. For example, there is direct evidence, in at least one case, for a disk around a massive star. Furthermore, there have been observations that are most simply interpreted as infall onto individual OB stars or subgroups and, in a few cases, evidence of residual infall onto entire massive cloud cores, suggesting that large scale collapse may the mechanism by which whole OB associations are formed at the same time. In a minority of cases, there is evidence of progressive star formation. The most ambiguous results have come from the large scale studies that are based on observations which average over large regions. For example, conflicting conclusions have been drawn about whether the rate of star formation is proportional to the density of molecular material or its square and about what is the role of spiral arms in star formation.

Many of the basic questions mentioned in the Introduction, such as what determines stellar masses and the IMF, still have no answers. The ambiguities from the studies that average over large scales argue that detailed high- resolution observations of many particular regions are more apt to bring firm results. The recent availability of many pixel IR cameras and high resolution synthesis arrays at millimeter wavelengths offer the means for the necessary detailed studies. We anticipate a flood of new data in the coming years from which answers to the basic questions may emerge.

REFERENCES

Allen, R. J., Atherton, P. D., and Tilanus, R. P. J. 1986, *Nature*, **319**, 296.
Barvainis, R. 1984, *Astrophys. J.*, **279**, 358.
Blaauw, A. 1964, *Ann. Rev. Astron. Astrophys.*, **2**, 213.
Blaauw, A. 1985, in IAU Symp., *The Milky Way Galaxy*, IAU Symp., ed H. van Woerden. (Dordrecht:Reidel) 335.
Blitz, L., and Thaddeus, P. 1980, *Astrophys. J.*, **241**, 676.
Dame, T. M., Ungerechts, H., Cohen, R. S., de Geus, E. J., Grenier, I. A., May, J.,Murphy, C. D., Nyman, L.-A., and Thaddeus, P. 1987, *Astrophys. J.*, **322**, 706.
Duerr, R., Imhoff, C. L., and Lada, C. J. 1982, *Astrophys. J.*, **261**, 135.
Elmegreen, B. 1987, in IAU Symp. 115, *Star Forming Regions*, eds. M. Peimbert and J. Jugaku (Dordrecht: Reidel).
Elmegreen, B. G., and Lada, C. J. 1977, *Astrophys. J.*, **214**, 725.
Garmany, C. D., Conti, P. S., and Chiosi, C. 1982, *Astrophys. J.*, **263**, 777.
Guelin, M., Garcia-Buretto, S., Blundell, R., Cernicharo, J., Despois, D., and Steppe, H. in *Highlights of Astronomy*, ed. D. McNally (Dordrecht:Kluwer, 1989), 8, 575.
Herbig, G. H. 1962, *Adv. Astron. Astrophys.*, **1**, 47.
Ho, P. T. P., and Hashick, A. D., 1986, *Astrophys. J.*, **304**, 501.
Humphreys, R. M. 1984, in, IAU Symp. 105, *Observational Tests of the Stellar Evolution Theory*, eds. A. Maeder and A. Renzini (Dordrecht:Reidel) 279.
Hunter, D. A., Gallagher, J. S., III, Rice. W. L., and Gillett, F. C. 1989, *Astrophys. J.*, **336**, 152.
Israel, F. 1989, in *Millimetre and Submillimetre Astronomy*, eds. R.D. Wolstencroft and W.B. Burton (Dordrecht:Kluwer), 423.
Keto, E. R., Ho, P. T. P., and Hashick, A. D. 1987, *Astrophys. J.*, **318**, 712.
Lada, C. J., Blitz, L., and Elmegreen, B. G. 1978, in *Protostars and Planets* ed. T. Gehrels (Tucson:University of Arizona Press), 341.
Lada, C. J. 1985, *Ann. Rev. Astron. Astrophys.*, **23**, 267.
Lin, C. C. , and Shu, F. H. 1964, *Astrophys. J.*, **140**, 646.
Mathieu, R. D. 1983, *Astrophys. J. Lett.*, **267**, 97.
Myers, P. C., and Benson, P. J. 1983, *Astrophys. J.*, **166**, 309.
Plambeck, R. L., Wright, M. C. H., and Carlstrom, J. E. 1990, *Astrophys. J Lett.*, in press.
Pratap, P., Batrla, W., and Snyder, L. E. 1990, *Astrophys. J.*, in press.
Rank, D. M., Townes, C. H., and Welch, W. J. 1971, *Science*, **174**, 1101.
Roberts, W. W. 1969, *Astrophys. J.*, **158**, 123.
Rudolph, A., Welch, W. J., Palmer, P., and Dubrulle, B. 1990, *Astrophys. J.* in press.
Salpeter, E. E. 1955, *Astrophys. J.*, **121**, 161.
Sargent, A. I., 1979, *Astrophys. J.*, **249**, 607.
Scoville, N. Z., and Solomon, P. M. 1975, *Astrophys. J.*, **199**, L105.
Shu, F. 1977, *Astrophys. J.*, **214**, 488.
Shu, F. H., Adams, F. C., and Lizano, S. 1987, *Ann. Rev. Astron. Astrophys.*, **25**, 23.
Snell, R. L. 1987, in, IAU Symp. 115, *Star Forming Regions*, (Dordrecht:Reidel) 213.
Snell, R. L., Loren, R. B., and Plambeck, R. L. 1980, *Astrophys. J. Lett.*, **239**, L17.
Solomon, P. M., and Sanders, D. B. in, *Protostars and Planets II*, ed. D. C. Black and M. S. Matthews (Arizona Press, 1985).
Straw, S. M., Hyland, A. R., and McGregor, P. J. 1989, *Astrophys. J. Suppl.*, **69**, 99.
Terebey, S., and Fich, M., in *Proceedings of Kerr Symposium on the Outer Galaxy*, ed. L. Blitz and F. J. Lockman (Berlin: Springer-Verlag, 1987), 192.
Vogel, S. N., Kulkarni, S. R., and Scoville, N. Z. 1988, *Nature*, **334**, 402.
Welch, W. J. 1988, *Astrophys. Lett. and Comm.*, **26**, 181.
Welch, W. J., Dreher, J. W., Jackson, J. M., Terebey, S., and Vogel, S. N., 1987, *Science*, **238**, 1550.
Wilson, T. L., and Walmsley, C. M. 1989, *Astron. Astrophys. Rev.*, **1**, 141.
Wood, D. O., and Churchwell, E. 1989, *Astrophys. J. Suppl.*, **69**, 831.
Young, J. S., Xie, S., Kenney, J. D. P., and Rice, W. L. 1989, *Astrophys. J. Suppl.*, **70**, 699.

Section V

Posters

Observational Constraints on an Embedded Cloud Model for
the Soft X-ray Diffuse Background

D. N. Burrows
Department of Astronomy, Pennsylvania State University
525 Davey Lab, University Park, PA 16802

Jakobsen and Kahn (1986) have constructed a model of the
soft X-ray diffuse background (SXRB) that uses absorption by clouds embedded in the hot X-ray emitting gas to
produce the observed anticorrelation between soft X-ray
intensity and neutral hydrogen column densities. The
model has two parameters: the ratio R of the scale
heights of the absorbing and emitting gas, and the
clumping parameter α_ν. For values of α_ν compatible
with HI observations of high-latitude small-scale
structure (Jahoda et al. 1985; Jahoda, McCammon, and
Lockman 1986), I show that this model is incompatible
with the observed B/C band ratio unless R ≥ 50, corresponding to local, unabsorbed emission (see Fig. 1).
The addition of a local, unabsorbed component allows the
model to fit the C band data, but it still cannot
simultaneously provide a good fit to the B band
data unless α_ν is small, in conflict with 21 cm data.
Figure 2 shows this model fit to the B band data, with
α_B =0.54 (solid line, best fit value: χ^2=63.6 for
39 points) and α_B =0.80 (dashed line, lowest value
compatible with 21 cm data: χ^2=41.8). Furthermore,
this fit requires R ≤ 0.1, which is equivalent to a
slab absorption model, in conflict with shadowing
experiments (Burrows et al. 1984).

REFERENCES

Burrows, D.N., McCammon, D., Sanders, W.T., and
Kraushaar, W.L. 1984, Ap. J., <u>287</u>, 208
Jahoda, K., McCammon, D., Dickey, J.M., and Lockman,
F.J. 1985, Ap. J., <u>290</u>, 229
Jahoda, K., McCammon, D., and Lockman, F.J. 1986,
Ap. J. (Letters), <u>311</u>
Jakobsen, P., and Kahn, S.M. 1986, Ap. J., <u>309</u>, 682

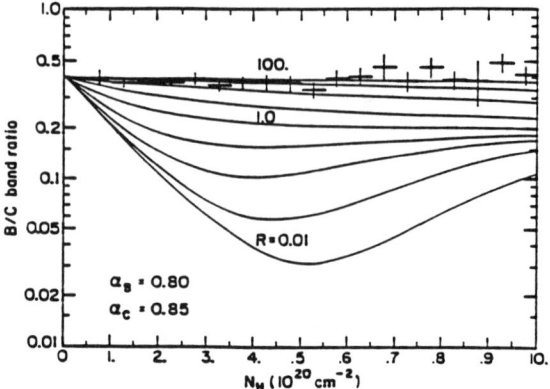

Figure 1

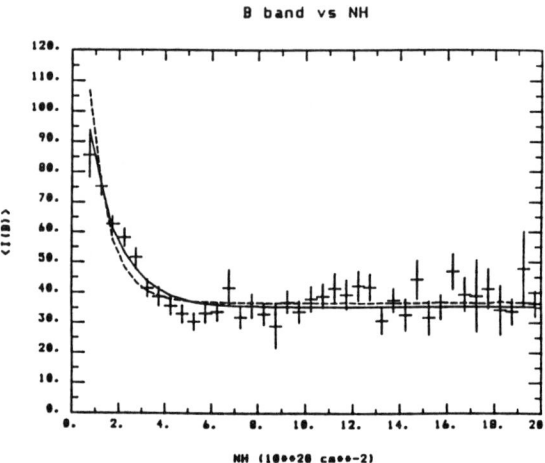

Figure 2

PROBING THE ISM WITH NEUTRON STARS

DIETER HARTMANN
Institute of Geophysics and Planetary Physics, Lawrence Livermore National Laboratory, Livermore, CA 94550

ABSTRACT The possibility to study the large-scale structure of the ISM using soft X-ray observations of neutron stars as a statistical probe is investigated. Spatial distribution and kinematic properties of neutron stars are determined from orbit simulations and the X-ray brightness distribution is calculated using Bondi accretion from a 3-phase ISM.

INTRODUCTION
High space velocities cause neutron stars to migrate far from the galactic plane and their radial distribution resembles that of major components of the ISM. This could make neutron stars the perfect tool to study the large-scale structure of the ISM if we are able to observe the interaction of a complete sample of galactic neutron stars with their environment. Accretion from the ISM is the underlying process by which moving neutron stars sample the properties of their surrounding medium and X-ray observations can be used to extract this information (Ostriker, Rees, and Silk 1970).

CALCULATIONS
The spatial distribution and kinematics of Pop I neutron stars is simulated by calculating stellar orbits in a realistic galactic potential using initial conditions obtained from radio observations of pulsars (Hartmann et al. 1990). Here we consider neutron stars with high rms space velocities (~ 200 km s^{-1}). Ostriker et al. (1970) investigated the observational consequences of accretion onto low velocity neutron stars (v ~ 10 km/s). We use a standard 3-phase model for the ISM including the cold intercloud medium (ICM), the warm ionized medium (WIM), and the hot ionized medium (HIM). We assume a spherically symmetric accretion flow, neglecting possible effects of strong magnetic fields and rotation. The standard estimate (Bondi 1952) of the accretion rate for a 1 M$_\odot$ neutron star moving through the ISM with relative velocity v (km/s) results in an unabsorbed X-ray flux at earth of

$$F_x \sim 4 \times 10^{-8} \, \eta_x \, D^{-2} \, (v^2 + \sigma^2)^{-3/2} \sum_i f_i^{1/3} \, n_i \, e^{-|z|/H_i} \, , \quad \text{ergs cm}^{-2} \text{ s}^{-1}$$

where η_x is the fraction of the radiated energy that emerges in the spectral window of the detector, D (kpc) is the distance to the neutron star and z its height above the galactic plane. ISM properties are density n_i, scale height H_i, and filling factor f_i of each component i ($\sigma = 15$ km s^{-1} was assumed).

RESULTS

Using the results of Hartmann et al. (1990) for the spatial distribution and kinematics of neutron stars we calculate their brightness distribution in the X-ray band accessible to detectors aboard the ROSAT satellite (0.1 - 2 keV; e.g., Trümper 1984) assuming $\eta_x = 0.1$ (Figure 1). Interstellar absorption was neglected. Sensitivity limits for the ROSAT all-sky survey and for pointed observations are shown. For an exposure of 10^6 s ROSAT should detect $\sim$ 0.01 of all neutron stars. The large scale structure of the ISM affects the angular distribution, so that multipole- and correlation analysis of detected point sources can be used to constrain parameters of ISM components. Most of the emission is due to accretion from the ICM phase, so that the distribution of detectable sources is most sensitive to H_{ICM}. For the standard ISM model adopted here most sources are located at galactic latitudes smaller than $\sim 5°$ (Figure 1). An increase of H_{ICM} to 0.8 kpc leads to a latitude distribution that extends to $\sim 10°$. Therefore, analysis of the angular distribution of unidentified X-ray point sources observed with ROSAT (Trümper 1984) or AXAF (Wilson 1987) should be useful to study the large scale structure of the ISM.

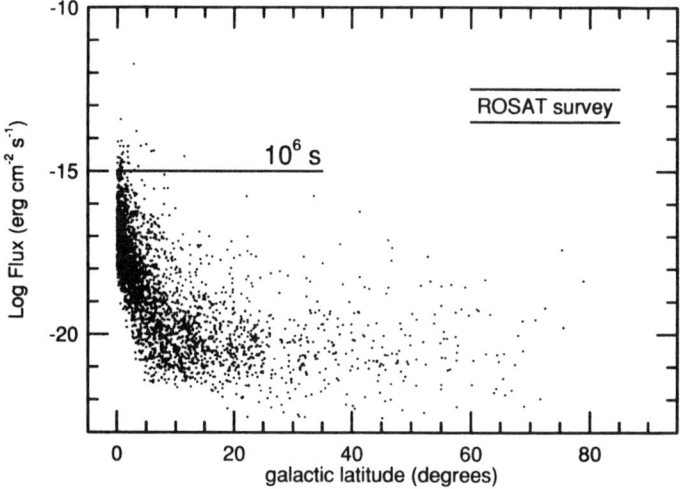

Fig. 1. Expected X-ray flux distribution as a function of galactic latitude from defunct pulsars accreting from the ISM. The mean detection threshold for the ROSAT all-sky survey and for 10^6 s pointed observations are shown.

REFERENCES

Bondi, H. 1952, *M.N.R.A.S.*, **112**, 195.
Hartmann, D. H., Epstein, R. I., and Woosley, S. E. 1990, *Ap. J.*, , in press.
Ostriker, J. P., Rees, M. J., and Silk, J. 1970, *Ap. J. (Letters)*, , **6**, 179.
Trümper, J. 1984, *Adv. Space Res.*, **3**, (10-12), 483.
Wilson, A. S. 1987, *Astr. Lett. Comm.*, **26**, 99.

LOW FREQUENCY OBSERVATIONS OF GALACTIC SNRS AND THE DISTRIBUTION OF LOW DENSITY IONIZED GAS IN THE ISM

NAMIR E. KASSIM
Naval Research Laboratory
Washington, DC 20375-5000

ABSTRACT We use new low frequency radio continuum observations of 47 Galactic SNRs to constrain the distribution of low density ionized gas in the ISM.

VALIDITY OF SNR SPECTRA AS PROBES OF THE ISM

Galactic SNR continuum spectra are characterized by a nonthermal spectral index α ($S \propto f^{+\alpha}$) which appears to be intrinsically constant throughout the radio range (i.e. 10 MHz - 10 GHz). The spectra shown in Fig. 1 (right) are typical examples, and the points with error bars are new low frequency measurements (30.9 or 57.5 MHz) made using the Clark Lake TPT telescope.

Our results for 47 Galactic SNRs (see Kassim 1989 and references therein) indicate that all SNRs at either high Galactic latitude or located towards the outer Galaxy show no low frequency turnovers in their spectra. Therefore any such turnovers can be attributed to free-free

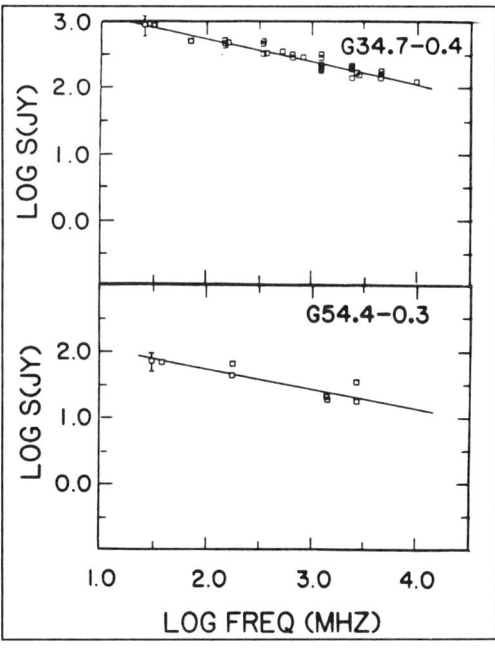

Fig. 1. Typical radio continuum spectra of 2 SNRs. Points with error bars are new data obtained with the Clark Lake TPT telescope.

absorption by ionized gas in the ISM located along the line-of-sight to the SNRs.

PATCHINESS IN THE ABSORPTION TOWARDS DISTANT SNRS IN THE INNER GALAXY AND CONSTRAINTS ON THE PHYSICAL PROPERTIES OF THE WIM

Towards the inner Galaxy, some distant SNRs show no turnovers in there spectra while many others (~2/3) do. Examples of both appear in Fig. 2 (below). This implies a patchy or inhomogeneous absorbing medium located along the lines-of-sight. The lack of absorption to some distant SNRs sets an upper limit to the electron density of any widely distributed Warm Ionized Medium (WIM) if an electron temperature is assumed.

Optical metastable lines of OII and NII superimposed upon a widespread diffuse Hα emission imply a WIM with an electron temperature T_e~3000-8000 K (see Kulkarni and Heiles 1987 and references therein). The lack of absorption ($\tau_{57.5} \leq 0.2$) towards the SNR G349.7+0.2 with a good distance estimate of 18.3 Kpc sets an upper limit on the electron density of the WIM of $n_e \leq 0.26$ cm^{-3} if T_e=8000 K or $n_e \leq 0.13$ cm^{-3} if T_e=3000 K.

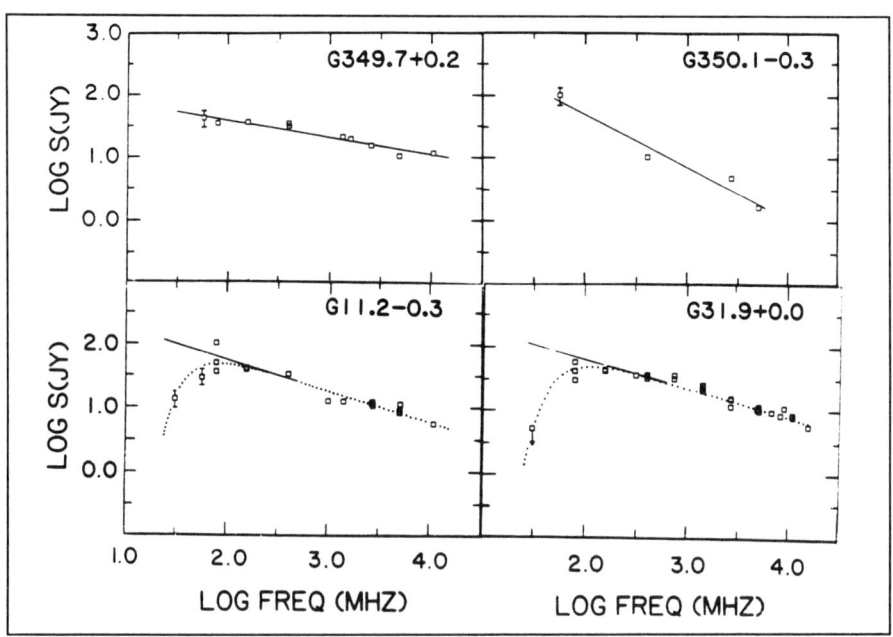

Fig. 2. Towards the inner Galaxy, some SNRs show low frequency turnovers while others do not. This implies absorption by a widespread but inhomogeneous absorbing medium.

EVIDENCE FOR EXTENDED HII REGION ENVELOPES (EHEs)

Low Frequency (~325 MHz) Galactic Ridge Recombination Lines (GRRLs) imply halos around normal HII regions (Anantharamaiah 1986). These halos or Extended HII Envelopes (EHE) have the following properties:

$T_e \sim$ 3000-8000 K
$n_e \sim$ 0.5-10 cm^{-3}
EM $\sim$ 500-3000 cm^{-6}pc^{-1}
Sizes $\sim$ 50-200 pc
Gal. Filling Factor $\leq$1%

We find that such EHEs can completely explain the low frequency continuum optical depths we measure towards Galactic SNRs. The EHE geometry and suspected association with "normal" HII regions is illustrated in Fig. 3 (right).

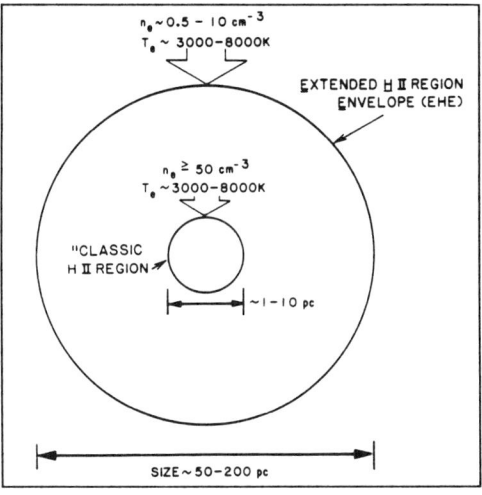

Fig. 3. Suspected EHE/HII Region geometry.

Conclusions

1) In the absence of absorbing gas along the line-of-sight, SNR spectral indices are intrinsically constant to dekametric wavelengths.
2) The absorbing gas in the Galactic plane is patchy.
3) Any widely distributed WIM must have $n_e \leq 0.26$ cm^{-3} if T_e= 8000 K or $n_e \leq 0.13$ cm^{-3} if T_e= 3000 K.
4) The absorbing gas has properties which are consistent with those of extended HII envelopes (EHE) surrounding classical HII regions as postulated by Anantharamaiah (1986) to explain his 325 MHz recombination line survey results.

REFERENCES

Anantharamaiah, K. R. 1986, J. Ap. Astr., 7, 131.
Kassim, N. E. 1989, Ap. J., 347, (in press).
Kulkarni, S. R. and Heiles, C. 1987, in *Interstellar Processes*, eds. D. J. Hollenbach and H. A. Thronson, Reidel:Dordrecht, p. 87.

FAR-IR OBSERVATIONS OF THE N/O RATIO IN INTERSTELLAR GAS

J. P. SIMPSON[1,2], S. W. J. COLGAN[1,3], E. F. ERICKSON[1],
M. R. HAAS[1], and R. H. RUBIN[1,4]
1. NASA/Ames Research Center, Moffett Field, CA 94035 2. UCB,
3. UCSC, 4. UCLA

Oxygen is produced in high mass stars by helium burning (primary nucleosynthesis), but nitrogen is produced both by massive stars and in the CNO cycle in intermediate mass stars (secondary nucleosynthesis). It has been observed that in halo stars and extragalactic H II regions of very low metallicity the N/O ratio is constant, indicating primary nucleosynthesis, but as the O/H ratio increases to solar values, the N/H ratio increases faster, indicating secondary nucleosynthesis. In H II regions in our galaxy it is well established that O/H and N/H decrease with increasing galactocentric radius R_G, but the presence of an N/O gradient is unclear. Rubin et al. (1988) concluded that the N/O ratio is flat in the outer galaxy ($R_G \geq 6$ kpc, $R_\odot = 8.5$ kpc) in agreement with some (but not all) of the optical observers, but that the infrared observations of H II regions with $R_G < 6$ kpc may show an increased N/O ratio in the 4-5 kpc ring of enhanced star formation. Their chief problem was correcting the low excitation H II regions for unseen ionization states. (Optical observers measure the ratio N^+/O^+, whereas infrared observers measure N^{++}/O^{++}.)

Using the Cooled Grating Spectrometer on the Kuiper Airborne Observatory we have made new observations of the [O III] 51.8 and 88.4 μm and [N III] 57.3 μm lines in the inner galaxy H II regions G1.1−0.1, G23.96+0.15, G29.9−0.0, W 43, G25.4−0.2SE, and G333.6−0.2 and the outer galaxy H II regions G298.2−0.3 and RCW 57. We emphasized high excitation H II regions where possible so that the corrections for the unseen ionization states of O^+ and N^+ would be small. Rubin et al. (1988) described how the ionization differences in N^{++} and O^{++} could be calculated by consideration of models of differing chemical abundances, ionizing luminosities, and effective temperatures, T_{eff}, of the exciting stars. Following the same procedure, we estimated the stellar effective temperatures from the ionization of helium, He^+/H^+, as measured by the radio recombination lines with the assumption that He/H = 0.1. From the measured abundances of neon and sulfur relative to hydrogen, we selected the appropriate model series from the compendium of H II region models of Rubin (1986). With the T_{eff} required of the models to give the observed helium ionization, we then calculated the ionization correction factor $icf = \langle O^{++}/O \rangle / \langle N^{++}/N \rangle$, which is multiplied by the observed N^{++}/O^{++} to give N/O. These values plus the N/O ratios derived by Rubin et al. (1988) are plotted in Figure 1.

One should not assume a constant He/H ratio if there is an abundance gradient in helium. For this case, we estimated the effective temperatures from the ionization of neon and sulfur; here we used our observations of Ne^{++} at 36.0 μm and S^{++} at 18.7 and 33.5 μm and took observations of Ne^+, and

S^{+3} from the literature and from IRAS LRS spectra (Simpson and Rubin, in preparation). The observed O^{++}/H^+ ratios are all smaller than the O/H ratio for the Orion Nebula ($= 4.0 \times 10^{-4}$, Simpson et al. 1986) even though the neon and sulfur abundances are of similar size or larger. A possible explanation is that the oxygen ionizations are uniformly much lower than one would expect from the observed He^+/H^+ ratios. The icf's for these oxygen T_{eff} result in N/O ratios larger than those in Figure 1, especially for the inner galaxy H II regions. Even though there is this uncertainty in the model and the final derived N/O, we see that the outer galaxy H II regions have N/O larger than solar (N/O = 0.126) (as pointed out by Rubin et al.), and the inner galaxy H II regions have N/O an additional factor of ~ 2 or more larger. The most reasonable explanation is that there is secondary nitrogen production in regions of enhanced star formation.

For a given effective temperature of the exciting star, the level of ionization of an H II region decreases as the heavy element abundances increase. Thus observations of lower ionization species can help establish the excitation. We report here the first detection of [N II] at 121.9 µm in any source. The line was measured in G333.6−0.2 and Saturn under conditions of exceptionally low water vapor overburden. The line flux is 4×10^{-18} W cm^{-2}. If the electron density is the same in the N^+ volume as in the S^{++} volume, $N^+/N^{++} \approx 1$. but if the electron density is much lower in the N^+ volume, as is found in the core-halo models of Rubin et al.(1983), the N^+/N^{++} ratio is less than 1.

Rubin, R. H. 1985, *Ap. J. Suppl.*, **57**, 349.
Rubin, R. H., Hollenbach, D. J., and Erickson, E. F. 1983, *Ap. J.*, **265**, 239.
Rubin, R. H., Simpson, J. P., Erickson, E. F., and Haas, M. R. 1988, *Ap. J.*, **327**, 377.
Simpson, J. P., Rubin, R. H., Erickson, E. F., and Haas, M. R. 1986, *Ap. J.*, **311**, 895.

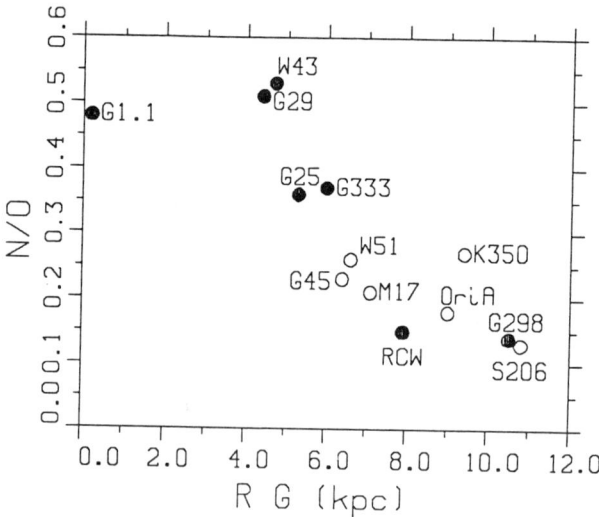

Figure 1. N/O ratios plotted against R_G with R_G for the Sun of 8.5 kpc. The open circles are the data from Rubin et al. (1988).

IRAS OBSERVATIONS OF A LARGE CIRCUMSTELLAR DUST SHELL AROUND W HYDRAE

GEORGE W. HAWKINS
University of California at Los Angeles, Dept. of Astronomy, Math Science Bldg., U.C.L.A., Los Angeles, CA 90024 U.S.A.

Extended 60 and 100 μm IRAS emission, indicating cool 25 K - 60 K dust grains, may reveal the true diameter and total mass of circumstellar envelopes better than molecular emission which is limited by photodissociation. IRAS observations reveal a large 30 - 40 arc minute ($\sim$ 1 parsec) diameter dust envelope surrounding the oxygen-rich red giant W Hya at $\geq 6\sigma$ above the noise, greatly extended beyond IRAS detector sizes 1.5'x4.5' and 3'x4.5' at 60 and 100 μm (figure 1). The emission is attributed to W Hya's mass loss because: 1) the emission is symmetrical and centered on W Hya with similar 60 and 100 μm diameters, 2) there are no 100-μm-only sources within a square degree of W Hya that would indicate infrared cirrus, and 3) there are no associations with objects other than W Hya in the catalogs searched and listed in the IRAS point source catalog. Additionally, the faint extended emission is not a detector memory effect from W Hya's strong point source flux densities (190 Jy at 60 μm, 70 Jy at 100 μm) as other sources with similar or greater flux densities (eg. IRC +10216: 900 Jy at 100 μm, IRC +10011: 70 Jy at 100 μm) do not show similar extended emission.

The dust mass loss rate of order 4×10^{-9} $M_\odot yr^{-1}$ derived from the region 3' - 5' from the star is the same as derived from the point source flux densities. The total dust mass of W Hya's envelope is $\dot{M}_{dust} \simeq 1\times10^{-4}$ $M_\odot$, assuming grains at 40 K, with grain opacity $\kappa(100$ $\mu m)=60$ cm^2 gm^{-1}, distance D=130 pc to W Hya with luminosity $L_{WHya} = 1\times10^4$ $L_\odot$ and 100 μm dust flux density of $\sim$100 Jy within 15' of W Hya.

Observations at 60 and 100 μm are consistent with a model of dust with emissivity $Q(\lambda) \propto \lambda^{-1.2}$ heated by W Hya to 50 K at 3', and 40 K at 10'. At 10' from the central star, the heating of dust grains, with $Q(\lambda)$ obtained from the dirty silicate model of Jones and Merril (1976), is $\sim$20 times greater from W Hya's radiation field than from the interstellar radiation field. In contrast, Gillett et al. (1986) suggested heating by interstellar ultraviolet radiation for the constant temperature dust envelope (18' or 8 pc diameter) around R CrB.

W Hya's dust envelope is much larger in apparent diameter than envelopes around stars with larger $M_{dust} = 10^{-7} - 10^{-6}$ $M_\odot yr^{-1}$. This may result from: 1) proximity of W Hya to Earth, which can give rise to higher dust temperatures at a given angle compared to stars farther away, 2) about 3 times greater mass loss at 1×10^4 years ago (determined from 100 μm intensity at $\theta = 15'$) compared to the present mass loss rate, 3) low background noise from low infrared cirrus emission at W Hya's 33° galactic latitude, and 4) low $1 \times 10^{-3} \leq n \leq .1$ gas densities in the local interstellar medium bubble, allowing the dust envelope to expand to a large radius.

Several other nearby giants also exhibit $\geq$ 20' diameter extended IRAS emission, tentatively attributed to mass loss. These include 30' – 40' diameter extended emission at 60 and 100 μm at the 3-6 σ level around R Dor (D=80 pc) and extended 20' diameter emission at 100 μm at the 3σ level around O Cet (D=77 pc) and RX Boo (D=200 pc).

ACKNOWLEDGEMENTS

This research was supported under the ADP program by NASA contract 957267 to U.C.L.A.

REFERENCES

Gillett, F. C. et al. 1986, Ap. J., 310, 842.
Jones, T. W., and Merril, K. M. 1976, Ap. J., 209, 509.

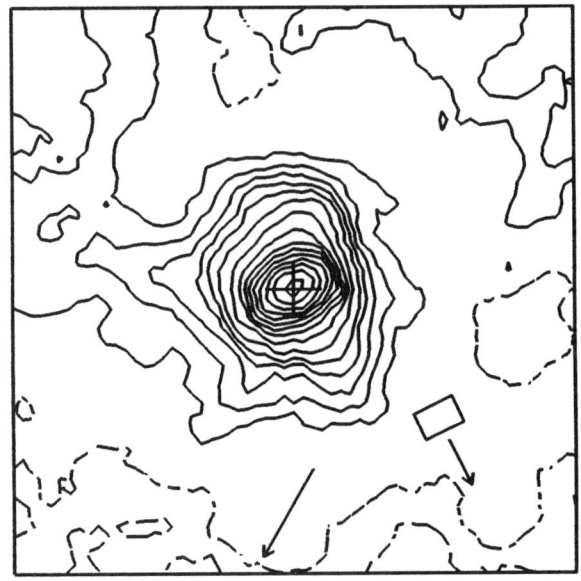

Fig. 1. IRAS 100 μm survey coadd intensities of a region one by one square degree centered on W Hydrae. The cross marks the optical position of W Hya. Solid line contours are 3, 6, 9, 12, 15, 20, 28.3, 40, 56.6, 80, 113.1, 160, 226.3, 320, and 452.5 times the background noise level of 8.0×10^4 Jy/sr. Dashed line contours are -3 and -6 times the noise level. The 100 μm detector size (3' x 4.5') is shown to scale, with the short arrow indicating the direction of the IRAS scans. The longer arrow points toward the galactic plane. North is at the top, East to the left.

HIGH RESOLUTION ^{12}CO (J=1→0) OBSERVATIONS OF NGC 7027

DAVID J. WILNER, JOHN H. BIEGING
Radio Astronomy Laboratory, U. California, Berkeley, CA 94720.

HARLEY A. THRONSON, JR.
Department of Physics and Astronomy, U. Wyoming, Laramie, WY 82071.

ABSTRACT We have combined observations from the Hat Creek interferometer and the NRAO Kitt Peak 12 m telescope to construct images of the molecular material surrounding the young planetary nebula NGC 7027. The ^{12}CO (J=1→0) emission reveals an expanding shell structure with diameter $\lesssim 20''$ and expansion velocity ~ 17 km s^{-1} exterior to the ionized region. There is also a roughly circular emission component that reaches a maximum diameter of $\sim 60''$ at the noise level of our maps. Both components exhibit considerable clumpiness. Position-velocity maps indicate symmetry in the neutral envelope along the same axes as the much more compact ionized nebula.

INTRODUCTION

Planetary nebulae are formed from the ejected atmospheres of low-to-moderate mass stars, whose ejecta are rich in molecular material. NGC 7027, however, is one of a very few planetary nebulae in which strong CO emission has been detected. Observations of the neutral gas in this object provide an important opportunity to examine the poorly understood stellar mass-loss process. Previous observations of the CO emission in NGC 7027 have been insufficient to determine the detailed morphology of the circumstellar molecular cloud: single dish measurements (e.g. Thronson 1983) lack the spatial resolution required to reveal significant structure, while interferometric observations (Masson et al. 1985) have missed more than half of the source flux density. Our combined Hat Creek interferometer and Kitt Peak single dish maps have spatial resolution $\sim 5''$, velocity resolution ~ 1 km s^{-1} and are not missing flux density at low spatial frequencies.

DISCUSSION

The line shape of the single dish spectrum is characteristic of an unresolved, expanding, optically thick shell. Our $\sim 5''$ beam clearly resolves this structure. The map at 25 km s^{-1}, the line center velocity, which shows a cross section of the CO emission in the plane of the sky, reveals the extent and complexity of the circumstellar material (Fig. 1a). The ionized nebula, as seen in 6 cm continuum emission (Masson 1986), is distinctly elliptical (with dimensions $10'' \times 14''$). A bright shell of CO emission closely surrounds the ionized gas. The expansion of the molecular material is clearly evident in a position-velocity diagram at a position angle along the minor axis of the ionized region (Fig. 1b). The expansion velocity is ~ 17 km s^{-1}, and the peak brightness temperature is $\sim 18°$K. Other position-velocity diagrams indicate that the neutral envelope possesses axial symmetry commensurate with that of the ionized nebula. The morphology of the CO emission implies that: (1) the molecular material is constraining the development of the ionized nebula; (2) mass loss from the central star has not been a simple, spherically symmetric, constant velocity outflow, but rather more axisymmetric; (3) the molecular envelope exhibits considerable structure on scales $\lesssim 5''$ (=0.03 pc).

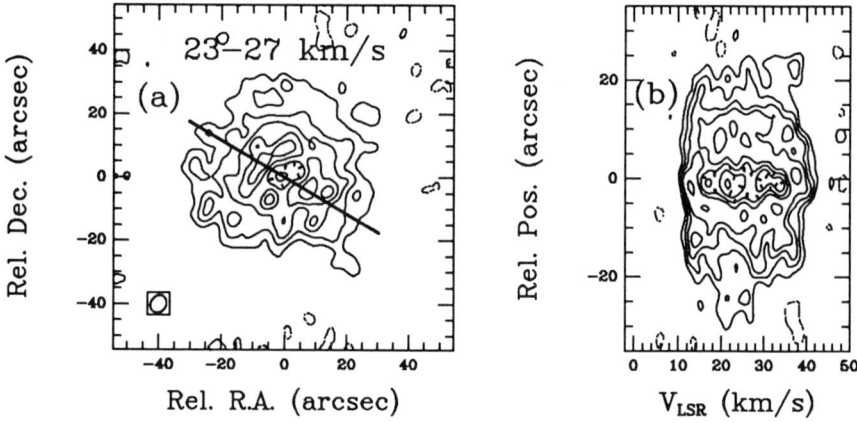

Fig. 1. (a) Intensity map of the CO emission averaged over the velocity range 23-27 km s^{-1}, which spans the spectrum's central velocity. This velocity range shows a cross section of the CO envelope in the plane of the sky. The map center is ($\alpha = 21^h 39^m 09\overset{s}{.}4$, $\delta = 42°02'03\overset{''}{.}0$). Contours are in linear steps of 2.5°K. The inset illustrates the 50% contour of the $5.7'' \times 4.8''$ clean beam. (b) A position-velocity map at position angle 60° (see Fig. 1a) shows the expansion of the molecular material.

ACKLOWLEDGEMENTS

Research at the U.C. Berkeley Radio Astronomy Lab is supported by NSF grant AST 87-14721. D.J.W. acknowledges support from a NSF Graduate Fellowship.

REFERENCES

Masson, C. et al. 1985, Ap. J., **292**, 464.
Masson, C. 1986, Ap. J. (Letters), **302**, L27.
Thronson, H. 1983, Ap. J., **264**, 599.

SYNTHESIS OBSERVATIONS OF J=1-0 HCO⁺ IN DR21

Rognvald P. Garden & Dan Grolemund
Dept. of Physics, U.C. Irvine, Irvine, CA 92717.

John Carlstrom
Dept. of Astronomy, U.C. Berkeley, Berkeley, CA 94720.

1. INTRODUCTION

The DR21 molecular outflow has been studied in great detail at both infrared and millimeter wavelengths by Garden et al. (1986, 1989a, 1989b, in press). From these observations it was determined that DR21 is probably the largest (extends over a projected distance of 5 pc), most massive (> 1000 $M_\odot$) and energetic ($> 3 \times 10^{48}$ ergs) outflow yet known. In addition, infrared images of the shock-excited H_2 v=1-0 S(1) line emission (*figure 1*), obtained using an infrared camera with an angular resolution of ~ 1 arcsec on the 4-m United Kingdom Infrared Telescope (Garden et al. 1989a), clearly show that the DR21 outflow is composed of two highly-collimated bipolar jets that are extremely clumpy and become more focused with increasing distance from the outflow origin.

2. RESULTS

A better understanding of the nature of the outflow phenomenon may result from a comparison of the H_2 image with observations of other molecular emissions at comparable angular resolution. Thinking that HCO⁺ would be a good choice for such a comparison, we have used the Berkeley-Illinois-Maryland Millimeter Array (BIMA) to obtain velocity-channel maps of the HCO⁺ J=1-0 line emission at an angular resolution of 8 arcsec over the entire length of the DR21 H_2 emission-line jets. Three fields were observed, centered on (i) the east H_2 jet, (ii) the west H_2 jet, and (iii) the DR21 cloud core (outflow center), with a 30 arcsec overlap between adjacent fields. The results of these observations are shown in *figure 2* in the form of contours of total integrated intensity superposed on contours of H_2 line emission.

3. DISCUSSION

The main results derived from the BIMA observations can be summarized as follows:

(1) We detect bright HCO⁺ emission not only from the central cloud core but also *from several dense, high-velocity clumps located within the outflow lobes.* Because both the east and west high-velocity outflow lobes lie along an axis that bisects the DR21 cloud core, we are inclined to believe that the HCO⁺ emission arises from ambient molecular gas that has been swept up by the passage of

a collimated wind that emanates from a powerful young stellar object located somewhere within the DR21 cloud core.

(2) Unlike most outflow sources observed to date, the DR21 outflow *does not appear to be bipolar in HCO^+ emission*; both lobes are b̲lue-shifted by ~ 15 km s^{-1} from the DR21 rest velocity. This agrees with earlier results on CO J=1-0 emission (Garden *et al.* 1989b). The high-velocity CO and HCO$^+$, however, peak at different spatial locations along the outflow lobes (see *figure 2*).

(3) The high-velocity HCO$^+$ emission associated with both outflow lobes is extremely bright and *bears a striking resemblance to the distribution of shock-excited H_2 line emission*. This close association may result if the abundance of HCO$^+$ has been enhanced in the post-shock gas.

Garden, R.P., Geballe, T.R., Gatley, I. and Nadeau, D. 1986,
    M.N.R.A.S., 220, 203.
Garden, R.P., Russell, A.P. and Burton M.G. 1989a, *Ap. J.*, *submitted*.
Garden, R.P., Geballe, T.R., Gatley, I. and Hayashi, T. 1989b,
    Ap. J. Suppl., *submitted*.

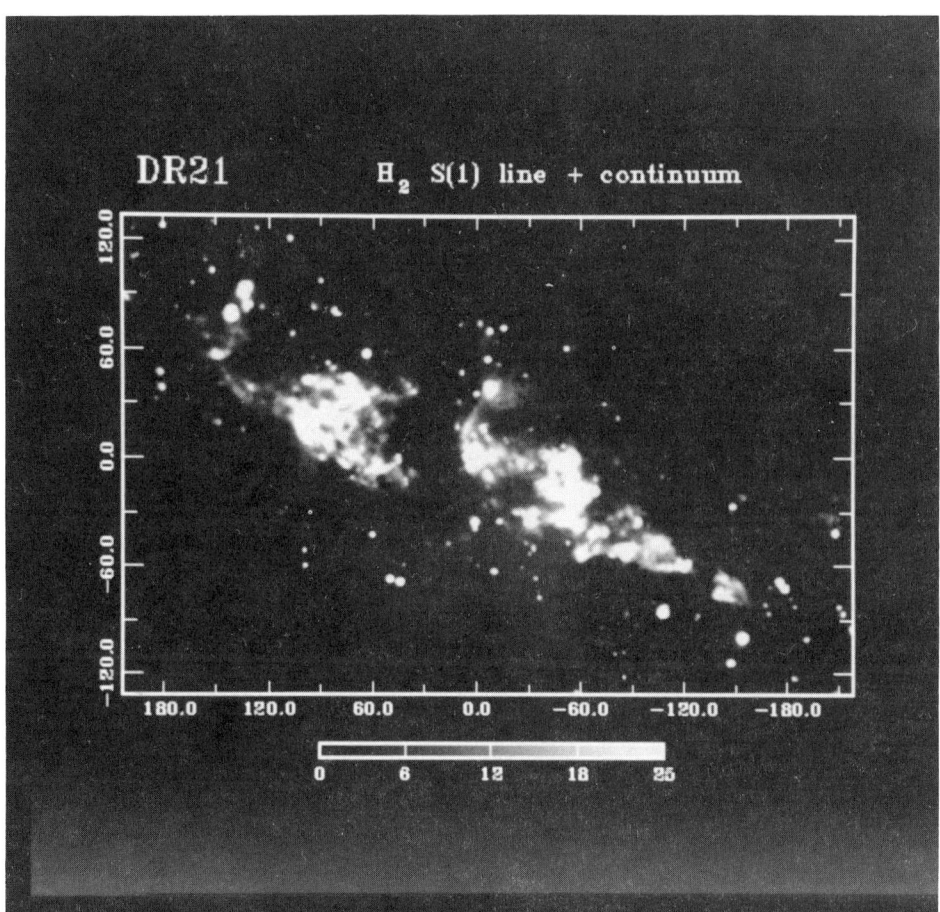

FIGURE 1. Grayscale image of integrated intensity in the v=1-0 S(1) line of molecular hydrogen at 2.12 μm.

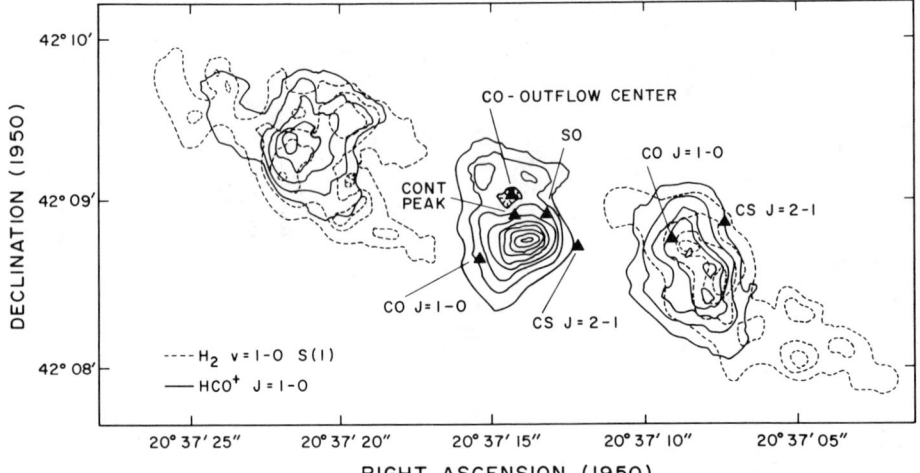

FIGURE 2. Contours of integrated HCO^+ J=1-0 emission superposed on H_2 line emission.

THE 3.3 MICRON FEATURE, H_2, AND IONIZED GAS IN THE ORION BAR

K. SELLGREN, A. T. TOKUNAGA
Institute for Astronomy, 2680 Woodlawn Dr., Honolulu, HI 96822 USA

Y. NAKADA
Department of Astronomy, University of Tokyo, Tokyo, JAPAN

INTRODUCTION

We present observations of the spatial distribution of the 3.3 μm feature, H_2 emission, and ionized gas in the Orion Bar. The Orion Bar is an ionization front 2' SE of θ^1C Ori, the star which ionizes the Orion Nebula. We also present results on the 3.3 μm feature width in the Orion Bar. Our results are presented in more detail in Sellgren, Tokunaga, and Nakada (1990).

THE SPATIAL DISTRIBUTION

We have used the IRTF to measure the spatial distribution of the 3.3 μm feature, Br α, P α, and the Q-branch of H_2, along a line perpendicular to the ionization front, and passing through Orion Position 4, 45" W 12" S of θ^2A Ori. The ionized gas traced by Br α and P α peaks 10" NW of Position 4, at the ionization front, while the H_2 brightness peaks 15" SE of Position 4 (Fig. 1). The 3.3 μm brightness drops dramatically inside the H II region, suggesting rapid destruction of the emitting material in the H II region. The 3.3 μm brightness also decreases at the H_2 peak. The 3.3 μm spatial distribution appears to be due to destruction of the emitting material within the H II region and extinction of the exciting radiation between the edge of the H II region and the H_2 peak, resulting in a maximum between the ionization front and the H_2 peak. The H_2 line ratios and H_2 surface brightness at the H_2 peak are consistent with either being due to a shock front or being due to UV-pumped fluorescence from dense clumps of H_2.

THE 3.3 MICRON FEATURE WIDTH

We have also used the IRTF's Cooled Grating Array Spectrometer at high spectral resolution ($\lambda/\Delta\lambda = 1400$), to observe the shape of the 3.3 μm feature in Orion in regions of varying UV intensity: at Position 4, inside the H II region, and at the H_2 peak. Previous measurements of the 3.3 μm feature profile (Nagata et al. 1988; Tokunaga et al. 1988; Geballe et al. 1989) find that the central wavelength is constant but that the wavelength of the half power point on the blue side of the 3.3 μm feature is variable. We observe that the blue half power point in Orion is constant at all positions, indicating a constant feature width. Geballe *et al.* have suggested that variations in 3.3 μm feature width are related to the intensity of the UV field. We suggest instead that the 3.3 μm feature is initially formed with a composition showing a narrow feature, and then is rapidly converted to a composition showing a broad feature.

REFERENCES

Geballe, T. R., Tielens, A. G. G. M., Allamandola, L. J., Moorhouse, A., and Brand, P. W. J. L. 1989, *Ap. J.*, **196**, 179.

Nagata, T., Tokunaga, A. T., Sellgren, K., Smith, R. G., Onaka, T., Nakada, Y., and Sakata, A. 1988, *Ap. J.*, **326**, 157.

Sellgren, K., Tokunaga, A. T., and Nakada, Y. 1990, *Ap. J.*, in press.

Tokunaga, A. T., Nagata, T., Sellgren, K., Smith, R. G., Onaka, T., Nakada, Y., Sakata, A., and Wada, S. 1988, *Ap. J.*, **328**, 709.

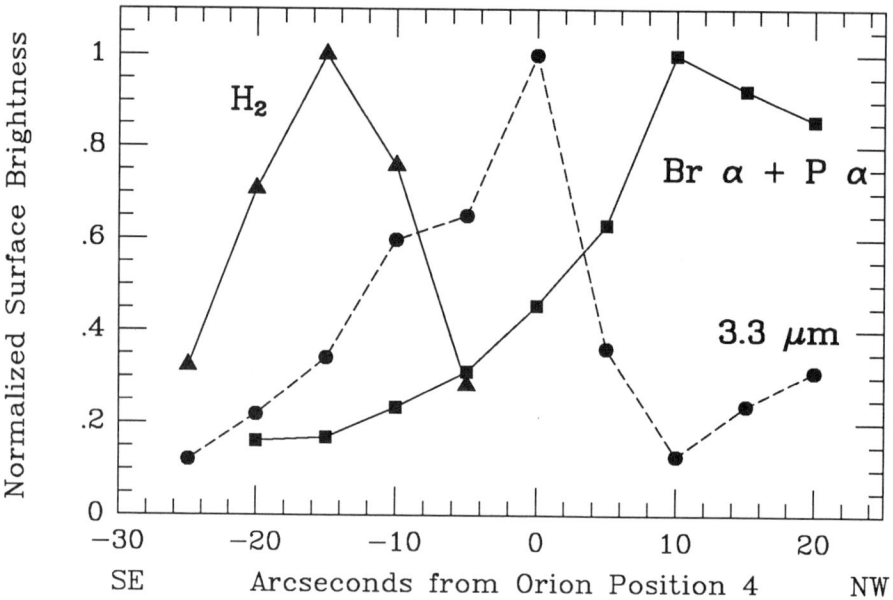

Fig. 1. The normalized surface brightness of the 3.3 μm feature, ionized gas traced by a combination of Br α and P α, and the 2.42 μm Q-branch emission of H_2, observed along a line passing through Orion Position 4 and perpendicular to the Orion ionization front.

OBSERVATIONS OF THE GALACTIC PLANE BY THE ZODIACAL INFRARED PROJECT

L. J RICKARD
E. O. Hulburt Center for Space Research, Naval Research Laboratory, Washington, DC 20375

S. W. STEMWEDEL
Applied Research Corporation, Landover, MD 20785

S. D. PRICE
Air Force Geophysical Laboratory, Hanscom AFB, MA 01731

The two rocket flights of the Zodiacal Infrared Project (ZIP; Murdock and Price 1985, *Astr. J.*, **90**, 375), flown 18 August 1980 and 31 July 1981, were intended to provide data on the near-infrared thermal emission of the interplanetary dust cloud over a broad range of ecliptic coordinates (latitudes -60° to +85°, solar elongation angles 22° to 90° and 140° to 180°). In addition, their multiple crossings of the Galactic plane provided low resolution spectral data ($\delta\lambda/\lambda$ ranging from 1. to 0.1, for effective wavelengths from 3 to 30 μm) for most of the first quadrant (longitudes 30° to 100°).

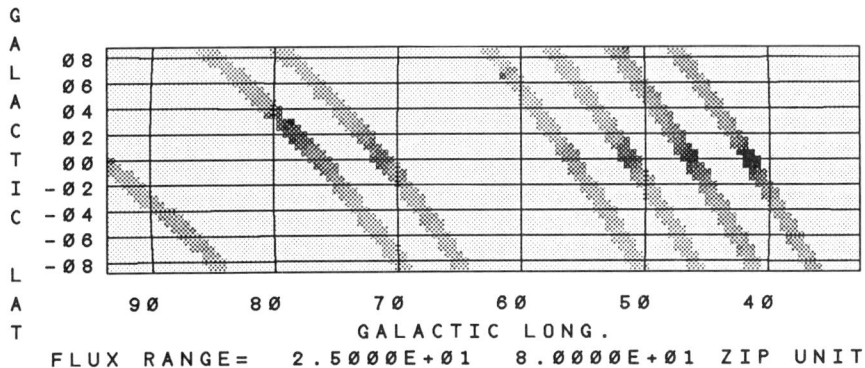

Figure 1. Example of Galactic crossing data from ZIP flight 1, detector 10 (11 μm).

Having made a thorough reanalysis of the calibration of the ZIP database, we present the salient features of the Galactic plane as observed by ZIP.

The binned, in-plane data, corrected for zodiacal emission, generally show an exponential decrease with increasing longitude.

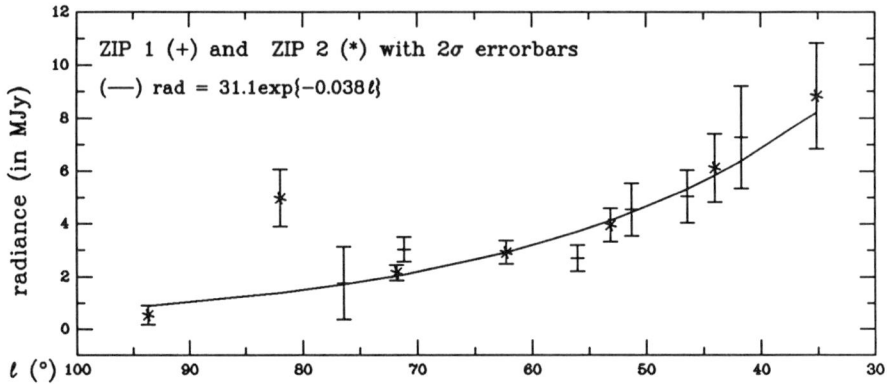

Figure 2. Variation of radiance with longitude, for detector 10 (11 μm), fitted by an exponential with scale-length 0.038 deg^{-1}.

The fitted exponential scale-length can be inverted to derive a radial density profile. Note as well the appearance of excess emission at 83° arising from material associated with the Cyg-X region.

Channel ratios are converted to temperatures by using model spectra in which thermal emitters, here with emissivity $\sim \lambda^{-1}$, are convolved with the filter responses.

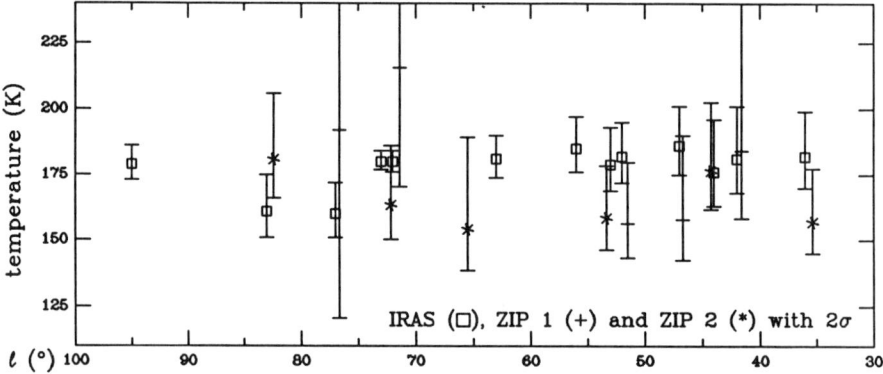

Figure 3. Color temperatures for channels 5 (11 μm) and 12 (21 μm), along with similarly derived temperatures from IRAS 12 μm and 25 μm data.

The ZIP data show little variation of temperature with longitude, consistent with IRAS results, both indicating a dust temperature of about 180° for this region of the Galactic plane.

A narrow spectral feature at 13 μm appears consistently in data for the plane (uncorrected for zodiacal emission). However, this is strongly contaminated by calibration problems for channel 8. We suggest that residual emission at 13 μm arises from the [Ne II] line at 12.8 μm.

STARS AND INTERSTELLAR MATTER IN OPHIUCHUS/SCORPIUS

E.J. DE GEUS
ASTRONOMY PROGRAM
UNIVERSITY OF MARYLAND COLLEGE PARK

INTRODUCTION

The early-type stars in the Sco Cen OB association cover the 4th quadrant at latitudes $-15° < b < +25°$. The morphology of the neutral hydrogen in the 4th quadrant is dominated by loop-like structures (Colomb, Pöppel, and Heiles 1980) extending far away from the plane. Outside the galactic plane, for $\ell < 340°$, little molecular material is present, except for a number of high-latitude clouds, which appear to be related to the large HI-structures (Gir and Blitz 1989).

In the region $360° > \ell > 340°$ at positive latitudes, a large amount of molecular material is present in the Ophiuchus clouds which are related to Upper-Scorpius stars, the youngest subgroup of the Sco-Cen OB association. The neutral hydrogen in this area is mainly situated in a loop surrounding the early-type stars.

INTERSTELLAR MATTER IN OPHIUCHUS

A large-scale CO map of the Ophiuchus clouds (de Geus, Bronfman, and Thaddeus 1989) shows that most molecular gas resides in strongly elongated structures, the total molecular mass of the region is 10^4 $M_\odot$. The IRAS 100 μm dust map is well correlated with the CO, except near hot stars, where the CO-emission lags the infrared. This discrepancy can be understood, because the regions around the massive stars will be hotter than average, which enhances the infrared emission although the dust column density may be as low as that of the molecular gas. The HI loop is part of an HI shell, expanding with approximately 10 km/s. The shell has a mass of 8×10^4 $M_\odot$, and thus a mechanical energy of approximately $2 \pm 1 \times 10^{50}$ ergs. Observations of CH and CH^+ absorption lines towards stars in Upper Scorpius indicate the presence of a slow ($\approx$ 15 km/s) shock along the line of sight, which is most likely due to a collision of -10 km/s gas with $+5$ km/s gas. The largest elongated CO clouds all seem to "radiate" outward, away from the dense ρ Oph cloud.

STARS IN UPPER SCORPIUS

Physical parameters of the stars in Upper Scorpius were determined using the Walraven five-colour photometric system and atmosphere models by Kurucz (1979), see de Geus, de Zeeuw, and Lub (1989). Based on their luminosities, the total mechanical energy output of the massive stars is estimated to be 10^{48} ergs. Fitting an IMF (Miller and Scalo 1979) to the present day mass function of Upper Scorpius revealed the possibility that one more massive star may have been a member of the subgroup. The most likely mass of the star is 40 $M_\odot$ (i.e. spectral type O7). The output of mechanical energy of such a star during its lifetime is $4 \pm 2 \times 10^{50}$ ergs, with roughly equal contribution by the stellar wind and the supernova explosion.

CONCLUSIONS

The HI shell is formed most likely by the stellar wind and supernova explosion of one massive star in the Upper-Scorpius OB subgroup. The expanding shell is moving into the surrounding medium, causing a slow shock in which CH and CH^+ molecules are formed. The encounter of the shell with the dense ρ Oph cloud caused part of the molecular material to be stripped off, and deposited in elongated structures in the wake of the shock. The Ophiuchus cloud is clearly a remnant of the molecular cloud from which the early-type stars in Scorpius were formed.

REFERENCES

Colomb, F.R., Pöppel, W.G.L., and Heiles, C. 1980, *Astr. Ap. Suppl.*, , **40**, 47.
de Geus, E.J., Bronfman, L., and Thaddeus, P. 1989, *Astr. Ap*, , in press.
de Geus, E.J., de Zeeuw, P.T., and Lub, J. 1989, *Astr. Ap*, , **216**, 44.
Gir, B.Y., and Blitz, L. 1989, *Ap. J.*, , in preparation.
Kurucz, R. 1979, *Ap. J. Suppl.*, , **40**, 1.
Miller, G.E., and Scalo, J.M., 1979, *Ap. J. Suppl.*, , **41**, 513.

HIGH SPECTRAL AND SPATIAL RESOLUTION IMAGING OF THE SHOCK WAVES IN ORION

MICHAEL BURTON[1,2], JOSS BLAND[3], D. AXON[4], P. BRAND[5],
R. GARDEN[2], T. GEBALLE[6], D. HOLLENBACH[1], J. HOUGH[7],
I. McLEAN[6], A. MOORHOUSE[5]
[1]NASA Ames, MS:245-6, Moffett Field, CA 94035, [2]U.C. Irvine, [3]Rice University, [4]Jodrell Bank Radio Observatory, [5]University of Edinburgh, [6]Joint Astronomy Centre Hilo, [7]Hatfield Polytechnic

ABSTRACT   Orion has been imaged in the v=1–0 S(1) line of molecular hydrogen, with 0.6" pixel scale and $12\,\mathrm{km\,s^{-1}}$ spectral resolution, and the most detailed picture yet of the shocked emission obtained.

Despite considerable effort over the past decade, the molecular hydrogen line emission from the OMC–1 in Orion has remained an enigma. It is clear that shocks are responsible for the phenomenom (*e.g.,* Beckwith *et al.* 1978, *Ap. J.*, 223, 463), and the velocity profiles of the excited lines, with widths of up to $150\,\mathrm{km\,s^{-1}}$ (*e.g.,* Nadeau *et al.* 1982, *Ap. J.*, 253, 154), imply widespread supersonic motions. Accounting for the shape and width of the profile has remained a problem since Kwan (1977, *Ap. J.*, 216, 713) established that H_2 would be dissociated in jump-shocks moving faster $25\,\mathrm{km\,s^{-1}}$ (or $\sim 40\,\mathrm{km\,s^{-1}}$ in magnetically-mediated shocks, Draine 1980, *Ap. J.*, 241, 1021). At the highest spectral and spatial resolution the line emission has been measured ($12\,\mathrm{km\,s^{-1}}$ and 5", Brand *et al.* 1989, *M.N.*, 237, 1009) the profile has remained smooth, relatively symmetrical, has a strong blue wing, and shows no hint of breaking into narrow emission features. The most plausible model seems to be that it arises from a conglomeration of discrete clumps, each individually accelerated to different velocities by a wind from the source IRc2, and being shocked. However, as Brand *et al.* (1989) discuss, this requires a minimum of 30–40 cloudlets along *every* line of sight to the source (a cylinder with cross-section $\sim 0.01\mathrm{pc}$, small compared to the source size of $\sim 0.2\mathrm{pc}$, yet with a velocity dispersion $\pm 75\,\mathrm{km\,s^{-1}}$). There are considerable difficulties in explaining the redirection of momentum from the source of the wind which ultimately drives the shocks.

The observations were obtained on the UKIRT telescope in January 1989 using the IR array camera IRCAM with 0.6" pixel scale. A Fabry–Perot interferometer (FP) with resolution $12\,\mathrm{km\,s^{-1}}$ was placed in the beam, and scanned in steps of $5\,\mathrm{km\,s^{-1}}$, over the profile of the H_2 1–0 S(1) line, at $2.1218\,\mu\mathrm{m}$, covering a 36" × 38" region at the peak of the H_2 line emission from OMC–1. The surface of constant phase (a paraboloid) was determined, and the phase shift across each frame deconvolved from the data (see *e.g.,* Bland *et al.* 1987, *M.N.*, 228, 595). In the figure are shown profiles of the line emission for one quadrant of the image. It must be cautioned that the reduction is still preliminary and the wavelength scale is non-uniform. Further work remains to linearise the scale and remove some artifacts from the data. Nevertheless, the results clearly demonstrate that H$_2$ line profiles with excellent S/N can now be obtained, with both high spectral and spatial resolution.

The first impression from the data is, despite the increased spatial resolution, that the profiles remain wide, have extended blue wings, and are generally smooth, all over the source. The problem of redirection of the wind momentum is exacerbated. There is, however, some structure in the profiles. Line splitting is seen for the first time in several locations, with the secondary component being a high-velocity blue-shifted feature. Examination of the images reveals that this arises from several filamentary, or 'jet-like' features, which emanate from the vicinity of IRc2. It is tempting to speculate that they delineate the conduits along which a high-velocity wind from IRC2 is flowing, with the H$_2$ emission arising from accelerated molecular gas swept-up and entrained along them. This then poses the question of what kind of collimation mechanism might lead to more than one jet escaping from the source?

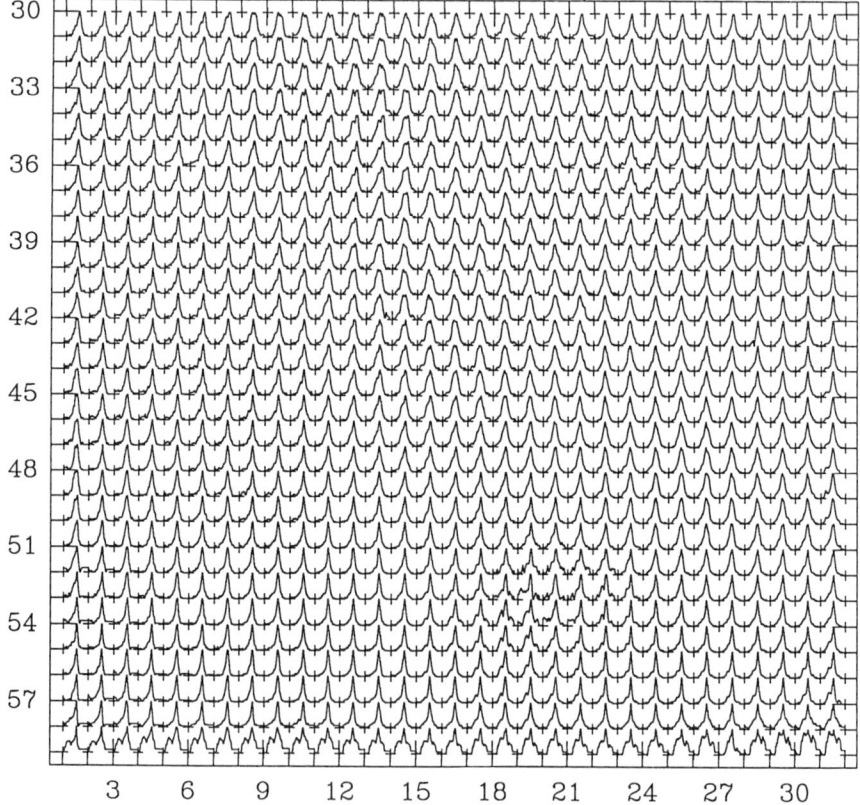

Figure 1: Line profiles of the H$_2$ 1–0 S(1) line for one quarter of the data, covering a portion of the peak emission region of OMC–1. Fluxes are scaled to the peak intensity in each pixel. N is to the bottom and E to the left.

GALACTIC STRUCTURE IN THE FOURTH QUADRANT DEDUCED FROM
INTERSTELLAR LINES, H II REGIONS, AND MOLECULAR HYDROGEN

J. J. RICKARD
P.O. Box 777, Borrego Springs, CA 92004, USA

ABSTRACT   Galactic structural features in the fourth
quadrant, L = 285-345, are examined using velocity and
distance data of interstellar Ca II K lines, optical H II
regions, young clusters and OB stars, and molecular CO
emission. Both the stellar and K line data of Rickard
(1974) are revised to reflect improved spectral classification, absolute magnitudes, and a new Local Standard of
Rest.

A new rotation curve is derived for the fourth quadrant
using young open cluster data of Hron (1987) and OB star
groups together with the 21-cm H I data (Sinha 1978) using
Ro = 8.5 Kpc and Vc = 220 Km/s.

Two large-scale structural features are seen: a portion of
the Sagittarius-Carina spiral arm located 1.5 Kpc from the
sun; and an interior feature at a distance of 3 Kpc.
The velocities of the Ca II K lines and H II regions in
this interior feature are closely correlated with the CO
Molecular Ring. This feature is closer than the spiral
pattern previously proposed by Georgelin and Georgelin
(1976).

Peculiar motions seen in the K line velocities toward
L = 328 may be confined to the H II complex RCW 96-99.

V-L DIAGRAM FOR CA II AND H II GAS

The revised velocity-longitude diagram derived from
interstellar K lines (Rickard 1974) and optical H II regions
(Georgelin and Georgelin 1976) is shown in Figure 1. Note that
the velocity features are not randomly distributed, but cluster
into three bands: low velocity, 0 to -25 km/s; medium velocity,
-35 to -50 km/s; and a few high velocity features -65 to -95
km/s.

STRUCTURAL FEATURE
(Distance from sun)

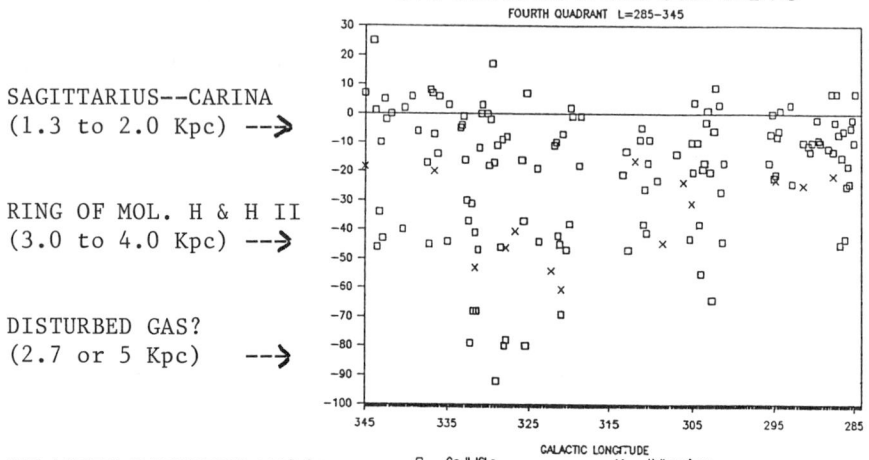

SAGITTARIUS--CARINA
(1.3 to 2.0 Kpc) -->

RING OF MOL. H & H II
(3.0 to 4.0 Kpc) -->

DISTURBED GAS?
(2.7 or 5 Kpc) -->

GALACTIC ROTATION MODEL

Figure 2. is the composite rotation model used to determine both the distance to the interstellar Ca gas and several H II regions with no or poor distances. The model is based upon the study of young open clusters (Hron,1987), H II regions with good distances, and the OB stars which were the background for the interstellar K lines. The optical model is valid to $R-R_o = -1.7$ Kpc. Interior to this the model was derived by rescaling the H I 21-cm tangent point velocity data of Sinha (1978).

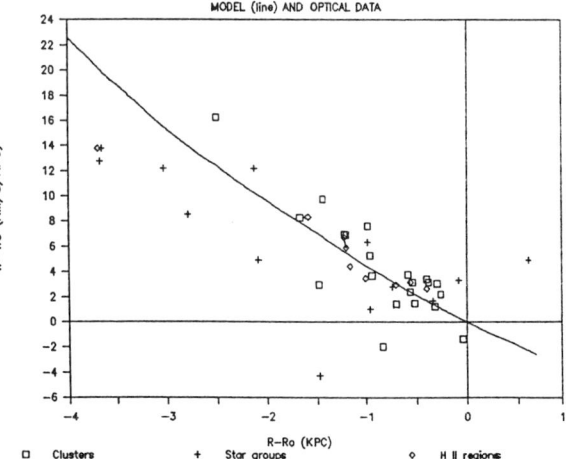

REFERENCES

Georgelin Y.M., Georgelin, Y.P. 1976, Astron. Astrophys. <u>49</u>, 57
Hron, J. 1987, Astron. Astrophys. <u>176</u>, 34
Rickard, J. J. 1974, Astron. Astrophys. <u>31</u>, 47
Sinha, R.P. 1978, Astron. Astrophys. <u>69</u>, 227

COMPARISON OF NH$_3$ AND CS DISTRIBUTION IN NGC 2071

Shudong Zhou[1] and Neal J. Evans II
Department of Astronomy, Univ. of Texas at Austin, Austin, TX 78712
[1] JCMT Fellow, Institute for Astronomy, 2680 Woodlawn Dr., Honolulu, HI 96822

Lee G. Mundy
Astronomy Program, Univ. of Maryland, College Park, MD 20742

ABSTRACT We have observed the NGC 2071 star forming region in the (J,K) = (1,1) and (2,2) lines of NH$_3$ and the J=2→1, 5→4, 6→5, and 7→6 lines of CS. The observations reveal a striking difference in spatial distribution between the NH$_3$ emission and the CS emission.

The NGC 2071 star forming region is in the northern part of the Orion B molecular cloud complex, with a distance of 390 pc (Anthony-Twarog 1982). A large scale map of the CS(J=2→1) emission shows a well defined structure 10' by 30' in size with a position angle of 75° (Lada et al. 1990). In the middle of this structure, there is a powerful, well-collimated outflow (P.A. = −45°) seen in many molecular lines. At least three infrared point sources (IRS 1-3) have been observed in the center of the outflow. The infrared source IRS 1 is the brightest object at 10μm (Persson et al. 1981). We use NH$_3$ and CS to probe the core structure on $\sim 10''$ scales near IRS 1.

THE NH$_3$ OBSERVATIONS

The NH$_3$(1,1) and (2,2) lines were observed simultaneously in 1988 September using the VLA D-array. The velocity resolution was 0.6 km s^{-1} and the best synthesized beam was 4''. To cover an area of 3' by 3', we cycled through four overlapping fields and combine the data using mosaicking techniques. The contour lines in Figure 1 shows the integrated intensity map for the NH$_3$(1,1) main hyperfine component with 10'' resolution.

The structure is very elongated with an aspect ratio of 3 to 1, which suggests a disk-like structure viewed edge-on. More detailed analysis show that most of the emission must come from a ring of gas approxiamtely 0.1 pc from IRS 1. Supporting evidence comes from both the low temperature ($\lesssim 20$ K) of the emitting gas and the lack of velocity gradient across the structure. As shown later, this ring-like structure of NH$_3$ emission does not reflect the overall gas and dust distribution in the region and is probably an artifact of the NH$_3$ abundance variation.

THE CS OBSERVATIONS

The CS data were obtained using three different telescopes: the IRAM 30 meter (J=2→1 and 5→4), the NRAO 12 meter (J=6→5), and the CSO 10.4 meter telescopes. We have mapped a region of 90" by 90" in all transitions.

In contrast to the elongated NH_3 distribution, the CS emission peaks strongly at the IRS 1 postion (gray scale in Figure 1). We have analyzed the data with spherically symmetric, microturbulent cloud models to derive the density structure. The gas temperature is assumed to be the same as dust temperature derived from far-infrared dust emission (Butner *et al.* 1990). For a uniform CS abundance, the best fit to data gived a density distribution of $n = 2 - 7 \times 10^6 (r/0.01 pc)^{-1} cm^{-3}$.

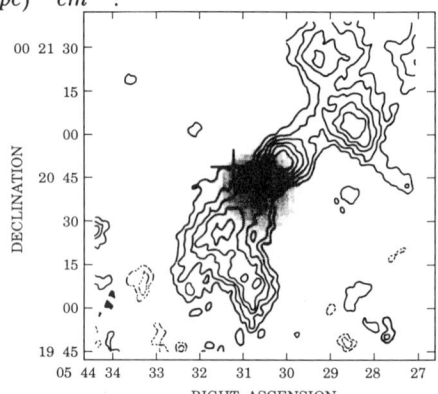

Fig. 1. The overlay of integrated intensity maps of the $NH_3(1,1)$ main line (contours, 10" beam) and the CS(5→4) line (grey scale, 11" beam).

Figure 1 shows a striking difference between the NH_3 and CS distribution. Comparison shows that CS mophorlogy agrees well with dust continuum emission at 800μm (G. Sandell, private comm), which indicates that NH_3 is not tracing the dust distribution well. The present result adivses caution in choosing molecular tracers near young stellar objects.

ACKNOWLEDGEMENTS

This work was supported in part by NSF grants AST 88-15801 and the W.M. Keck Foundation.

REFERENCES

Anthony-Twarog, B.J. 1982, *A.J.*, **87**, 1213.
Butner, H.M., Evans, N.J. II, Harvey, P.M., Munday, L.G., and Natta, A. 1990, *Ap.J.*, in prep.
Lada, E.A., Bally, J., and Stark, A. 1990, *Ap.J.*, in prep.
Persson, S.E., Geballe, T.R., Simon, T., Lonsdale, C.J., and Baas, F. 1981, *Ap.J. Lett.*, **251**, L85.

METHANOL MASERS AND STAR FORMATION: VLA OBSERVATIONS OF THE NGC 6334 REGION

K. M. MENTEN AND M. J. REID
Harvard–Smithsonian Center for Astrophysics, 60 Garden Street,
Cambridge MA 02138

ABSTRACT   The VLA was used to map five different methanol (CH_3OH) transitions and the H_2O maser line toward the prominent star forming regions NGC 6334–I/I(N). These two regions are in very different stages of early stellar evolution. Four of the CH_3OH lines show maser emission in the immediate neighbourhood of the cool dust continuum source I(N) and thermal emission in the hot molecular core associated with the infrared source I and the compact HII region F. Maser action in the other methanol transition is observed from the interface region between the HII region F and the hot core, which also gives rise to OH maser emission. H_2O masers are found toward I and I(N) Our results corroborate the theory that two different types of methanol masers exist that probe very different regions in the environment of newly formed massive stars.

INTRODUCTION

NGC 6334 is an extended molecular cloud/HII region complex at a distance of 1.7 kpc that forms a ridge of dimensions 10 × 3 pc parallel to the galactic plane. The large number of embedded infrared and sub–millimeter sources and compact HII regions make this region one of the most active sites of high–mass star formation in the Galaxy.

The (sub–)millimeter continuum sources I and I(North) are found toward the northern end of the NGC 6334 complex (Gezari 1982). While NGC 6334–I is associated with a compact HII region (also called NGC 6334–F) and OH masers (Gaume and Mutel 1987), no centimeter–wave continuum emission has been observed toward I(North), a cool massive dust emission peak $\approx 2'$ north of I. Observations of these sources provide the opportunity to simultaneously study two objects at distinctly different stages in the early evolution of massive stars toward the same molecular cloud region.

A large number of different methanol (CH_3OH) maser transitions have been detected toward NGC 6334–I and I(N) (Batrla *et al.* 1987, Menten and Batrla 1989, Haschick *et al.* 1989, 1990). Since methanol masers are interesting probes of regions of massive star formation, we used the VLA to study several methanol transitions toward both I and I(N).

RESULTS AND DISCUSSION

To study the small scale distribution of the methanol masers found in this region and to compare their positions with other features, we used the VLA to conduct $\approx 5''$ resolution observations of the 22 GHz H_2O maser line, and four different methanol (CH_3OH) transitions from the $J_2 \rightarrow J_1 E$ series ($J = 2, 3, 4$, and 6) near 25 GHz toward *both* sources. Additionally, we observed the 23 GHz $9_2 \rightarrow 10_1 A^+$ CH_3OH maser transition toward the southern source and obtained maps of the 1.3 cm continuum emission and the H66α recombination line.

We have determined accurate positions for the two 25 GHz CH_3OH maser components observed toward the I(N) source and find (1) that are located at two distinct positions separated by $17''$ and (2) that they are both offset by $9''$ from the H_2O maser position.

In contrast, *thermal* emission in the 25 GHz methanol lines is observed toward the hot dense molecular core $6''$ north–west of the HII region. This hot core also has associated water masers. The spatial and radial velocity distribution of the H_2O maser spots indicate that they are part of a bipolar outflow emerging from the center of the core. The strong maser emission from the 23 GHz $9_2 \rightarrow 10_1 A^+$ transition and the OH maser emission seem to be emitted from a narrow dense interface region of hot gas between the HII region and the molecular core.

Methanol maser emission from various other transitions has been detected toward both I and I(N). It is interesting to note that none of the seven CH_3OH transitions that have been found to show maser action toward I(N) is masing toward I, and none of the six lines (among them the 12 GHz $2_0 \rightarrow 3_{-1} E$ transition) masing toward I is detected toward I(N). Our observations strongly support the hypothesis that one type of methanol masers (*Class A*) is found near regions in a very early (possibly protostellar) evolutionary state, in our case I(N). In contrast, *Class B* methanol masers are associated with compact HII regions and OH masers, both of which are found towards NGC 6334–I. All of the methanol maser transitons hitherto found can be ascribed to one of these two classes.

REFERENCES

Batrla, W., Matthews, H. E., Menten, K. M., Walmsley, C. M.: 1987, *Nature*, **326**, 49.
Gaume, R. A., Mutel, R. L.: 1987, *Ap. J. Suppl.*, **65**, 193.
Gezari, D. Y.: 1982, *Ap. J. (Letters)*, **259**, L29.
Menten, K. M., and Batrla, W.: 1989 *Ap. J.*, **341**, 839.
Haschick, A. D., Baan, W. A., Menten, K. M. 1989, *Ap. J.* (in press).
Haschick, A. D., Menten, K. M., Baan, W. A. 1990, *Ap. J.* (submitted).

STELLAR MASS LOSS RATES IN MAGELLANIC CLOUD STARS

CATHARINE D. GARMANY
Joint Institute for Laboratory Astrophysics, University of Colorado and National Institute of Standards and Technology, Boulder, Co. 80309

ABSTRACT Mass loss rates determined from high resolution IUE spectra are presented for seven Magellanic cloud stars.

INTRODUCTION

Stellar winds from massive stars are an important step in the galactic recycling process. In the giant molecular clouds the winds are second only to supernovae as sources of energy to the interstellar medium (Abbott, 1982b). This phenomenon is well studied in galactic stars, and has spurred the development of radiation driven wind theory (Castor, Abbott and Klein 1975). One particularly interesting prediction is the dependence on chemical composition. According to Abbott (1982a) the mass loss scales directly with chemical composition Z; Pauldrach, Puls and Kudritzki (1987) predict that it should scale with $Z^{1/2}$. Not only would the resolution of this be useful for the refinement of wind theory, but it has important implications for stellar evolution in other galaxies with different chemical enrichment histories. The range of Z for Galactic stars is too small to provide a suitable test, but the Magellanic Clouds are well suited to the problem, and this is the issue we have addressed.

MAGELLANIC CLOUD STARS

Attempts have been made to determine stellar mass loss (Prinja 1987, Garmany and Fitzpatrick 1988) by using low resolution (6 A) IUE spectra to study the wind lines, but the inability of this method to model the shape of the line and account for circumstellar components makes the results uncertain. However, the results suggested rates that differ little from galactic stars of corresponding spectral types, although the wind terminal velocity is significantly lower. Leitherer (1988a) has found the same thing from Hα emission.

In an attempt to resolve the issue, we have obtained high resolution (0.25 A) spectra of 7 Magellanic Cloud stars. The C IV, Si IV and N V profiles were extracted, flux calibrated and then normalized, utilizing low dispersion spectra of the same star as an aid in the normalization process. The wind profiles were fit with theoretical profiles to determine the wind column density. Table 1 lists the stars, their stellar parameters and the preliminary mass loss rates. We have arbitrarily adopted carbon and nitrogen abundances to be down by a factor of 8 in the SMC and down a factor of 4 in the LMC. For the SMC stars we also list the rates determined from the Hα profiles by Leitherer (1988b): the agreement is excellent. It appears that the mass loss rates for the Magellanic Cloud stars are only a factor of 2-3 lower than the galactic stars of the same luminosity, although a more complete analysis using stellar parameters derived from high dispersion optical spectra is in progress.

TABLE I   The Program Stars

Star	Sp. Type	T_{eff}(K)	M bol	Wind ($km\,s^{-1}$)	Log $\dot{m}$ UV	Log $\dot{m}$ Hα
AV 26	O7III	38,100	-10.4	2100	-5.54	-5.65
AV 75	O5III	42,300	-10.5	2200	-5.08	-5.43
AV 243	O6III	40,200	-9.2	1500:	-6.20	-6.29
AV 388	O4V	46,400	-9.4	2100	-5.90	-5.97
SK-66 100	O6III	40,200	-9.1	2300	-5.54	
SK-66 172	O5V	44,300	-9.8	3300	-5.92	
SK-67 51	O6.5III	39,200	-9.8	2700	-5.39	

What are the predictions of radiation wind theory for these Magellanic Cloud stars? Convential wisdom says that the rates should be significantly lower than in corresponding galactic stars by a factor of three to ten. However, this may not be correct. Leitherer (1988b) has explored the scaling relations presented by Friend and Abbott (1986) and finds excellent agreement for galactic stars, but theoretical rates that are systematically lower than observed in the Clouds. These questions will be explored in a later paper.

This research was supported by NASA grant NAG5-78 to the University of Colorado.

REFERENCES
Abbott, D. C. 1982, *Ap. J.*, **259**, 282.
Abbott, D. C. 1982, *Ap. J.*, **263**, 723.
Castor, J. I., Abbott, D. C. and Klein, R. I. 1975, *Ap. J.*, **195**, 157.
Friend, D. B. and Abbott, D. C. 1986, *Ap. J.*, , **311**, 701.
Garmany, C. D. and Fitzpatrick, E. L. 1988, *Ap. J.*, , **332**, 711.
Leitherer, C. 1988a, *Ap. J.*, , **326**, 356.
Leitherer, C. 1988b, *Ap. J.*, , **334**, 626.
Prinja, R. K. 1987, *M.N.R.A.S.*, , **228**, 173.
Pauldrach A., Puls J. and Kudritzki, R. P. 1986, *Astr. Ap.*, , **164**, 86.

DIFFUSE IONIZED GAS IN THE ANDROMEDA GALAXY

RENE A.M. WALTERBOS
Astronomy Department, University of California
Berkeley, CA 94720

ROBERT BRAUN
NRAO, Socorro and NFRA, P.O. Box 2
7990 AA Dwingeloo, The Netherlands

ABSTRACT   We have detected diffuse ionized gas in the spiral arms of M31 in Hα and [SII] emission lines, using deep CCD imagery. The [SII]/Hα line ratio in the diffuse gas is systematically higher in the diffuse medium than in the discrete HII regions, similar to the situation in our Galaxy.

OBSERVATIONS

Current observational techniques allow studies of the interstellar medium in nearby galaxies at a level that is hard or, due to our unfavorable position in the disk, impossible to reach in our own Galaxy. We have completed a large radio and optical survey of the interstellar medium and young stars in the NE half of the nearby spiral M31. The VLA was used to map the 20-cm radio continuum and 21-cm atomic hydrogen emission at high sensitivity and spatial resolution, 5" to 10", corresponding to 17 to 34 pc (Braun, 1989a,b). The No1 36-inch telescope at Kitt Peak was used to obtain deep CCD-images of the Hα and [SII] line emission, and broadband B and R exposures of the stellar associations in the main spiral arms. The line images reach emission measures as low as a few pc cm^{-6} and provide accurate absolute fluxes for about 1000 HII regions, supernova remnants and planetary nebulae (Walterbos and Braun, in prep.).

RESULTS

Our CCD images show widespread diffuse emission in the spiral arms of M31. The emission is characterized by a high ratio of [SII]/Hα emission, typically 0.5. It can be traced to very faint levels, below emission measures of a few pc cm^{-6}. Similar properties have been found for diffuse ionized gas (DIG) in the solar neighborhood (e.g. Reynolds, 1988 and references therein), but this is the first clear measurement of this important component of the interstellar medium in another galaxy. Observations of other galaxies allow us to determine the spatial distribution of the DIG. The diffuse emission in M31 is present across the

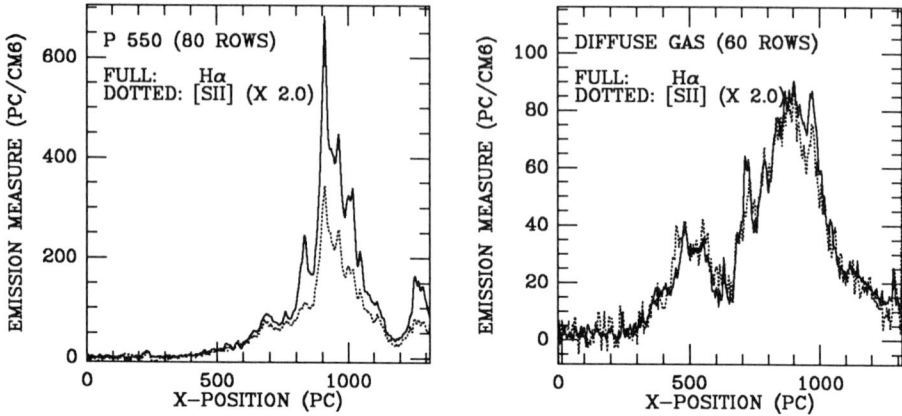

Fig. 1. Cross-cuts in Hα and [SII] emission lines through the 10-kpc spiral arm on the NE side of M31. The plots were obtained by averaging the indicated number of rows in the continuum-subtracted CCD images. The [SII] intensities have been multiplied by two. The left diagram includes the large HII region Pellet 550, while the right diagram was obtained for a region with mainly diffuse ionized gas.

spiral arms. It is more extended and reaches higher brightnesses in places where star formation activity is high. We observe an increase in the ratio of [SII]/Hα emission immediately outside the discrete HII regions (see Fig. 1), different from the situation in our Galaxy, where Reynolds (1988) does not detect
emission with ratios of [SII]/Hα as high as 0.5 in the immediate surroundings of HII regions. However, this may be due to the limited extent of the region that can be observed in our Galaxy. We also observe the ratio of [SII]/Hα to be fairly constant in the DIG, which seems to indicate that the gas is photo- rather than shock-ionized, since the ratio is critically dependent on shock velocity. Photo-ionization models can reproduce the [SII]/Hα line ratios in the diffuse medium (Mathis, 1986). The association of the DIG with regions of star formation is consistent with both photo- or shock-ionization.

We have extended our observations to a sample of some 20 nearby galaxies to investigate the properties of the interstellar medium as a function of Hubble type.

Kitt Peak National Observatory and the National Radio Astronomy Observatory are operated by Associated Universities, Inc., under contract with the National Science Foundation.

REFERENCES

Braun, R. 1989a, *Ap. J. Suppl.*, in press.
Braun, R. 1989b, submitted to *Astroph. J. Suppl.*
Mathis, J.S. 1986, *Ap. J.*, **301**, 423.
Reynolds, S. 1988, *Ap. J.*, **333**, 341.

DENSE GAS IN THE STARBURST GALAXY NGC 253

J. E. CARLSTROM
Radio Astronomy Lab., University of California, Berkeley, CA 94720

ABSTRACT   The Hat Creek millimeter array was used to obtain high resolution maps of HCO^+, HCN, and $\lambda\,3.3$ mm emission toward the starburst galaxy NGC 253. The molecular emission, which traces gas with density greater than $10^4\,cm^{-3}$, is distributed along a $\sim 35''$ bar–like structure. The continuum emission is confined to the inner region of the molecular gas distribution. The data suggests the presence of a torus of dense gas surrounding the nucleus with a projected diameter of $9''$, 140 pc at the assumed distance of 3.25 Mpc.

INTRODUCTION

The energetic activity in the nuclear barred region of NGC 253 is attributable to a burst of star formation. NGC 253 is in many ways a twin of the 'prototypical' starburst galaxy M82; the strong non-thermal radio continuum, high far-infrared luminosity, and bright molecular emission of the central 1 Kpc parallel the morphology of the M82 starburst. Furthermore, the filamentary low ionization optical emission and extended X-ray emission along the minor axis in NGC 253 is similar to a scaled down version of the well developed galactic bipolar wind in M82. The infrared luminosity of NGC 253, $3 \times 10^{10} L_\odot$, is comparable to M82 but is emitted from a smaller region (Telesco and Harper 1980). This suggests that the NGC 253 starburst may be more intense and at an earlier evolutionary stage than M82. However, the presence of a non-stellar AGN in NGC 253 may complicate the comparison (Turner and Ho, 1985).

Here we report on recent observations of the $J = 1 \to 0$ transitions of HCN and HCO^+, and of $\lambda\,3.3$ mm continuum toward NGC 253, obtained with the Hat Creek millimeter array[1].

DISCUSSION

The HCN and HCO^+ emission trace high density molecular gas ($n(H_2) > 10^4\,cm^{-3}$). Therefore the similar distributions of the HCO^+ emission (see Fig. 1), the HCN emission (not shown here), and the bright CO emission

---

[1] Operated by the University of California at Berkeley, the University of Illinois, and the University of Maryland, with support from the National Science Foundation.

(see Fig. 1 in Canzian *et al.* 1988) suggests that the molecular gas associated with the starburst is dense. Although not resolved in the integrated emission map, the spectra (see Fig. 1) and a position velocity map (not shown) reveal two components of dense gas. The two velocity components are separated by approximately 9″ ($\sim$ 140 pc in projection). The emission centroid is coincident with the nucleus and the base of the X–ray emission (see Fig. 1). The two components are most easily explained as limb brightened emission from a torus of dense gas similar to the molecular circumnuclear ring inferred for M82. However, the molecular ring in NGC 253 is much smaller than the ring in M82 (140 pc vs. 400 pc).

The 3.3 mm continuum emission, resolved by the interferometer observations (see Fig. 2), is confined to the heart of the starburst region. As in M82 (Carlstrom 1988), the 3.3 mm emission is distributed similar to emission at longer wavelengths which are dominated by nonthermal emission. The continuum emission in both galaxies peaks at the base of the optical filamentary emission and X–ray emission and appears to be 'bracketed' by dense molecular gas. The 3.3 mm flux (0.35±0.05 Jy) sets an upper limit to the thermal free–free emission from HII regions. Although thermal emission from dust is a small fraction of the 3.3 mm flux, the contribution from non-thermal emission is not presently known. The ratio of the infrared luminosity to Lyman

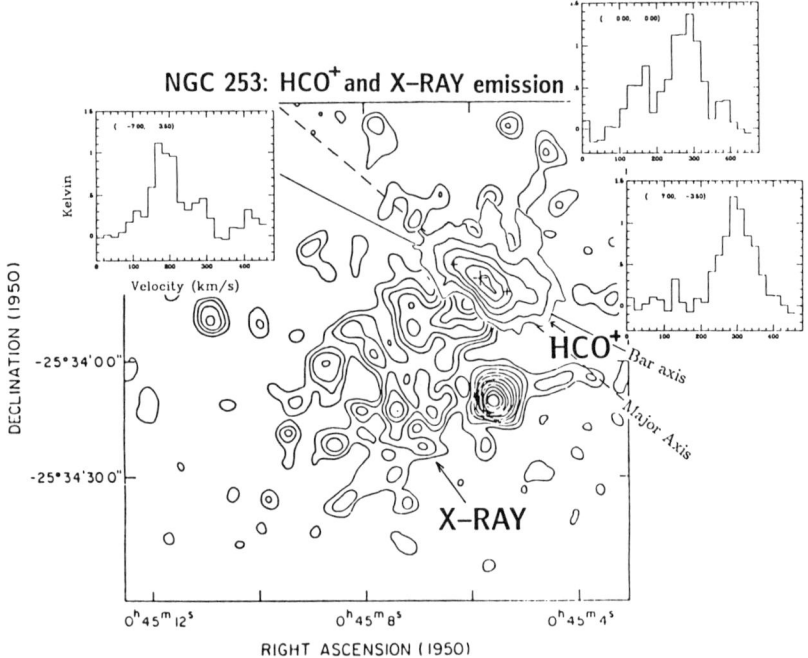

Fig. 1. The HCO$^+$ integrated emission plotted on the X–ray map of Fabbiano and Trinchieri (1984). The HCO$^+$ emission contours are multiples of 13.5 Jy km s^{-1} per beam. The synthesized beam is 10.8″ × 7.0″. The dashed line marks the position angle of the major axis and the solid line marks the position angle of the stellar bar (Scoville *et al.* 1985).

continuum luminosity ($N_L = 4.7 \times 10^{53} s^{-1}$ for $S_{ff} = 0.35$ Jy at 3.3 mm) for NGC 253 is twice the ratio measured for M82 (Carlstrom 1988). This suggests that the HII regions may be less evolved in NGC 253 with dust competing effectively for the ionizing photons. Alternatively, lower mass stars may be relatively important in NGC 253, supplying more non-ionizing luminosity.

The observations also support the evolutionary scenario of starbursts suggested by several authors. In this scenario NGC 253 is at an earlier evolutionary stage than M82. Superbubbles of the thermalized ejecta of numerous supernovae have broken out of one of the poles of the galaxy disrupting the molecular clouds. Within the plane of the galaxy, a portion of the kinetic energy has compressed and pushed the molecular clouds out of the very center forming a dense molecular torus.

This work was supported by NSF grant AST87-14721.

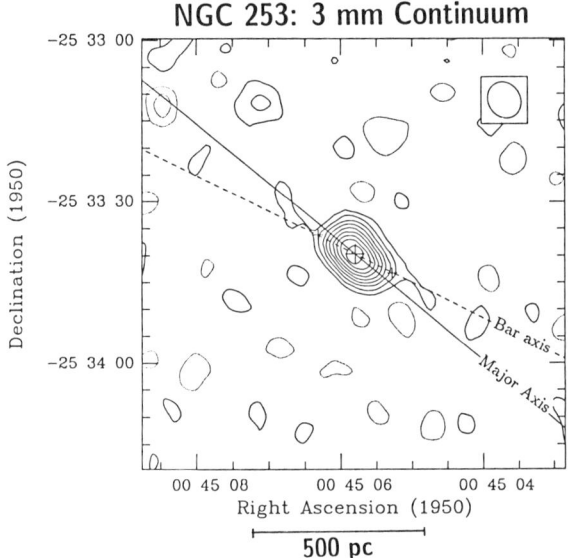

Fig. 2. The 3.3 mm continuum map. The contour interval is 18 mJy per beam. The beam, 7" × 6", is shown in the upper right corner. The cross marks the position of the nucleus. The total flux is 0.35 ± 0.05 Jy.

REFERENCES

Canzian, B., Mundy, L. G., Scoville, N. Z. 1988, *Ap. J.*, **333**, 157.
Carlstrom, J. E. in "Galactic and Extragalactic Star Formation" ed. R. E. Pudritz and M. Fich, Dordrecht: Reidel, 1988.
Carlstrom, J. E., 1988, Ph.D. Thesis, University of California, Berkeley
Fabbiano, G., and Trinchieri, G. 1984, *Ap. J.*, **286**, 491.
Scoville, N. Z., Soifer, B. T., Neugebauer, G., Young, J. S., Matthews, K., and Yerka, J. 1985, *Ap. J.*, **289**, 129.
Telesco, C. M. and Harper, D. A. 1980, *Ap. J.*, **235**, 392.
Turner, J. T., and Ho, P. T. P. 1985, *Ap. J. (Letters)*, **299**, L77.

HIGH RESOLUTION OBSERVATIONS OF NGC 7538 IRS 1

PREETHI PRATAP
University of Illinois, Astronomy Dept., 1011 W. Springfield,
Urbana, Illinois 61801

WOLFGANG BATRLA
N.R.A.O., P.O. Box 2, Green Bank, WV 24944

LEWIS E. SNYDER
University of Illinois, Astronomy Dept., 1011 W. Springfield
Urbana, Illinois 61801

ABSTRACT   High resolution observations of NGC 7538 IRS 1 in the J=1-0 transition of HCO^+ show an outflow which originates 15″ south of IRS 1. The molecular cloud appears very clumpy and there is a cavity in the material surrounding the H II region.

INTRODUCTION

NGC 7538 IRS 1, IRS 2, and IRS 3 are a group of compact H II regions embedded in the dense molecular material at the interface between a molecular cloud and a visible H II region. High resolution observations (3″.9 x2″.7) of the molecular cloud surrounding this source have been done with the Berkeley- Illinois-Maryland Array (BIMA) in the J=1-0 transitions of HCO^+ (at 89.18852 GHz) and HCN (at 88.6318 GHz).

RESULTS

The channel maps across the HCO^+ line (Figure 1) show that the molecular cloud is very clumpy. The channel map corresponding to a velocity of -57.5 $km\,s^{-1}$ shows an elongated structure extending from the northeast to the southwest. This structure is more evident in the map with the bold contours which has a resolution of 8″.5x6″.7. The elongated structure is perpendicular to the direction of a molecular outflow which has been observed in CO (Campbell and Thompson 1984; Scoville *et al.* 1986). The outflow is also detected in the HCO^+ line and can be seen in the channel maps at -59.5 and -60.0 $km\,s^{-1}$. The HCO^+ outflow is in the same direction as the CO outflow but is less extended since HCO^+ traces the denser material closer to the origin of the outflow. The

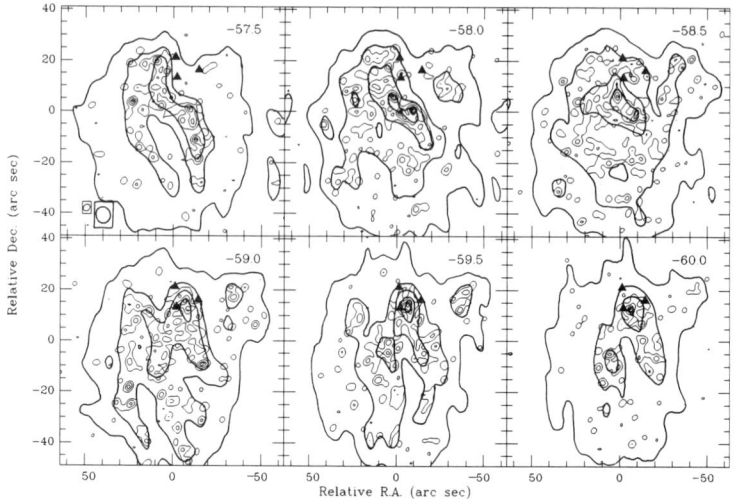

Fig. 1. Bold contours correspond to a resolution of 8".5x6".7 while the high resolution contours correspond to a resolution of 3".9x2".7. The three triangles correspond to the positions of IRS 1, IRS 2 and IRS 3. The relative R.A. and Dec. are with respect to $\alpha(1950)=23^h11^m37^s$; $\delta(1950)=61°11'35"$.

The map at -58.0 km s^{-1} shows that the emission is strung around IRS 1 apparently outlining a cavity in the molecular cloud. This cavity is also seen in the HCN line (Pratap, Batrla, and Snyder 1989a,b). The cavity could be caused by the expansion of the H II region into the molecular cloud. In the context of the maser emission in the source, the clumps around the cavity can provide the different conditions required for the excitation of different molecular masers. The cavity can be interpreted as being due to a lack of high density material around the H II region since both HCO$^+$ and HCN trace H$_2$ densities greater than $\sim 10^5$ cm^{-3}.

REFERENCES

Batrla, W., Pratap, P., and Snyder, L. E. 1988, *Ap. J. (Letters)*, **330**, L67.
Campbell, B., and Thompson, R. I. 1984, *Ap. J.*, **279**, 650.
Pratap, P., Batrla, W., and Snyder, L. E., 1989a, *Ap. J.*, **341**, 832.
Pratap, P., Batrla, W., and Snyder, L. E., 1989b, accepted for publication to the *Ap. J.*
Scoville, N. Z., Sargent, A. I., Sanders, D. B., Claussen, M. J., Masson, C. R., Lo, K. Y., Phillips, T. G. 1986, *Ap. J.*, **303**, 416.

EVOLUTION OF A SUPERBUBBLE BLASTWAVE IN A MAGNETIZED MEDIUM

KATIA M. FERRIERE, ELLEN G. ZWEIBEL
APAS Dept., University of Colorado, Boulder, CO 80309-0391

MORDECAI-MARK MACLOW
NASA Ames Research Center, Moffet Field, CA 94035

ABSTRACT We investigate the effects of interstellar magnetic fields on the evolution and structure of interstellar "superbubbles," using both analytic and numerical MHD calculations.

Superbubbles result from the combined actionof stellar winds and supernova explosions in OB associations. They consist of a cavity of hot gas, surrounded by a shell of cold dense material preceded by a shock wave.

If the medium in which a superbubble goes off is homogeneous and unmagnetized, the blast wave expands isotropically. As the interstellar gas flows through the shock, it cools significantly and gets strongly compressed such that thermal pressure remains approximately equal to ram pressure. Hence, the swept up material is confined to a very thin shell.

However, if the ambient medium is permeated by a uniform magnetic field $B_0 \sim 3\mu G$ (typical value for the ISM), the configuration loses its spherical symmetry, and, due to magnetic pressure, the shell of swept up material does not remain thin. We find the following qualitative differences:

1. Except in the immediate vicinity of the magnetic poles, the shell is supported by magnetic pressure. Its thickness, which is now determined by flux conservation, increases continuously form the poles to the equator.

2. The refraction of field lines at the shock and the thermal pressure gradient along the shell both contribute to accelerating the gas toward the equator. The resulting mass flux considerably decreases the column density at the magnetic poles.

3. Away from the poles, magnetic tension in the shell causes the field lines (particularly the inner boundary) to elongate in the direction of $\vec{B_0}$. In contrast, the shock wave radius increases with increasing θ, as would be the case if the wave were linear. At late times, however, the shock surface tends to flatten near the equator.

4. The reduced inertia of a parcel in the polar neighborhood makes it easier to decelerate, and accounts for the dimple which appears at the poles in numerical simulations. This dimple also results form the necessity to call on intermediate shocks in order to insure a smooth transition between a purely thermal shock at the poles and a magnetic shock in the rest of the shell.

5. The shock wave propagates faster than in the absence of magnetic field, except near the poles where the reduced mass of the shell allows it to be more efficiently decelerated. The shell inner boundary travels slower in a magnetized medium, regardless of the value of θ.

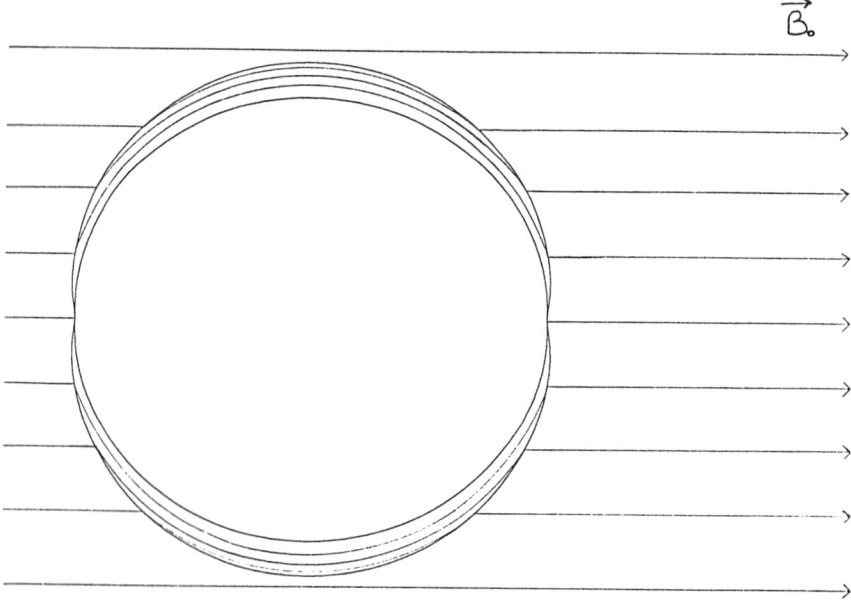

Shape of a superbubble in a medium with $n_0 = 0.324\ cm^{-3}$, $T_0 = 8000K$, $B_0 = 3\mu G$, at $t = 3 \times 10^6\ yr$.

ACKNOWLEDGMENTS

This research was supported by the NASA Astrophysical Theory Program, NAGW-766, at the University of Colorado.